Lecture Notes in Physics

Volume 953

The Lecture Notes in Physics

The series Lecture Notes in Physics (LNP), founded in 1969, reports new developments in physics research and teaching-quickly and informally, but with a high quality and the explicit aim to summarize and communicate current knowledge in an accessible way. Books published in this series are conceived as bridging material between advanced graduate textbooks and the forefront of research and to serve three purposes:

- to be a compact and modern up-to-date source of reference on a well-defined topic
- to serve as an accessible introduction to the field to postgraduate students and nonspecialist researchers from related areas
- to be a source of advanced teaching material for specialized seminars, courses and schools

Both monographs and multi-author volumes will be considered for publication. Edited volumes should, however, consist of a very limited number of contributions only. Proceedings will not be considered for LNP.

Volumes published in LNP are disseminated both in print and in electronic formats, the electronic archive being available at springerlink.com. The series content is indexed, abstracted and referenced by many abstracting and information services, bibliographic networks, subscription agencies, library networks, and consortia.

Proposals should be sent to a member of the Editorial Board, or directly to the managing editor at Springer:

Lisa Scalone
Springer Nature
Physics Editorial Department
Tiergartenstrasse 17
69121 Heidelberg, Germany
Lisa.Scalone@springernature.com

More information about this series at http://www.springer.com/series/5304

Werner Ebeling • Thorsten Pöschel

Lectures on Quantum Statistics

With Applications to Dilute Gases and Plasmas

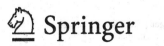 Springer

Werner Ebeling
FB Physik
Humboldt Universität zu Berlin
Berlin, Germany

Thorsten Pöschel
Friedrich-Alexander-Universität
Erlangen-Nürnberg
Erlangen, Germany

ISSN 0075-8450 ISSN 1616-6361 (electronic)
Lecture Notes in Physics
ISBN 978-3-030-05733-6 ISBN 978-3-030-05734-3 (eBook)
https://doi.org/10.1007/978-3-030-05734-3

Library of Congress Control Number: 2019933558

This Springer imprint is published by the registered company Springer Nature Switzerland AG
The registered company address is: Gewerbestrasse 11, 6330 Cham, Switzerland

Preface

In spite of the dominating importance of solids and liquids in our life, it is a fact that most of the matter in our universe occurs in the state of gas and plasma. The overwhelming part of this matter exists under extreme conditions. Therefore, the physics of gases and plasmas is a field of growing interest. We believe that we are just at the beginning of the exploration of the world beyond the narrow window of experience given by the state of our planet. Thus, the deeper reason for the growing interest in the physics of gases and plasmas is the dominating role of this state of matter in the universe. The subject of research is fundamental physics and in particular the quantum-statistical thermodynamics of states of matter, starting from rare gases and ending with plasma states. We are deeply convinced that the thermodynamics and quantum statistics are still the solid foundation on which even the most advanced research is operating. It remains our conviction that in spite of a flood of modern concepts, the traditional concepts of thermodynamics and statistics with terms such as temperature, pressure, energy, and entropy and the theory developed by Planck, Einstein, Nernst, Saha, and other pioneers are fundamental to the field. The body of these notes is based on lectures at universities and presentations at seminars and conferences. In particular, these lecture notes are based on the lectures on quantum statistics and plasma physics given by Werner Ebeling at the University of Rostock between 1970 and 1979 and at the Humboldt University of Berlin between 1980 and 2001, including shorter guest lectures at the Universities Paris VI 1977, Minneapolis 1986, Moscow 2005, and Krakow 2008. The body of the text is based on the last full two-semester course on quantum statistics held jointly by Werner Ebeling and Thorsten Pöschel at the Humboldt University of Berlin, assisted by several former coworkers such as Andreas Förster, Burkhard Militzer, Lutz Molgedey, and Waldemar Richert and by several students such as Jörn Dunkel, Hendrik Hache, Stefan Hilbert, Dirk Holste, Thomas Pohl, Michael Spahn, and others. This text is based on the notes including of the complete lecture written down by Thorsten Pöschel as well as several later revisions.

Following the personal interests of the authors and the traditions of quantum statistical thermodynamics in Berlin and Rostock, we concentrate on the fundamentals and applications to gases and plasmas. This is in contrast to most textbooks and monographs on quantum statistics which reveal some bias to condensed matter and in particular to solid state systems. We believe that the focus on gases and plasmas corresponds to trends in research evidenced, e.g., by several recent large international research projects and conferences.

This book is written on an intermediate level, that is, most of its contents should be accessible for undergraduate and graduate students, while a few results might be on a more advanced level and, thus, addressed to young researchers in the field. Readers interested in additional information, in particular also on numerical methods and extreme states of matter, are referred to a monograph by one of the current authors (W.E.) together with V.E. Fortov and V. Filinov which appeared recently in Springer's "Series in Plasma Science and Technology" (Ebeling et al. 2017).

On many occasions, the text contains historical remarks on the roots of quantum statistics. There are two reasons why we decided to include this historical material:

- The first reason is related to the *genius loci*: In fact, a large part of the book goes back to lectures given in Berlin where many fundamentals of quantum statistics were developed between 1900 and 1932.
- We are convinced—and this is based on a long experience in lecturing—that students are strongly motivated by looking back to the history of the development of our science.

The spirit of this textbook is influenced by meetings and discussions with several pioneers of quantum statistical thermodynamics, such as Alexander A. Abrikosov, Berni Alder, Nikolay N. Bogolyubov, Alexander S. Davydov, Hans Falkenhagen, Michael Fisher, Vitali L. Ginzburg, Günter Kelbg, Hugh DeWitt, and in particular Yuri L. Klimontovich. Further, we have to thank our colleagues in Berlin, in particular Dietmar Ebert, Wolfgang Muschik, and Lutz Schimansky-Geier, and in Rostock, in particular Klaus Kilimann, Wolf-Dietrich Kraeft, Dietrich Kremp, and Gerd Röpke, as well as Peter Hänggi from Augsburg for collaboration, advice, and discussions.

Berlin, Germany   Werner Ebeling
Erlangen, Germany   Thorsten Pöschel
Summer 2018

Reference

Ebeling, W., V.E. Fortov, and V.S. Filinov. 2017. *Quantum Statistics of Dense Gases and Nonideal Plasmas. Springer Series in Plasma Science and Technology.* Springer.

Contents

Chapter 1
Basic Physics of Gases and Plasmas

1.1 Matter in Form of Gases and Plasmas

Matter appears on our planet, in the solar system and in the rest of Universe in rather different forms. The focus of this text is the quantum physics of gases and plasmas leaving plasma-like matter of high energy density apart. While for our everyday life solid state physics is of large interest, one should keep in mind that less than 1% of the visible matter in the Universe is in condensed states but 99% of our world occurs in form of gases, plasmas or extreme states. This includes not only the atmosphere of our planet, many gases in technological applications, gases in the outer space of Earth and in the interstellar space, but also the stars from the big giants to the sun-like stars, white dwarfs up to the neutron stars and other exotic stars as well as many other still less known forms of matter such as dark matter (Fortov 2011).

In contrast, condensed states of matter which dominate the education of physicists are a relatively seldom form of matter on the scale of the Universe which however is central for life in our Universe. The condensed state is a rather special state of matter based on bound states of electrons and nuclei, forming atoms and molecule, which exists only in a small region of density and temperature.

By happy chance, our planet Earth assumes a state in this relatively small region of density and temperature. Only under the conditions of planets the evolution of life was possible (Feistel and Ebeling 1989, 2011). Life needs atoms and molecules, that is, bound states of electrons and nuclei. Thus, one of the subjects of this book is indeed to elaborate the conditions for the formation of bound states as atoms and molecules. We will show that only in a small region of density and temperature the nuclei and electrons are able to form such bound states and subsequently liquids and solids. From our more global point of view, these are entities based on strong correlations between particles and we speak therefore about the region of bound states or in other context about the *corner of correlations* (CoC), see Fig. 1.1.

In general we consider in detail mostly hydrogen as model system, which is the simplest and at the same time the most abundant kind of matter in the Universe. We

© Springer Nature Switzerland AG 2019
W. Ebeling, T. Pöschel, *Lectures on Quantum Statistics*,
Lecture Notes in Physics 953, https://doi.org/10.1007/978-3-030-05734-3_1

Fig. 1.1 Region of bound
states (atoms, molecules etc.)
in hydrogen plasmas, also
called *corner of correlations*
(CoC)

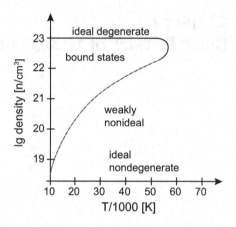

will study hydrogen in form of plasma, atomic, and molecular gas. In the first part
of the book, we study in particular the equilibria between hydrogen plasma, atomic
hydrogen gases and molecular hydrogen gases. It will become clear that the behavior
of hydrogen is in most respect quite universal and typical for all substances. In order
to understand the principal aspects of the formation of these states of matter and
the transitions between them, alongside, we develop the tools of thermodynamics
and statistical physics, and explain the rôle of Coulomb forces, interatomic, and
intermolecular forces. Central to this book is the equation of state, and various
transport properties. Out of the scope of this book are the details of the formation of
liquids and solid states.

1.2 Basic Physics of Gases

1.2.1 Classical Gas Laws

Scientific studies of the properties of gases go back to the seventeenth century
when Robert Boyle (1627–1691) experimented with gases and found that inverse
pressure is proportional to volume, $p \sim 1/V$, at constant temperature, T. A
few years later, Edme Mariotte found the same law, nowadays known as Boyle-
Mariotte law of gases. Another perfect gas law says that at constant volume,
pressure is proportional to temperature, $p \sim T$. An application of this law is the
hydrogen thermometer. When heated, a constant volume of hydrogen responds by
increased pressure. Therefore, for measurements of volume of a gas, it is important
to specify temperature and pressure of the gas. Standard temperature and pressure is
defined as 273.15 K (0 °C) and 101.325 kPa (760 mm Hg). According to the Boyle-
Mariotte gas law, the state of a gas changes when energy (heat) is transferred to
or from the system. The state of a gas can also be changed without changing
its heat, called adiabatic change. A gas responds by decreased temperature when

expanded adiabatically, demonstrated by the experiments by Joule and Kelvin. Further, Dalton's law of partial pressures states that in a mixture of gases, the pressure each gas exerts is the same as if it would be the only gas in the volume. Finally, due to Avogadro's law, equal volumes of gases at the same temperature and pressure contain equal numbers of molecules. A mole is defined as the number of atoms in 0.012 kg of Carbon 12, that is approximately 6.022×10^{23}. At standard temperature and pressure, a mole of gas occupies a volume of 22.4 l. Combining the perfect gas laws and Avogadro's law we obtain the universal ideal gas law,

$$pV = \frac{\nu RT}{V} \quad \text{or} \quad p = nk_BT ; \qquad n = \frac{N}{V} , \tag{1.1}$$

where ν is quantifies the number of atoms, N, measured in moles, R is Avogadro's constant, and k_B is Boltzmann's constant. This universal law is connected with the work of Joseph Gay-Lussac (1778–1850). The energy density of an ideal gas is given by

$$\rho_E(T) = c_\nu nT , \tag{1.2}$$

with the specific heat, c_v, assuming the value $c_v = 3/2 k_B$ for dilute molecular gases. Of some importance is the equation of state for adiabatic processes, that is, for processes where no heat or matter is transferred between the system and its environment,

$$p \sim n^\gamma ; \qquad \gamma = \frac{c_p}{c_v} . \tag{1.3}$$

where $\gamma = 5/3$ for molecular gases.

Historically, temperature scales were defined by the universality of the ideal gas law and gauged by the triple point of water, by definition, to $T_c = 273.16$ K, where pure ice, pure water and pure water vapour can coexist at equilibrium.

The next big step in the understanding of gases was the gas kinetic theory, based on a simplified particle description of a gas. August Krönig (1822–1879) and Rudolf Clausius (1822–1888) in Berlin started from the simple idea that the atoms are like elastic balls moving stochastically. From this simple idea many gross properties of the gas could be derived—for an historical review see (Ebeling and Hoffmann 1991; Rompe et al. 1987). A few years later, James Clerk Maxwell (1831–1879) and Ludwig Boltzmann (1844–1906) developed then the kinetic theory of gases, which became one of the most important concepts of physics of the nineteenth century. At the end of the nineteenth century, Johannes van der Waals (1837–1923) from the Netherlands introduced more realistic interactions between the particles, resulting in a reliable theory, in particular at moderate density and higher temperature. An independent and even more general approach to statistical thermodynamics, emphasizing the rôle of entropy was developed by the American physicists Josiah Willard Gibbs (1839–1903). Gibbs developed the ensemble approach, the entropy functional and was the first to understood the rôle of the maximum entropy principle. His monograph on the principles of statistical physics belongs to the most important books in the field of Statistical Physics. Other important contributions to the body of literature in this field are the book on the Kinetic Theory by Chapman and Cowling (1941) and the thick volume covering *everything* in the theory of gases and liquids by Hirschfelder et al. (1954).

In the course of this book we concentrate on the development of the quantum theory of gases, leaving out many interesting applications as, e.g., the physics of granular gases (Brey et al. 2009; Brilliantov and Pöschel 2004; Pöschel and Brilliantov 2003; Pöschel and Luding 2001) and the physics of gases at extremely low temperature.

The quantum statistics of gases was born in Berlin, starting with the work of Max Planck (1858–1947) who was around 1900 interested in the foundations of statistical thermodynamics and in developing a theory of radiation (Fig. 1.2). His work was based on the famous relation between entropy and probability (Ebeling et al. 1990; Feistel and Ebeling 1989, 2011),

$$S = k \log W , \tag{1.4}$$

which he first wrote down explicitly. By finding the quantum radiation laws in 1900, Planck was the founder of the first quantum theory of gas. The quantum statistics of gases was formulated by Planck for the photon gas which is a very particular case. It took nearly a quarter of a century to find the generalization for particles with rest mass and integer spin by Einstein (1924) and for particles with half-integer spin by Fermi (1926). The track to these fundamental laws was not easy and the third law of thermodynamics played an essential rôle.

Fig. 1.2 Max Planck (left) and right Walther Nernst, after a portrait by Max Liebermann 1912, property family Nernst, lost 1945. Source: (left) http://www. roro-seiten.de/physik/extra/ planck/max_planck.html; (right) https://commons. wikimedia.org/wiki/File: Nernst,_Walther_1912.jpg

1.2.2 Third Law of Thermodynamics and the Roots of Quantum Statistics

The long way from Planck's radiation law to the Bose-Einstein and Fermi-Dirac statistics is connected with the third law of thermodynamics which paved the way to generalize the classical gas laws (see Ebeling and Hoffmann 1991; Simonyi 1990). In 1905, Nernst found relations which he considered as the "missing stone in thermodynamics", which is now called *third law of thermodynamics* (Nernst 1906). The idea which led Nernst to the new law arose from the critical analysis of experimental data on chemical and electrochemical reactions in the liquid phase at low temperature. Here, as known already for some time, appeared an unexpected and unexplained correspondence between the free energy and the internal energy. Bertholet had already hypothesized the identity of these quantities, and Nernst found that the correspondence improved at lower temperature. This led him to suggest that the difference between the internal energy and free energy vanishes asymptotically when approaching $T \rightarrow 0$. Some years later, Planck formulated Nernst's new principles as *The entropy of all bodies which are in internal equilibrium vanishes at the zero point of temperature.*

After postulating his new theorem, Nernst and his collaborators worked hard to prove and develop further this new law of nature. The specific heat, being of particular importance, was determined for several substances at low temperature. This was a very difficult scientific problem which required new scientific instruments. On the other hand, the new theorem raised also fundamental theoretical problems. Nernst understood soon that the laws of ideal gases and other relations as, e.g., for the heat capacity of solids contradict the third law. However, Nernst was confident about his theorem such that he formulated the needs for a new theory of gases, a new theory of heat capacity etc., compatible with the third law of thermodynamics.

At the same time, 1905–1906, when Nernst and his group developed the third law and its experimental verification, a young theoretician working in Bern published three fundamental papers in volume 17 of the *Annalen der Physik*, one of them on a new quantum theory of the photo effect closely related to Planck's work. At that time, when Einstein was at the age of 26 years, he published in the same volume of the Annalen, also two other fundamental papers devoted to the theory of relativity and on the theory of Brownian motion.

The first theoretical work by Albert Einstein (1879–1955) was carried out during the time when he worked at the patent office in Bern. Einstein started his work on statistical physics in 1902/03 with two interesting papers on "Kinetische Theorie des Wärmegleichgewichtes und des zweiten Hauptsatzes der Thermodynamik",[1] published in the leading physical journal of that time, the *Annalen der Physik* (Einstein 1902). Here independently of Gibbs, Einstein developed the basic ideas of an ensemble theory and the statistics of interacting systems. In his dissertation,

---

[1] In English: "The kinetic theory of thermal equilibrium and the second law of thermodynamics."

presented in 1905 to the Zürich University (Einstein 1905a,b), he developed a first correct theoretical interpretation of Brownian motion. The first experimental check of Einstein's relations on Brownian motion was published already in 1906 by Svedberg (1906a,b) in *Zeitschrift für Elektrochemie*. In the same year appeared the first paper by Marian von Smoluchowski (1872–1917) who was at that time Professor at the University of Lemberg in the Austrian Empire (now Lviv, Ukraine). Smoluchowski also contributed many fundamental results to the theory of Brownian motion. But after this side steps let us come back to Einstein's papers on the photo effect "Über einen die Erzeugung und Verwandlung des Lichtes betreffenden heuristischen Gesichtspunkt"[2] (Einstein 1905c) and "Zur Theorie der Lichterzeugung und Lichtabsorption"[3] (Einstein 1906). In spite of the fact that Einstein's papers on the theory of light are so closely related to Planck's ideas, Planck was not excited about the 1905/06 papers, which he considered as too speculative. Today it looks strange to us, that Planck considered still in 1913 Einstein's 1905 hypothesis of light quanta as pure speculation (Pais 1982). However, Einstein continued to attribute physical reality to Planck's quanta. In 1907, Einstein attacked the unsolved problem of the specific heat at low temperature (Einstein 1907). He proposed, that quantum effects are responsible for the vanishing of the specific heat at zero temperature. His theory can be considered as the second origin of quantum statistics. In fact this work can be considered as the origin of the quantum statistics of phonon gases. This work was immediately recognized by Planck who stated *So war er vor Allem der Erste, der die Bedeutung der Quantenhypothese für die Atom- und Molekularbewegungen nachgewiesen hat, indem er aus dieser Hypothese eine Formel für die spezifische Wärme fester Körper ableitete.*[4] Einstein's work on the specific heat *Die Plancksche Theorie der Strahlung und die Theorie der spezifischen Wärme"*[5] (Einstein 1907) attracted the attention of Nernst and his collaborators and by 1910 the Nernst group succeeded in confirming this prediction. In this way the third law of thermodynamics as well as the young and still controversial quantum theory found one of its first experimental verifications. Through these investigations, Nernst became not only one of the earliest and most committed prophets of the quantum theory—he was the initiator of the first Solvay conference (1911)—but also a firm supporter of the young Einstein (for details see Ebeling and Hoffmann 1991). In 1913, together with Planck, he introduced the *new Copernicus* into the prominent circle of Berlin physicists. Of particular importance is Einstein's paper on *Strahlungs-Emission und -Absorption nach der Quantentheorie*[6] (Einstein 1916).

---

[2]In English: "On a heuristic point of view concerning the production and transformation of light."

[3]In English: "On the theory of light production and light absorption."

[4]In English: "Above all, he was the first who has demonstrated the importance of the quantum hypothesis for the motion of atoms and molecules, by deriving a formula for the specific heat of solid bodies from this hypothesis."

[5]In English: "Planck's Theory of Radiation and the Theory of Specific Heat.

[6]In English: "Emission and absorption of radiation in quantum theory."

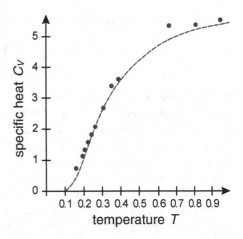

Fig. 1.3 Data for the specific heat as a function of temperature presented by Einstein on a conference of the Bunsengesellschaft 1912 in order to demonstrate his new law

This work provided a theory of radiative transitions, including a new kinetic derivation of Planck's law. Further, this paper by Einstein on spontaneous emission of light and induced emission and adsorption can be considered as the theoretical foundation of the nonlinear dynamics and stochastic theory of photon gases and modern LASERS (for details see Haken et al. 2016) (Fig. 1.3).

The next big step in the development of the quantum statistics of gases was done only in 1924 by the Indian physicist Satendranat N. Bose (1892–1974). Bose's method (Bose 1924) is based on a division of the phase space of the photons into cells of volume h^3 similar as suggested earlier by Sackur (1913), Planck (1916), and Tetrode (1912a,b) for the calculations of the Nernst constant in the entropy of pure gases. Otto Sackur (1880–1914), who was a young physical chemist from Breslau, and Hugo Tetrode (1895–1931) from the Netherlands who was a 17 years old student of physics at the University at Leipzig found independently and nearly at the same time a the relation between the unknown chemical constant in Nernst's theory of chemical equilibria and Planck's constant (Sackur 1911; Tetrode 1912b). They discussed the problem how to count the number of possible states of motion of the atoms of a monatomic gas. Using similar arguments, they found that the number of allowed states in a given energy range depends on the minimal distance of particles in the phase space of positions and momenta. If a lower limit existed for the distance representing two particles in phase space, this would give an upper limit to Planck's constant, W, and allow the explicit calculation of W and, thus, of the entropy.

Tetrode started his derivation with an equation from Gibb's classical statistical mechanics. He assumed, that the product of the elements in the phase space of momenta and positions, cannot be smaller than Planck's constant. Sackur was more close to the style of Max Planck's reasoning, he limited the spacing of the allowed states for the elements of energy. This way both scientists obtained an equation which by today is known as the *Sackur-Tetrode equation*,

$$S = k_B N \left[\ln \left(\frac{V}{N \Lambda^3} \right) + \frac{5}{2} \right].$$ (1.5)

This equation can be used to calculate the entropy of ideal monatomic gases. The Sackur-Tetrode equation is, in fact, a first quantum statistical result for the entropy of gases which, however, has only limited validity restricted by the condition $N \Lambda^3 \ll V$ (see Fig. 3.2). To extend the theory of gases beyond this limitation, entirely new methods had to be developed, as on the side of the theory, Bose's new method of counting states and Einsteins revolutionary ideas on condensation and on the side of the experiment entirely new methods for the study of the condensation of atomic gases at extremely low temperature. In spite of intensive search only in 1995 Petrich et al. (group of Eric Cornell) reported at the *International Conference of* LASER *spectroscopy* on the island Capri on experiments at the University of Boulder which show the properties of dense Bose-Einstein gases. Similar experiments were performed at the same time in the group of Wolfgang Ketterle at MIT (Davis et al. 1995). More details on these exciting systems are discussed in Chaps. 3 and 5.

1.3 Basic Physics of Plasmas

1.3.1 Coulomb Interactions and Ionization Equilibrium

About 1785, Charles Augustin de Coulomb recovered the law for the force between two charges, e_a and e_b. The force is a function of the distance, r, in radial direction (in Gaussian units)

$$F = -\frac{e_a e_b}{\varepsilon_r r^2}.$$ (1.6)

This force is repulsive for equal charges and attractive for opposite charges. The charge-carrying particles belong to the species a and b and ε_r is the relative dielectric constant of the imbedding medium. The corresponding potential is

$$V_{ab}(r) = \frac{e_a e_b}{\varepsilon_r r}.$$ (1.7)

In most cases, we will silently assume that the charges are in vacuum, $\varepsilon_r = 1$. The Coulomb potential is long-range. It was Joseph E. Mayer (1904–1983) who detected that the Fourier transform of the Coulomb potential is of primary importance for solving problems of screening and cluster theory (Mayer 1950). The Fourier transform of a potential is defined by

$$\tilde{V}_{ab}(\mathbf{t}) = \int_V d\mathbf{r}\, V_{ab}(r) e^{i\mathbf{t}\cdot\mathbf{r}}.$$ (1.8)

For infinite volume, the integral over the volume diverges for Coulomb interaction. We will show that this problem is of principal nature and, thus, the Coulomb potential in its original form needs some regularization. The great pioneer of the statistical theory of Coulomb systems, Joseph Mayer, proposed to introduce a decay factor $e^{-\alpha r}$ with some small α. Exploiting radial symmetry, we obtain

$$\tilde{V}_{ab}(t) = \int_0^\infty \frac{e_a e_b}{t} \sin(rt) e^{-\alpha r}\, dr = 4\pi \frac{e_a e_b}{t^2 + \alpha^2}. \tag{1.9}$$

For the limit $\alpha \to 0$ we get

$$\tilde{V}_{ab}(t) = 4\pi \frac{e_a e_b}{t^2}. \tag{1.10}$$

When we perform the limit $t \to 0$ after the limit $\alpha \to 0$, the integral (related to the self-energy of the Coulomb field $\tilde{V}_{ab}(t = 0)$) which is infinite for infinite volume. We should keep in mind that the Coulomb-Mayer model includes some small parameter, α, and that possibly a result can depend on the coarse of the calculations, where the limit $\alpha \to 0$ is performed. Alternative concepts based on discrete Fourier series for the Coulomb potential have been developed by Bohm and Pines (1951), Pines (1961), and Pines and Nozieres (1966) and more recently by Bobrov et al. (2015). We will come back to this problem in Chap. 6.

Ionization Equilibrium Between Atoms, Electrons and Ions

Plasmas are gases consisting of charged particles, mainly electrons and ions, created through ionization of atoms at high density or high temperature. The development of the quantum statistics of ionization and plasmas around 1920 was intimately related to the requirements of a new scientific discipline, namely astrophysics. Although astronomy is as ancient as recorded history itself, it was long separated from the study of physics. In the Aristotelian world view, the celestial world tended towards perfection. In this view the sky seemed to be organized in perfect spheres moving in perfectly circular orbits. Only at the end of the nineteenth century, it was discovered that, when decomposing the light from the Sun a multitude of spectral lines were observed (regions where there was less or no light). Experiments with hot gases showed that the same lines could be observed in the spectra of gases, specific lines corresponding to unique chemical elements. In this way it was proved that the chemical elements found in the Sun were also found on Earth. Indeed, the element helium was first discovered in the spectrum of the Sun and only later on Earth, which explains its name. During the twentieth century, spectroscopy and the study of these spectral lines advanced, particularly as a result of the advent of quantum physics that was necessary to understand the astronomical and experimental observations. In order to understand why spectra appeared under definite conditions, a theory of abundances was required. The theoretical basis for this was created by the German physico-chemist John Eggert (1891–1973) and the Indian physicist Meghnad Saha (1893–1956). The work of Eggert and Saha resides on two theoretical results obtained a few years earlier:

(1) Planck's work on the chemical constant in the entropy and the chemical potential of gases as expressed by Sackur and Tetrode
(2) Rutherford's new model of atoms.

The work by Sackur and Tetrode was already explained in the previous section, let us discuss Rutherford's new model. In May 1911, Rutherford (1911) came forth with a physical model for the structure of atoms. The Rutherford model provided a first understanding why electron scattering can so deeply penetrate into the interior of atoms, which was an unexpected experimental result. Rutherford explained scattering as a passage of rapid electrons through an atom having a positive central charge, Ne^+, surrounded by a compensating charge of N electrons.

In 1913, Bohr (1913a,b,c) (see also Bohr 1921) developed a dynamical model based on Rutherford's ideas, which describes in a semi-classical approximation the electron orbits and the corresponding energies. According to Bohr's theory, the radii of the orbits are

$$a_s = s^2 a_B \, ; \qquad a_B = \frac{\hbar^2}{\mu e^2} \, ; \qquad \hbar = \frac{h}{2\pi} \, ; \qquad \mu = \frac{m_e m_+}{m_e + m_+} \, , \tag{1.11}$$

where m_e is the mass of an electron. The corresponding energies are

$$E_s^* = -\frac{\mu e^4}{2\hbar^2 s^2} \, . \tag{1.12}$$

Here and in the following we denote the so-called main quantum number by s in order not to be mixed with the density n. According to the Bohr model there are infinitely many levels close to the series limit $s \to \infty$ (see Fig. 1.4). The existence of a lower bound for the energy of atoms provides the stability of atoms, however, surprisingly enough, it does not guarantee the stability of matter consisting of charges in bulk. Instead, its stability is assured through the fermionic character of the electrons (Lieb 1976).

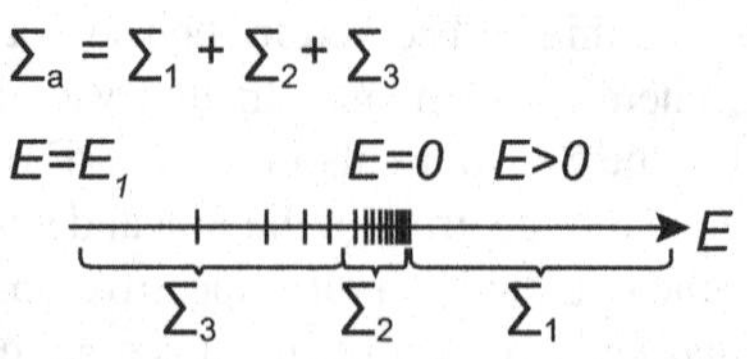

Fig. 1.4 Planck's schema of the atomic levels of hydrogen. Left the ground state and the lower bound states, followed by the weakly bound discrete states, the series limit and right by the continuous level

The first step towards a theory of the ionization of atoms was done by John Eggert who used the results on chemical equilibrium obtained in Nernst's group (where he worked as a scientific assistant) and the results on the entropy of gases obtained by Sackur and Tetrode. Eggert considered the ionization as a chemical equilibrium between electrons, ions, and atoms and used the Sackur-Tetrode results for calculating the chemical potential of the particles. Eggert did, however, not succeed to describe multiple ionization of iron atoms in the interior of stars. Based on the work by Eggert, this was achieved by the young Indian physicist Saha who realized the significance of the ionization potential, I, of atoms. The importance of I was apparent from the theoretical work by Bohr, and the experiments by Franck and Hertz.

For a brief derivation of the Eggert-Saha equation, we consider a binary Coulomb system with n_+ positive ions (cations) and n_- negative charges (electrons) per cubic centimeter, where $n_- = n_+$. The density of free atoms is n_a and the total density is $n_0 = n_+ + n_a = n_- + n_a$. In the following we will use the plasma notations, that is, we call the negative charges *electrons* and the positive charges simply *ions*. In he spirit of Nernst's and Eggert's ideas, we assume a chemical equilibrium between electrons, ions and atoms,

$$e^- + i^+ \to a_0; \qquad \mu_e + \mu_i = \mu_a.$$ (1.13)

Here the μ_k are the chemical potentials of the corresponding species. For atoms we assume

$$\mu_a = I + k_B T \ln n_a + k_B T \ln \Lambda_a^3$$ (1.14)

and for the charges

$$\mu_e = k_B T \ln n + k_B T \ln \Lambda_e^3; \qquad \mu_i = k_B T \ln n + k_B T \ln \Lambda_i^3.$$ (1.15)

The last terms in these expressions contain Planck's constant h, due to the work by Sackur and Tetrode:

$$\Lambda_k = \frac{h}{\sqrt{2\pi m_k k_B T}}.$$ (1.16)

From the condition of chemical equilibrium follows the Eggert-Saha formula

$$\frac{n_a}{n_e n_i} = \Lambda^3 e^{\frac{I}{k_B T}}; \qquad \Lambda = \frac{\Lambda_e \Lambda_i}{\Lambda_a}.$$ (1.17)

In the work by Saha, this simple reasoning was extended to more complicated ionization phenomena, including multiple ionization. Already after receiving his PhD in India in 1918, Saha became interested in astrophysics and in particular the relation between thermodynamics and astrophysics as in chromospheric and

Fig. 1.5 Meghnad Saha
(1893–1956) in Berlin, 1921.
According to Eddington his
work is a landmark in the
history of astronomy. Source:
https://commons.wikimedia.
org/wiki/File:SahaInBerlin.
jpg

stellar problems. Still working in India he came across with the mentioned paper
of Eggert on the dissociation of gases in stars published 1919 in Physikalische
Zeitschrift (Eggert 1919). By boldly applying the results from thermodynamics
and Bohrs theory to stellar matter, Saha derived a formula by which the degree of
ionization in a very hot stellar gas could be expressed in terms of its temperature and
electron pressure. He published his results in several papers in 1920/21 (Saha 1920,
1921). For the quantitative astronomy, the work by Saha was a true breakthrough.
Saha's greatest contribution is the further development of Eggerts ionization theory
and the correct application of ionization theory to stellar atmospheres. Using the
language of physical chemistry he called his theory "the equation of the reaction-
isobar for ionization". Saha travelled for the first time to Europe in the 20th visiting
centers in England and Germany (see Fig. 1.5). The work of Eggert and Saha has
shown that an analysis of stellar spectra provides a rich information about the
stellar atmospheres, such as the chemical composition and temperature. Further,
the deviations from thermal equilibrium, the density distribution of the various
elements, and the value of gravity in the atmosphere and its state of motion could
be studied. The impetus given to plasma physics and astrophysics by the work of
Eggert and Saha can scarcely be overestimated, as nearly all later progress in this
field has been influenced by it.

   After this brief discussion of the treatment of ionization phenomena in astrophysical and normal plasmas by means of ionization theory let us briefly discuss
different other types of plasma states.

1.3.2   Different Plasma States on Earth

By definition, a plasma is a neutral system of free electrons and ions. We consider
here mainly gaseous plasmas but plasmas exist also in the solid state. For example,
the conduction electrons in metals and semiconductors are a special form of plasma.
The behavior of plasmas depends on density and temperature. Degenerate and

Table 1.1 Ionization
energies of the first two outer
electrons, I_1 and I_2 for
several gases

Element	I_1	I_2
Hydrogen	13.595	–
Lithium	5.399	75.62
Sodium	5.138	47.29
Potassium	4.339	31.81
Rubidium	4.176	27.50
Cesium	3.893	25.10
Beryllium	9.306	18.187
Helium	24.60	54.38
Neon	21.6	–
Argon	15.8	–
Krypton	14.0	–
Xenon	12.1	–

non-degenerate plasmas behave in quite different way. In the logarithmic density-temperature-plane (see Fig. 3.2) the line $n_e \Lambda_e^3 = 1$ is the interface between degenerate and non-degenerate regions. Here $\Lambda_e = h/\sqrt{2\pi m_e k_B T}$ denotes the de Broglie wave length of electrons of thermal momentum and n_e is the density of free electrons. Free electrons are formed in equilibrium with bound electrons, e.g. the core electrons of atoms. This equilibrium depends crucially on the binding energies. As a rule of thumb, we consider bound states as relevant if $(I/k_B T) > 10$, that is, $T < (I/10k_B)$. Here I is the ionization energy. For example, in the case of hydrogen, $I = |E_{10}|$ and $I/k_B \approx 156,000$ K. Measuring T in electron Volt ($1\,\text{eV} \approx 10^4$ K), according to our rule we find for hydrogen (with $I = 13.6\,\text{eV}$) that bound states are relevant for $T < 15,000$ K. In Table 1.1 we present the ionization energy of several elements, where I_1 and I_2 denote the energies needed to remove the first and the second electron, respectively. In this book, we mainly deal with hydrogen and helium plasmas taking into account their abundances in the Universe. Our Sun and all stars are made of plasma, much of interstellar space is filled with a plasma, albeit a very sparse one, and intergalactic space as well. We mention also astrophysical jets of luminous ejected plasma. In our solar system, interplanetary space is filled with the plasma of the Solar Wind that extends from the Sun out to the heliopause. The planet Jupiter accounts for most of the non-plasma matter within the orbit of Pluto (about 0.2% by mass, or 10–15% by volume). On Earth the non-plasma state of matter dominates while the plasma state is rather exotic. And among the plasmas, Hydrogen and Helium are not as dominant as in the Universe but still of some interest and importance.

Aiming to fusion power plants, research is focused to hydrogen and deuterium plasmas in two different types of experiment, the Tokamak and the stellarator (Fortov 2009, 2011). Today, most devices are of Tokamak type, which is best investigated and comes closest to the ignition conditions. Large amounts of money are spent on the development of fusion reactors, due to the hope to find a solution of the energy problem. However, during recent decades, also a considerable amount of experimental and theoretical effort has been devoted to alkali and noble gas plasmas. Special interest found the behavior of alkali metals in the liquid and plasma states produced by their heating (Bonzel et al. 1989; Ebeling and Hoffmann 1991; Fortov and Yakubov 1994; Hensel et al. 1985; Sizyuk et al. 2006) and in particular Li^+, Na^+, K^+, Rb^+, Cs^+ and Be^{2+} plasmas. Of special interest has been the liquid-vapor critical point (Hensel et al. 1985). In recent years hot alkali plasmas are also widely employed in many technical projects. For instance, lithium plasmas are planned to be used in inertial confinement fusion devices, in solar power plants, for the electrochemical energy storage, in magnetohydrodynamic power generators and many other applications. Recent advances in the field of extreme ultraviolet lithography have revealed that laser-produced Li plasmas are source candidates for next generation microelectronics (Sizyuk et al. 2006). For this and other reasons, the investigation of static and dynamic properties of alkali plasmas are certainly of large interest.

Another field of great interest is semiconductor plasmas (Ashcroft and Mermin 1976; Blakemore 1962; Kittel 2005). In solid state materials, the conducting band and the valence band are separated by a nearly empty gap. Near to conducting or valency band one find energy states due to impurities (doping). The Fermi edge of semiconductors is within the band. We have to differ between donor and acceptor semiconductors. Both consist of a host lattice with some imbedded impurity atoms. Donors are impurity atoms, which have one excess electron in comparison with the atoms of the host lattice. An example is the semiconductor germanium doted with arsen atoms which have in the outer shell one electron more as the Ge-lattice. This electron is only loosely bound (see Table 1.2). The loosely bound excess electron can escape from the donor moving freely in the lattice and behaving similar as the free electron in a plasma. In particular it can form hydrogen-like states. Acceptors are impurity atoms in a lattice having one valence electron less than the host lattice The missing electron behaves like a hole, that is, like a positive electron. In real semiconductors there can be doping with donors as well as with acceptors. If the donor density is larger than the acceptor density we say the semiconductor is of n-type, otherwise we say its of p-type. Since the relative dielectric constant of typical semiconductors $\varepsilon_r \approx 10 \ldots 30$, the Coulomb interaction is weaker than in vacuum.

Table 1.2 Ionization energies (in eV) for donor and acceptor states in Ge and in Si

Host lattice	Ge	Si
P	0.012	0.045
As	0.013	0.050
Sb	0.010	0.039
B	0.0104	0.05
Al	0.0102	0.06
Ga	0.0108	0.07
In	0.0112	0.16

According to a simple estimate, the interaction law is

$$V(r) \sim \frac{e^2}{\varepsilon_r r} \tag{1.18}$$

and the masses are

$$m_e^* \approx 0.1 \ldots 1 \, m_e \, ; \qquad m_h^* \approx 0.1 \ldots 1 \, m_e \, . \tag{1.19}$$

The effective relative mass for electron-hole pairs is

$$m^* = \frac{m_e^* m_h^*}{m_e^* + m_h^*} \, . \tag{1.20}$$

For the Bohr radius and the ground state energy we get

$$a_B^* = \varepsilon_r \frac{\hbar^2}{m^* e^2} \, ; \qquad E_{10}^* = -\frac{m^* e^4}{2 \, (\varepsilon_r \hbar)^2} \, . \tag{1.21}$$

Since $\varepsilon_r^2 \sim 100 \ldots 1000$, the ground state energy is in the range $E_{10}^* \approx -(0.01 \ldots 0.1) \, \text{eV}$ and the Bohr radius $a_B^* \approx 10 \ldots 100 \, \text{Å}$.

We wish to mention that there are some fields of plasma physics which are not well presented in this book. The first example is *ultracold plasmas*. These plasmas which found a lot of interest in the last decade have temperature below 1 mK. Ultracold plasmas are created in magneto-optical traps (MOT) by trapping and cooling neutral atoms and then using another laser to ionize the atoms by giving each of the outermost electrons just enough energy to escape the electrical attraction of its parent ion. One special property of ultracold plasmas is that they are well characterized and their parameters are well tuneable as the initial conditions, their size and electron temperature. By adjusting the wavelength of the ionizing laser, the kinetic energy of the free electrons can be tuned as low as 0.1 K. This limit is set by the presently available frequency bandwidth of the laser pulse. The ion temperature are in the milli-Kelvin range as well as the temperature of the neutral atoms. The ions are, however, quickly heated through a process known as disorder induced heating. The physics of ultracold plasmas is related to the physics of so-called *Rydberg matter*. As Rydberg matter we denote one of the metastable states of a strongly non-ideal plasma, which forms upon condensation of excited atoms.

The second example is *dusty plasmas* and *grain plasmas*. A dusty plasma contains tiny charged particles of dust as for example found in interplanetary and interstellar space but also under many conditions on Earth (Morfill and Ivlev 2009). The dust particles acquire high charges and interact with each other. A plasma that contains larger particles is called grain plasma. Under laboratory conditions, dusty plasmas are also called complex plasmas.

References

Ashcroft, N., and N.D. Mermin. 1976. *Solid State Physics*. Boston: Cengage Learning.

Blakemore, J.S. 1962. *Semiconductor Statistics*. Oxford: Pergamon Press.

Bobrov, V.B., A.G. Zagorodny, and S.A. Trigger. 2015. Coulomb Potential of Interaction and Bose-Einstein Condensate. *Low Temperature Physics* 41: 1154–1163 (in Russian).

Bohm, D., and D. Pines. 1951. A Collective Description of Electron Interactions I. Magnetic Interactions. *Physics Review* 82: 625–634.

Bohr, N. 1913a. On the Constitution of Atoms and Molecules. *Philosophical Magazine* 26: 1–25.

Bohr, N. 1913b. On the Constitution of Atoms and Molecules. Part II. – Systems Containing only a Single Nucleus. *Philosophical Magazine* 26: 476–502.

Bohr, N. 1913c. On the Constitution of Atoms and Molecules. Part III. – Systems Containing Several Nuclei. *Philosophical Magazine* 26: 857–875.

Bohr, N. (1921). *Abhandlungen über Atombau*. Braunschweig: Vieweg.

Bonzel, H.P., A.M. Bradshaw, and G. Ertl, eds. 1989. *Physics and Chemistry of Alkali Metal Adsorption. Materials Science Monographs*. Vol. 57. Amsterdam: Elsevier.

Bose, S.N. 1924. Plancks Gesetz und Lichtquantenhypothese. *Zeitschrift für Physik* 26: 178–181.

Brey, J.J., I. Goldhirsch, and T. Pöschel, eds. 2009. *Granular Gases: Beyond the Dilute Limit. European Physical Journal - Special Topics*. Vol. 179. Berlin: Springer.

Brilliantov, N.V., and T. Pöschel. 2004. *Kinetic Theory of Granular Gases*. Oxford: Oxford University Press.

Chapman, S., and T.G. Cowling. 1941. *The Mathematical Theory of Non-Uniform Gases. Cambridge Mathematical Library*. Cambridge: Cambridge University Press.

Davis, K.B., M.O. Mewes, M.R. Andrews, N.J. van Druten, D.S. Durfee, D.M. Kurn, and W. Ketterle. 1995. Bose-Einstein Condensation in a Gas of Sodium Atoms. *Physical Review Letters* 75: 3969–3973.

Ebeling, W., and D. Hoffmann. 1991. The Berlin School of Thermodynamics Founded by Helmholtz and Clausius. *European Journal of Physics* 12: 1–9.

Ebeling, W., A. Engel, and R. Feistel. 1990. *Physik der Evolutionsprozesse*. Germany: Akademie-Verlag.

Eggert, J. 1919. Über den Dissoziationzustand der Fixsterngase. *Physikalische Zeitschrift* 20: 570–574.

Einstein, A. 1902. Kinetische Theorie des Wärmegleichgewichtes und des zweiten Hauptsatzes der Thermodynamik. *Annalen der Physik* 9: 417–433.

Einstein, A. 1905a. Eine neue Bestimmung der Moleküldimensionen. PhD thesis. Universität Zürich, 138–148.

Einstein, A. 1905b. Über die von der molekularkinetischen Theorie der Wärme geforderte Bewegung von in ruhenden Flüssigkeiten suspendierten Teilchen. *Annalen der Physik* 17: 549–560.

Einstein, A. 1905c. Über einen die Erzeugung und Verwandlung des Lichtes betreffenden heuristischen Gesichtspunkt. *Ann. Phys.* 17: 138–148.

Einstein, A. 1906. Zur Theorie der Lichterzeugung und Lichtabsorption. *Annalen der Physik* 20: 199–206.

Einstein, A. 1907. Die Plancksche Theorie der Strahlung und die Theorie der spezifischen Wärme. *Annalen der Physik* 22: 180–190.

Einstein, A. 1916. Strahlungs-Emission und -Absorption nach der Quantentheorie. *Verh. Dt. Phys. Ges.* 18: 318–323.

Einstein, A. 1924. Quantentheorie des einatomigen idealen Gases. *Sitzungsber. Preuss. Akad. Wiss. Phys.-math. Kl.* 22: 261–267.

Feistel, R., and W. Ebeling. 1989. *Evolution of Complex Systems*. Dordrecht: Kluwer.

Feistel, R., and W. Ebeling. 2011. *Physics of Selforganization and Evolution*. Weinheim: Wiley-VCH.

Fermi, E. 1926. Über die Wahrscheinlichkeit der Quantenzustände. *Zeitschrift für Physik* 26: 54–56.

Fortov, V.E. 2009. *Extreme States of Matter*. Moskva: FizMatGis (in Russian).

Fortov, V.E. 2011. *Extreme States of Matter: On Earth and in the Cosmos*. Berlin: Springer.

Fortov, V.E., and I.T. Yakubov. 1994. *Nonideal Plasmas*. Moskva: Énergoatomizdat (in Russian).

Haken, H., P. Plath, W. Ebeling, and Yu. M. Romanovsky. 2016. *Beiträge zur Geschichte der Synergetik*. Berlin: Springer.

Hensel, F., S. Jüngst, F. Noll, and R. Winter. 1985. Metal-Nonmetal Transition and the Critical Point Phase Transition in Fluid Cesium. In *Localisation and Metal Insulator Transitions*, ed. D. Adler and H. Fritsche, 109–117. Berlin: Springer.

Hirschfelder, J.O., C.F. Curtis, and R.B. Bird. 1954. *Molecular Theory of Gases and Liquids*. New York: Wiley.

Kittel, C. 2005. *Introduction to Solid State Physics*. New York: Wiley.

Lieb, Elliott H. 1976. The Stability of Matter. *Reviews of Modern Physics* 48: 553–569.

Mayer, J.E. 1950. The Theory of Ionic Solutions. *The Journal of Chemical Physics* 18: 1426–1436.

Morfill, G.E., and A.V. Ivlev. 2009. Complex plasmas: An interdisciplinary research field. *Reviews of Modern Physics* 81: 1353–1404.

Nernst, W. 1906. Über die Berechnung chemischer Gleichgewichte aus thermischen Messungen. *Nachr. Kgl. Ges. Wiss. Gött.* 1: 1–40.

Pais, Abraham. 1982. *'Subtle is the Lord...', the science and life of Albert Einstein*. Oxford: Oxford University Press.

Petrich, W., M.H. Anderson, J.R. Ensher, and E.A. Cornell. 1995. Stable, tightly confining magnetic trap for evaporative cooling of neutral atoms. *Physical Review Letters* 74: 3352–3355.

Pines, D. 1961. *The Many Body Problem - A Lecture Note*. New York: Benjamin.

Pines, D., and P. Nozieres. 1966. *The Theory of Quantum Liquids*. New York: Benjamin.

Planck, M. 1916. Die physikalische Struktur des Phasenraumes. *Annalen der Physik* 355(12): 385–418.

Pöschel, T., and N.V. Brilliantov, eds. 2003. *Granular Gas Dynamics*. Berlin: Springer.

Pöschel, T., and S. Luding, eds. 2001. *Granular Gases*. Berlin: Springer.

Rompe, R., H.J. Treder, and W. Ebeling. 1987. *Zur Großen Berliner Physik*. Leipzig: Teubner.

Rutherford, E. 1911. The scattering of α and β particles by matter and the structure of the atom. *Philosophical Magazine* 21: 669–688.

Sackur, O. 1911. Zur kinetischen Begründung des Nernstschen Wärmetheorems. *Annalen der Physik* 339.3: 455–468.

Sackur, O. 1913. Die universelle Bedeutung des sog. elementaren Wirkungsquantums. *Annalen der Physik* 40: 67–86.

Saha, M.N. 1920. Ionization in the Solar chromosphere. *Philosophical Magazine Sr. VI* 40: 472–478.

Saha, M.N. 1921. Versuch Einer Theorie der Physikalischen Erscheinungen bei Hohen Temperaturen Mit Anwendungen auf die Astrophysik. *Zeitschrift für Physik* 6: 40–47.

Simonyi, K. 1990. *Kulturgeschichte der Physik*. Berlin, Frankfurt: Urania, Harri Deutsch.

Sizyuk, V., A. Hassanein, and T. Sizyuk. 2006. Three-Dimensional Simulation of Laser-Produced Plasma for Extreme Ultraviolet Lithography. *Journal of Applied Physics* 100: 103106.

Svedberg, T. 1906a. Über die Eigenbewegung der Teilchen in Kolloidalen Lösungen. *Z. f. Elektroch.* 12: 853–860.

Svedberg, T. 1906b. Ueber die elektrische Darstellung colloïdaler Lösungen. *Berichte der deutschen chemischen Gesellschaft* 39 (2): 1705–1714.

Tetrode, H. 1912a. Berichtigung zu Meiner Arbeit: Die Chemische Konstante der Gase und das Elementare Wirkungsquantum. *Annalen der Physik* 39: 255–256.

Tetrode, H. 1912b. Die Chemische Konstante der Gase und das Elementare Wirkungsquantum. *Annalen der Physik* 38: 434–442.

Chapter 2
Elements of Quantum Statistical Theory

2.1 Many-Body Quantum Theory

2.1.1 Quantum States

Quantum statistics is a many body theory describing macroscopic matter. Let us first summarize concepts of classical many body theory and subsequently concepts of many body quantum theory, just what we need in the following. After this we will proceed to the simplest quantum statistical ensembles.

In the classical mechanics of many-body systems, we consider systems of point masses with more than two particles described by sets of coordinates and momenta $(q_1 \ldots q_f \, p_1 \ldots p_f)$, including interactions of particle pairs. Examples are:

- one-dimensional chains of mass points and springs ($d = 1$; $f = N$),
- two-dimensional electron layers on the interface of semiconductors ($d = 2$; $f = 2N$),
- three-dimensional gases or condensed matter ($d = 3$; $f = 3N$).

Following the concepts of classical mechanics, the state of a system is described in a $2f$-dimensional space, called *phase space* or Γ-space. The actual state of the system is a point in this phase space, and the evolution of a classical system is described by a trajectory in the phase space,

$$\left[q_1(t) \ldots q_f(t) \, p_1(t) \ldots p_f(t) \right] \tag{2.1}$$

which is determined by Hamilton's equations.

In the quantum mechanical description of a many-body system, the state is also given by a point in a space. However in difference to the classical case, the state is now given as a vector in an ∞-dimensional space, the *Hilbert space*. Following the notation proposed by Dirac (1932), we introduce a Ket-Vector, $|\psi\rangle$. In most cases we will use a special version of the Hilbert space, namely the space of quadratic

© Springer Nature Switzerland AG 2019
W. Ebeling, T. Pöschel, *Lectures on Quantum Statistics*,
Lecture Notes in Physics 953, https://doi.org/10.1007/978-3-030-05734-3_2

integrable functions, $\mathcal{L}^2$. In the space $\mathcal{L}^2$ exist two representations of special interest, the coordinate space representation, $\Psi(q_1 \ldots q_f, t)$, and the momentum space representation, $\Psi(p_1 \ldots p_f, t)$. For the description of the quantum state one of them is sufficient. This is an essential difference to the classical representations which include always coordinates and momenta. In quantum mechanics we have an alternative and we will most often follow Schrödinger by using the coordinate representation.

Let us introduce now the concept of a *basis* by means of an expansion with respect to an arbitrary orthonormal system, $\psi_n(q_1 \ldots q_f)$, which spans the Hilbert space, or in the corresponding vector space of coefficients, $a_m(t)$:

$$\Psi(q_1 \ldots q_f, t) = \sum_{n=1}^{\infty} a_n(t)\psi_n(q_1 \ldots q_f) \tag{2.2}$$

with the set of *orthonormal basis functions* $\{\psi_n(q)\}$ obeying

$$\int dq \, \psi_m^*(q)\psi_n(q) = \delta_{mn}. \tag{2.3}$$

The time evolution of a quantum system is described by a trajectory in the Hilbert space,

$$\int dq \, \psi_m^* \Psi = \sum_{n=1}^{\infty} a_n(t) \int \psi_m^* \psi_n \, dq = \sum_{n=1}^{\infty} a_n \delta_{nm} = a_m(t). \tag{2.4}$$

The trajectory is determined by the Schrödinger equation. Alternatively, we can represent the system in the momentum space,

$$\Psi(p_1 \ldots p_f, t) = \sum_{n=1}^{\infty} b_n(t)\psi_n(p_1 \ldots p_f). \tag{2.5}$$

In classical mechanics, the physical quantities are functions of coordinates and momenta $f = f(q, p)$. An example is the classical *Hamilton function*, which for an N-particle system is given by

$$H(q, \hat{p}) = \sum_{i=1}^{N} \frac{\mathbf{p}_i^2}{2m_i} + \frac{1}{2} \sum_{i,j=1}^{N} \Phi(\mathbf{r}_i, \mathbf{r}_j). \tag{2.6}$$

Here, Φ is the interaction potential, e.g., a Lennard-Jones-potential or in the case of charged particles a Coulomb potential. The trajectory of the classical dynamics is then given by the Hamilton equations,

$$\dot{p}_k = -\frac{\partial H}{\partial q_k}, \qquad \dot{q}_k = \frac{\partial H}{\partial p_k}. \tag{2.7}$$

In quantum mechanics, the physical quantities correspond to Hermitian operators dexHermitian operator in a Hilbert space. The operators corresponding to coordinates and momenta are

$$q_k \rightarrow \hat{q}_k = q_k \hat{1}; \qquad p_k \rightarrow \hat{p}_k = \frac{\hbar}{i} \frac{\partial}{\partial q_k}. \tag{2.8}$$

The rule for the construction of operators of complex mechanical quantities is as follows: we write first the quantity as a function of coordinates and momenta and replace then in this function the momenta by the momentum operators,

$$f(q_1 \ldots q_f \, p_1 \ldots p_f) \rightarrow \hat{f} = f(q_1 \ldots q_f \, \hat{p}_1 \ldots \hat{p}_f). \tag{2.9}$$

Since this is not a unique prescription we have to add the request of Hermiticity. For example the *Hamilton operator* is defined by

$$H(q, p) \rightarrow \hat{H} = H(q, \hat{p}) = -\frac{\hbar^2}{2} \sum_{i=1}^{N} \frac{1}{m_i} \Delta_k + \frac{1}{2} \sum_{i,j=1}^{N} \Phi(\mathbf{r}_i, \mathbf{r}_j). \tag{2.10}$$

As physicists learned in the first decades of the twentieth century from experiments, in particular the Stern-Gerlach-experiment, the representation of quantum systems just by mechanical coordinates and corresponding operators is incomplete. For the complete description, we need an additional quantity, the *Spin*. The *spin hypothesis* was formulated by Pauli, Kronig, Uhlenbeck and Goudsmith. Evidently Wolfgang Pauli was the first to propose the concept of spin, but he did not name it. Then in 1925, Ralph Kronig, George Uhlenbeck and Samuel Goudsmit at Leiden University understood that the spin is a kind of angular momentum. They suggested the picture that the particles are spinning around their own axis. The mathematical theory for introducing the spin into quantum mechanics was worked out by Pauli in 1927. We can conditionally understand the spin as one of two types of angular momentum in quantum mechanics, the other being the orbital angular momentum. The orbital angular momentum operator is the quantum mechanical counterpart of angular momentum which arises when a particle executes a rotating trajectory. The spin is less easy to interpret, but the existence of spin angular momentum is evidenced by many experiments, such as the Stern-Gerlach experiment. In these experiments, particles are observed to possess an angular momentum that cannot be accounted for by orbital angular momentum alone. In some sense, spin is a kind of a vector quantity; it has a definite magnitude, and it has a direction. We have to understand the spin as a property of elementary particles. All elementary particles of a given kind have the same magnitude of spin angular momentum, the spin quantum which is $\hbar/2$ for an electron. Summarizing our findings for electrons, we state:

(1) Electrons have a magnetic moment, $M = \pm M_B$, with

$$M_B = \frac{e\hbar}{2m_e}. \tag{2.11}$$

(2) Electrons have also an intrinsic angular momentum which is the spin, **S**. The z-component of this angular momentum, S_z, can assume only two values.

For finding the actual value of the spin for a given particle one can perform the Einstein-de Haas experiment. In this experiment one can determine directly the spin:

$$\pm \frac{e\hbar}{2m_e c} = \frac{e}{m_e c} S_z .$$ (2.12)

For electrons, this leads to

$$S_z = \pm \frac{1}{2}\hbar .$$ (2.13)

According to Pauli, the spin state is an additional variable for the description of a quantum state. We have to define, therefore, the physical state by a wave function, $\Psi(x, y, z, \sigma, t)$, where we denoted $\sigma = S_z/\hbar$. For electrons, the spin variable has two possible values, $\sigma = \pm\frac{1}{2}$. In general, there exist three components of the spin, S_x, S_y, and S_z. According to the concepts of quantum mechanics, a physical variable, that is, an observable is associated with operators and exchange relations. This leads to the assumption of spin operators which will be discussed now. For the angular momentum, $\hat{L} = (\hat{L}_x, \hat{L}_y, \hat{L}_z)$, we have the relations

$$\hat{L}_x\hat{L}_y - \hat{L}_y\hat{L}_x = i\hbar\hat{L}_z$$

$$\hat{L}_y\hat{L}_z - \hat{L}_z\hat{L}_y = i\hbar\hat{L}_x$$ (2.14)

$$\hat{L}_z\hat{L}_x - \hat{L}_x\hat{L}_z = i\hbar\hat{L}_y$$

and in spherical coordinates

$$\hat{L}^2 = \hat{L}_x^2 + L_y^2 + L_z^2 = \frac{\hbar^2}{\sin\theta^2}\left[\sin\theta\frac{\partial}{\partial\theta}\left(\sin\theta\frac{\partial}{\partial\theta}\right) + \frac{\partial^2}{\partial\varphi^2}\right] .$$ (2.15)

For the z-component we have, in particular,

$$\hat{L}_z = \frac{\hbar}{i}\frac{\partial}{\partial\varphi} .$$ (2.16)

By solving the corresponding eigenvalue problem we find the eigenvalues

$$L^2 = \hbar^2 l(l+1) ; \qquad l = 0, 1, \ldots$$ (2.17)

$$L_z = m\hbar ; \qquad m = 0, \pm 1, \pm 2, \ldots, \pm l .$$ (2.18)

We transfer these properties to the spin, understanding that a simple transfer of the mechanical aspects makes no sense, since the spin has no direct analog in mechanics. Summarizing the spin hypothesis we note: there exists a physical property of particles with an operator $\hat{S} = (\hat{S}_x, \hat{S}_y, \hat{S}_z)$ satisfying

$$\hat{S}_x \hat{S}_y - \hat{S}_y \hat{S}_x = i\hbar \hat{S}_z$$
$$\hat{S}_y \hat{S}_z - \hat{S}_z \hat{S}_y = i\hbar \hat{S}_x \qquad (2.19)$$
$$\hat{S}_z \hat{S}_x - \hat{S}_x \hat{S}_z = i\hbar \hat{S}_y$$

and

$$\hat{S}^2 = \hat{S}_x^2 + \hat{S}_y^2 + \hat{S}_z^2 , \qquad (2.20)$$

with the eigenvalues

$$S^2 = \hbar^2 s(s+1); \qquad s = 0, \frac{1}{2}, 1, \frac{3}{2}, \ldots \qquad (2.21)$$

$$\Delta S_z = \pm \hbar , \quad \text{that is,} \quad \frac{S_z}{\hbar} = -s, \ -s+1, \ldots, \ s-1, \ s . \qquad (2.22)$$

Problem 1 Derive from the exchange relations for the operators the relations for the eigenvalues using:

$$\hat{S}_\pm = \hat{S}_x + i\hat{S}_y \qquad (2.23)$$
$$S^2 = S_- S_+ + S_z^2 + \hbar S_z \qquad (2.24)$$
$$\hat{S}_z(\hat{S}_\pm \phi_m) = (m \pm 1)\hbar (\hat{S}_\pm \phi_m) , \quad \text{if } \hat{S}_z \phi_m = m\hbar \phi_m . \qquad (2.25)$$

Problem 2 Find (2×2)-matrices which satisfy the given exchange relations for spin operators. Discuss whether this representation is unique.

Let us formulate now the *Pauli postulate*: the spin S is a fixed number for a given species of elementary particles. Examples are:

$S = 0$	for π-mesons, μ-mesons, photons
$S = \frac{1}{2};\quad S_z = \pm\frac{\hbar}{2}$	for electrons, protons, neutrons
$S = 1;\quad S_z = -\hbar, 0, \hbar$	for certain simple nuclei, and conditionally photons
$S = \frac{3}{2};\quad S_z = -\frac{3}{2}\hbar, -\frac{1}{2}\hbar, \frac{1}{2}\hbar, \frac{3}{2}\hbar$	for composite nuclei.

(Note that photons have rest mass zero and are, therefore, relativistic particles which cannot be correctly classified by the schema of quantum mechanics given here. This is possible only in the framework of relativistic quantum mechanics).

Our task is now to seek for explicit representations of spin operators (see also the exercises above). For $S = 1/2$, we have the eigenvalues $S_z = \pm\frac{\hbar}{2}$. This suggests to represent the associated operators by (2×2)-matrices which have two eigenvalues. We write $S_x = \frac{\hbar}{2}\sigma_x$, $S_y = \frac{\hbar}{2}\sigma_y$, $S_z = \frac{\hbar}{2}\sigma_z$ and try the Ansatz

$$\hat{\sigma}_z = \begin{pmatrix} a_{11} & a_{12} \\ a_{21} & a_{22} \end{pmatrix}. \tag{2.26}$$

With the conditions $\hat{\sigma}_z \chi_\pm = \chi_\pm$, and $\chi_+ = \binom{1}{0}$ and $\chi_- = \binom{0}{1}$, we find

$$\hat{\sigma}_z = \begin{pmatrix} 1 & 0 \\ 0 & -1 \end{pmatrix}. \tag{2.27}$$

2.1.2 Identity and Symmetry

In nature there exist about 10^3 species of elementary particles, such as electrons, e, protons, p, neutrons, n, positrons, e^+, and others. Typical macroscopic systems consist of many particles of a few species, for instance, a hydrogen gas consists of many electrons and the same number of protons. Looking at the electrons in a quantum system, e.g., in a metal they appear as identical objects which cannot be distinguished. Sometimes, our provider of electricity for households offers us expensive *green current* or cheap *nuclear current*. From the point of physics, this is problematic since all electrons are identical, we cannot give them labels.

The term *identity* is of fundamental importance for physics. A basic concept of classical mechanics is distinguishability, that is, each particle can be attached a label or a number. This is true even for particles of the same type. The first physicist who challenged this assumption was the great American pioneer of statistical physics Josiah Willard Gibbs. His doubts about the distinguishability of particles led him to the *Gibb's paradox*. His deep insight was that identical particles (which cannot be labeled) and non-identical particles (which can be labeled or numbered) lead to different physical properties of the corresponding many-particle systems. In particular as shown by Gibbs, the entropy of a system of identical particles is different from the entropy of a system of non-identical particles (*Gibb's paradox*). Identity is not a technical aspect but it has deep physical consequences. As a matter of fact, elementary particles such as electrons, protons, neutrons, etc. are identical. In the classical picture, all equal particles can be assigned a number, this is not possible for particles of the quantum world. If quantum particles are identical, they cannot be labelled, thus identity and non-identity are well defined. The deepest miracle of the quantum world, namely that systems of identical particles behave very different from systems of non-identical particles, was first understood by Einstein (1924, 1925a,b) in his papers on quantum statistics of atomic gases. This way, Einstein introduced once more a revolutionary concept, this time against the protest of his closest friends and colleagues such as Ehrenfest and Planck (2014). Note that Nernst and Schrödinger were among the first colleagues supporting the new idea by

Einstein. We underline that it was definitely Einstein but not Bose who understood the connection of the new statistics to the identity and indistinguishability of equal particles for the first time. This new view was a real revolution in physics as understood first by Planck and Ehrenfest who expressed serious protests. However, in 1924/25 Einstein's voice was still that of an early bird and only after several years and after the work of other pioneers like Pauli, Dirac, Pauli, Fermi, Wigner, Sommerfeld, Nordheim, Landau, and von Neumann it was generally acknowledged that identity and the corresponding indistinguishability of particles is one of the most fundamental new properties of quantum matter.

Let us repeat the new concept in different words: two particles are either identical or non-identical. Elementary particles like electrons are identical, protons are identical, neutrons are identical. An electron and a proton are different; an electron is different from a muon. The Gedanken experiment that we could change the properties of an elementary particle gradually and convert it gradually to another kind, e.g., by changing the mass to convert an electron to a myon is impossible. Processes which violate the identity principle do not occur and this new principle of impossibility is as serious as the second law.

The *principle of identity* is not only one of the most fundamental principles in physics but has also most serious consequences for the properties of matter. In Chaps. 4–6 we will show that the structure of our world like atoms, molecules and solid states is in fact deeply influenced by the principle of identity—even the thermodynamic stability of our world is guaranteed only by this principle as shown by Lenard, Dyson, Lieb, Lebowitz and others.

Let us now study systems of N identical particles, such as electrons, in detail. The particle with number k is described by coordinates and spin $q_k = x_k, y_k, z_k, \sigma_k$ and the wave function by $\Psi = \Psi(q_1 \ldots q_N, t)$. In the following, frequently we will use the short-hand notation $\Psi = \Psi(1, \ldots, N, t)$. Note that the fact that we numerate the particles by the index k does not mean that they are distinguishable. We will account for this fact in the calculations. The Hamiltonian operator of an N-particle system with potential energy U is

$$\hat{H} = \hat{H}(q, \hat{p}) = \sum_{k=1}^{N} \left(-\frac{\hbar^2}{2m_k} \nabla_k^2 + U(q_k) \right) + \sum_{1 \leq k < l \leq N} U(q_k, q_l), \qquad (2.28)$$

with $q_i = \{\mathbf{r}_i, \sigma_i\}$. The operation of permutation (permutation operator) is defined by

$$\hat{P}_{ij} f(q_1 \ldots q_i \ldots q_j \ldots q_N) = f(q_1 \ldots q_j \ldots q_i \ldots q_N). \qquad (2.29)$$

Application to the Hamiltonian yields

$$\hat{P}_{ij} \hat{H}(q_1 \ldots q_i \ldots q_j \ldots q_N) = \hat{H}(q_1 \ldots q_i \ldots q_j \ldots q_N). \qquad (2.30)$$

The Hamilton operator, $\hat{H}$, is, thus, invariant with respect to the permutation operator. We call $\hat{H}$ also *symmetric*. Applying the permutation operator to the wave function gives

$$\hat{P}_{ij}\Psi\left(q_1 \ldots q_i \ldots q_j \ldots q_N, t\right) = \Psi\left(q_1 \ldots q_j \ldots q_i \ldots q_N, t\right). \tag{2.31}$$

This means that the quantum state created by a permutation is the same as the original state, according to the principle of identity. If, however, the new state is the same as the original state the wave function can differ only by a constant factor:

$$\hat{P}_{ij}\Psi\left(q_1 \ldots q_i \ldots q_j \ldots q_N, t\right) = \Psi\left(q_1 \ldots q_j \ldots q_i \ldots q_N, t\right)$$
$$= \lambda\Psi\left(q_1 \ldots q_i \ldots q_j \ldots q_N, t\right). \tag{2.32}$$

Applying a permutation $\hat{P}_{ij}$ a second time leads then to

$$\hat{P}_{ij}^2\Psi = \lambda\hat{P}_{ij}\Psi = \lambda^2\Psi, \tag{2.33}$$

It follows $\lambda^2 = 1$ and $\lambda = \pm 1$. By definition, the wave function is symmetrical if $\lambda = 1$ and otherwise antisymmetrical.

Symmetry Postulate (Pauli's Postulate)

The symmetry properties of elementary particles are given and invariant. Particles with integer values of spin, $S = 0, 1, 2, \ldots$, have symmetrical wave functions and are called *bosons*. Particles with half integer spin, $S = \frac{1}{2}, \frac{3}{2}, \frac{5}{2}, \ldots$, have antisymmetrical wave functions and are called *fermions*.

Symmetry Principle (Pauli's-Principle)

The symmetry character of elementary particles reflects a law of nature which is based on the identity of equal elementary particles. Particles in an ensemble have wave functions which are symmetrical or anti-symmetrical with respect to the exchange of two identical particles.

This principle has serious consequences as seen only on a second glance. Let us look at one important consequence—the Pauli exclusion principle: we expand the wave functions of the N-particle problem with respect to 1-particle wave functions using a set of eigenfunctions $\int \psi_n^*(q)\psi_m(q)dq = \delta_{ij}$. For $N = 2$ we obtain

$$\Psi(q_1, q_2, t) = \sum_{n_1}\sum_{n_2} c(n_1, n_2, t)\psi_{n_1}(q_1)\psi_{n_2}(q_2). \tag{2.34}$$

Let us consider two fermions and study the probability to find the first fermion in state n_1 and the second in state n_2:

$$w(n_1, n_2) = |c(n_1, n_2, t)|^2 . \tag{2.35}$$

The symmetry principle requests

$$\Psi(q_2, q_1, t) = -\Psi(q_1, q_2, t). \tag{2.36}$$

This request leads to the equations

$$\sum_{n_1, n_2} c(n_1, n_2, t)\psi_{n_1}(q_1)\psi_{n_2}(q_2) = -\sum_{n_1, n_2} c(n_2, n_1, t)\psi_{n_1}(q_2)\psi_{n_2}(q_1), \tag{2.37}$$

and we conclude

$$c(n_1, n_2, t) = -c(n_2, n_1, t); \quad c(n, n, t) = -c(n, n, t). \tag{2.38}$$

It follows $w(n, n, t) = 0$, that is, the probability to find in a measurement the two particles in the same state is zero. We formulate this result as the *Pauli exclusion principle*: *The probability that two fermions are in the same one-particle quantum state is zero.*

The Pauli exclusion principle is a special formulation of the more general Pauli principle. There is no classical interpretation of this deep quantum property. Fermions behave in a way as if they knew about the quantum states of all other identical fermions in the system and avoid to occupy a state which is already occupied by another identical particle. Fermions can occupy only states which are anti-symmetrical with respect to the total wave function.

2.2   Quantum Dynamics of Many Particles

2.2.1   Schrödinger Equation

Let us consider the quantum dynamics of a system of N particles (Eyring et al. 1944; Slater 1939). In order to describe the system, we need $3N$ classical variables, which characterize the coordinates and N spins,

$$q = \{x_1 \, y_1 \, z_1 \, \sigma_1 \ldots x_N \, y_N \, z_N \, \sigma_N\} , \tag{2.39}$$

with $\sigma_k = S_{k_z}/\hbar$. The corresponding wave function reads

$$\Psi = \Psi(q, t) = \Psi(\mathbf{r}_1 \sigma_1 \ldots \mathbf{r}_N \sigma_N, t) , \tag{2.40}$$

with $\mathbf{r}_i = \{x_i, y_i, z_i\}$. In the simplest case, the coordinate dependence and the spin dependence are decoupled, that is,

$$\Psi(q, t) = \chi(\sigma_1 \ldots \sigma_N) \, \phi(\mathbf{r}_1 \ldots \mathbf{r}_N, t) . \qquad (2.41)$$

The corresponding classical Hamiltonian reads

$$H = \sum_{k=1}^{N} \left(\frac{p_k^2}{2m_k} + U_k(\mathbf{r}_k) \right) + \sum_{1 \leq k < l \leq N}^{N} U(\mathbf{r}_k, \mathbf{r}_l) , \qquad (2.42)$$

where $U(\mathbf{r}_k, \mathbf{r}_l)$ denotes the potential of the interaction between the particles k and l. In the Schrödinger picture, we postulate that the wave function is a solution of the partial differential equation

$$i\hbar \frac{\partial}{\partial t} \Psi(q, t) = \hat{H} \, \Psi(q, t) \qquad (2.43)$$

with natural boundary conditions, that is, the amplitude must disappear at infinity. In Eq. (2.43), $\hat{H}$ denotes the Hamilton operator of the N-particle system

$$\hat{H} = \hat{H}(\mathbf{p}_k, \mathbf{r}_k) = \sum_{k=1}^{N} \left(-\frac{\hbar^2}{2m} \nabla_k^2 + U_k(\mathbf{r}_k) \right) + \sum_{1 \leq k < l \leq N}^{N} U(\mathbf{r}_k, \mathbf{r}_l) . \qquad (2.44)$$

In the stationary case, we obtain the energy eigenfunction problem

$$\Psi(q, t) = e^{-\frac{i}{\hbar} E t} \, \psi(q) ; \quad \hat{H} \psi(q) = E \psi(q) \qquad (2.45)$$

with appropriate boundary conditions.

For an N-particle system this is so far still a postulate, supported by the following arguments:

(1) For $N = 1$ follows the standard Schrödinger equation.
(2) With the expansion with respect to the Planck parameter,

$$\Psi(q, t) = \chi(\sigma_1 \ldots \sigma_N) e^{-\frac{i}{\hbar} S(\mathbf{r}_1 \ldots \mathbf{r}_N, t)} \qquad (2.46)$$

$$S(\mathbf{r}_1 \ldots \mathbf{r}_N, t) = S^{(0)}(\mathbf{r}_1 \ldots \mathbf{r}_N, t)$$
$$+ i\hbar \, S^{(1)}(\mathbf{r}_1 \ldots \mathbf{r}_N, t) + (i\hbar)^2 \, S^{(2)}(\mathbf{r}_1 \ldots \mathbf{r}_N, t) + \ldots \qquad (2.47)$$

follows the classical Hamilton-Jacobi equation in the limit $\hbar \to 0$:

$$\frac{\partial}{\partial t} S^{(0)} (\mathbf{r}_1 \ldots \mathbf{r}_N, t) + H \left(\mathbf{r}_k, \frac{\partial S^{(0)}}{\partial \mathbf{r}_k} \right) = 0; \qquad k = 1 \ldots N . \qquad (2.48)$$

(3) Calculations based on the given N-particle Schrödinger equation agree with experiments as shown for atoms, molecules, solid materials.

(4) For small dispersion follows the N-particle Ehrenfest law:

$$\langle \mathbf{r}_i \rangle = \int |\Psi (\mathbf{r}_1 \ldots \mathbf{r}_N)|^2 \, \mathbf{r}_i \, d\mathbf{r}_1 \ldots d\mathbf{r}_N \qquad (2.49)$$

$$\left\langle \frac{\partial}{\partial \mathbf{r}_i} U (\mathbf{r}_1 \ldots \mathbf{r}_N) \right\rangle = \int |\Psi (\mathbf{r}_1 \ldots \mathbf{r}_N)|^2 \, \frac{\partial}{\partial \mathbf{r}_i} U (\mathbf{r}_1 \ldots \mathbf{r}_N) \, d\mathbf{r}_1 \ldots d\mathbf{r}_N$$

$$(2.50)$$

$$m_i \frac{d^2}{dt^2} \langle \mathbf{r}_i \rangle = - \left\langle \frac{\partial}{\partial \mathbf{r}_i} U (\mathbf{r}_1 \ldots \mathbf{r}_N) \right\rangle . \qquad (2.51)$$

We approximate the average on the right-hand side by assuming that we can go directly to the average of the arguments. This is certainly true only for linear functions. Since the potential is in general nonlinear, the replacement of the arguments of the potential by $\langle \mathbf{r}_i \rangle = r_i$ is a rough approximation which leads to

$$\left\langle \frac{\partial}{\partial \mathbf{r}_i} U (\mathbf{r}_1 \ldots \mathbf{r}_N) \right\rangle \approx \frac{\partial U (\mathbf{r}_1 \ldots \mathbf{r}_N)}{\partial \mathbf{r}_i} . \qquad (2.52)$$

This way follows also Newton's law as a classical limit for $\hbar \to 0$:

$$m_i \frac{d^2}{dt^2} \mathbf{r}_i \approx - \frac{\partial U (\mathbf{r}_1 \ldots \mathbf{r}_N)}{\partial \mathbf{r}_i} . \qquad (2.53)$$

Problem 3 Derive in the limit $\hbar \to 0$ the Hamilton-Jacobi equations for a one-dimensional system with the wave function $\Psi (x, t)$.

In order to check the new N-particle equations, let us first discuss the case $N = 2$ with a central potential $U = U(r)$ and $r = |\mathbf{r}_1 - \mathbf{r}_2|$. The wave function is $\Psi (q, t) = \Psi (\mathbf{r}_1, \sigma_1, \mathbf{r}_2, \sigma_2, t)$, and the dynamics is given by

$$i\hbar \frac{\partial \Psi}{\partial t} = \left(-\frac{\hbar^2}{2m_1} \Delta_1 - \frac{\hbar^2}{2m_2} \Delta_2 \right) \Psi + U(r)\Psi . \qquad (2.54)$$

With the transformation

$$M = m_1 + m_2; \qquad \mu = \frac{m_1 m_2}{m_1 + m_2} \tag{2.55}$$

$$x = x_1 - x_2; \qquad y = y_1 - y_2; \qquad z = z_1 - z_2 \tag{2.56}$$

$$X = \frac{\mu}{m_2} x_1 + \frac{\mu}{m_1} x_2; \quad Y = \frac{\mu}{m_2} y_1 + \frac{\mu}{m_1} y_2; \quad Z = \frac{\mu}{m_2} z_1 + \frac{\mu}{m_1} z_2 \tag{2.57}$$

we find

$$i\hbar \frac{\partial \Psi}{\partial t} = -\frac{\hbar^2}{2M} \Delta_{\mathbf{r}} \Psi - \frac{\hbar^2}{2\mu} \Delta_{\mathbf{R}} \Psi + U(r) \Psi . \tag{2.58}$$

By decoupling the motion of the center of mass and the spins follows

$$\Psi(q,t) = \chi(\sigma_1, \sigma_2) e^{-\frac{i}{\hbar} \mathbf{P} \mathbf{R}} e^{-\frac{i}{\hbar} Et} \psi(x, y, z) , \tag{2.59}$$

where $\mathbf{R}$ is the vector of the center of mass and $\mathbf{P}$ its momentum. The Schrödinger equation is decomposed into

$$E = \frac{P^2}{2M} + \varepsilon ; \qquad \frac{\hbar^2}{2\mu} \Delta_{\mathbf{r}} \psi(\mathbf{r}) + (\varepsilon - U(r)) \psi(\mathbf{r}) = 0 , \tag{2.60}$$

where ε is the energy of relative motion $\varepsilon = E - P^2/2M$. Thus, the two-particles problem can be reduced to the problem of relative motion of a particle with reduced mass μ in the field $U(r)$. For $N > 2$ many different approximations can be found in the literature. Most important is the *Born-Oppenheimer approximation* which is based on the idea that electrons and nuclei move on different time scales due to the large difference of their masses. This leads to the idea to decouple the motion of the heavy nuclei from the motion of the light electrons. Following this model, the dynamics of the system is decoupled into two steps:

(1) The nuclei are first frozen at given positions $\mathbf{R}_1 \ldots \mathbf{R}_M$ and the Schrödinger equation is solved in the relative coordinates $\mathbf{r}_i$ in the field of the fixed nuclei

$$\hat{H} = -\sum_{i=1}^{N} \frac{\hbar^2}{2m_e} \nabla_i^2 - \sum_{j=1}^{N} \frac{\hbar^2}{2M_j} \nabla_j^2 + V(\{\mathbf{r}, \mathbf{R}\}) . \tag{2.61}$$

Here the Laplacian is the derivative with respect to the electron coordinates, that is, $\nabla_i^2 = \partial^2/\partial \mathbf{r}_i^2$. In the limit $M_j \to \infty$, we neglect the operator character of the kinetic term for the center of mass motion and obtain the Schrödinger equation

$$\left[\hat{T}_e + V\left(\{\mathbf{r}, \mathbf{R}\}\right) \right] \psi_n\left(\{\mathbf{r}, \mathbf{R}\}\right) = E_n\left(\mathbf{R}_1 \ldots \mathbf{R}_N\right) \psi_n\left(\{\mathbf{r}, \mathbf{R}\}\right) . \tag{2.62}$$

(2) Next, the motion of the heavy nuclei is treated classically or semi-classically with effective interactions

$$\left[\hat{T}_n + V_n\left(\mathbf{R}\right) + V_n'\left(\mathbf{R}\right) \right] \psi = E\psi . \tag{2.63}$$

Here $V_N'\left(\mathbf{R}\right) \psi$ is the electronic ground state energy as derived in the first step.

Dynamics of the Symmetry Properties (Blochinzew 1953)

We assume that the wave function, Ψ, is a solution of the Schrödinger-equation

$$i\hbar \frac{\partial}{\partial t} \Psi\left(q_1 \ldots q_i \ldots q_j \ldots q_N, t\right) = \hat{H} \Psi\left(q_1 \ldots q_i \ldots q_j \ldots q_N, t\right) \tag{2.64}$$

with the Hamilton operator

$$\hat{H}(1 \ldots N) = \sum_{k=1}^{N} \left(-\frac{\hbar^2}{2m_k} \nabla_k^2 + U\left(q_k\right) \right) + \sum_{1 \le k < l \le N}^{N} U\left(q_k, q_l\right) . \tag{2.65}$$

Since the Hamilton operator is symmetrical, we have

$$\hat{P}_{ij} \hat{H}(1 \ldots i \ldots j \ldots N) = \hat{H}(1 \ldots j \ldots i \ldots N) \hat{P}_{ij} \tag{2.66}$$

and

$$i\hbar \frac{\partial}{\partial t} \hat{P}_{ij} \Psi(1 \ldots i \ldots j \ldots N, t)$$

$$= \hat{P}_{ij} \hat{H}(1 \ldots i \ldots j \ldots N) \Psi(1 \ldots i \ldots j \ldots N, t)$$

$$= \hat{H}(1, \ldots i \ldots j \ldots N) \hat{P}_{ij} \Psi(1 \ldots i \ldots j \ldots N, t) . \tag{2.67}$$

Consequently, the function

$$\Psi'(1 \ldots i \ldots j \ldots N, t) = \Psi(1 \ldots j \ldots i \ldots N, t) \tag{2.68}$$

should also be a solution. Consider now the dynamics of the wave function, $d\Psi = \frac{1}{i\hbar}\hat{H}\Psi dt$. For symmetrical wave functions, Ψ_s, we find by application of the exchange operator

$$dΨ_s = \frac{1}{i\hbar}\hat{H}\Psi_s dt \quad \rightarrow \quad \hat{P}_{ij}dΨ_s = \frac{1}{i\hbar}\hat{H}\Psi_s dt = dΨ_s. \tag{2.69}$$

This means that the changes are symmetric. Correspondingly, for the antisymmetric wave function we obtain

$$dΨ_a = \frac{1}{i\hbar}\hat{H}\Psi_a dt \quad \rightarrow \quad \hat{P}_{ij}dΨ_a = -\frac{1}{i\hbar}\hat{H}\Psi_a dt = -dΨ_a, \tag{2.70}$$

that is, the changes are antisymmetric. From this, we conclude that the symmetry of a wave function does not change in the course of time. The symmetry character is an invariant of the quantum dynamics. The symmetry character is an invariant property of elementary particles.

We distinguish between Bose particles (bosons) with integer spins, $S = 0, 1, 2, \ldots$, and Fermi particles (fermions) with non-integer spins, $S = \frac{1}{2}, \frac{3}{2}, \frac{5}{2} \ldots$. To the class of Bose particles belong photons, π-mesons, k-mesons, α-particles and the H atom. Electrons, positrons, protons, neutrons and their anti-particles are fermions. The photon does not fit completely into this schema since photons do not have a rest mass. In some context, e.g., in the context of spin matrices, the value $S = 1$ provides a better description of the observed properties of photons.

2.2.2 Pure and Mixed Ensembles

For the proper description of macroscopic systems, besides many-body quantum mechanics we need tools which relate the microscopic variables and the macroscopic properties, such as entropy, the thermodynamic functions and further macroscopic properties. In order to perform a transition from quantum mechanics of many bodies to quantum statistics let us remind, how the transition from classical mechanics to statistical thermodynamics is performed (see, e.g., Dirac 1932; Fetter and Walecka 1971; Feynman 1972; Fick 1981; Kadanoff and Baym 1962; Kremp et al. 2005; Landau and Lifshitz 1976, 1990; Martin and Schwinger 1959; Reichl 1980; ter Haar 1995; Tolman 1938).

In classical mechanics, a many-particle system is described by the set coordinates and momenta as a function of time, $q_1(t) \ldots q_f(t)$, $p_1(t) \ldots p_f(t)$. This set corresponds to a point in a $2f$-dimensional space which is called the Γ-space or phase space. For a macroscopic system, f is of the order of 10^{23} which is an extremely large number. In classical statistics, the state is described by a probability density for the points in the phase space

$$\rho\left(q_1 \ldots q_f\, p_1 \ldots p_f\right). \tag{2.71}$$

In the classical case, for sharp values of coordinates and momenta the density is a product of delta functions,

$$\rho = \prod_{i=1}^{f} \delta\left(q_i - q_i(t)\right) \delta\left(p_i - p_i(t)\right) . \tag{2.72}$$

The dynamics of the classical density is given by the Liouville equation

$$\frac{\partial \rho}{\partial t} + \{H, \rho\} = 0 ; \qquad \{H, \rho\} = \sum_{i=1}^{f} \left(\frac{\partial H}{\partial p_i} \frac{\partial \rho}{\partial q_i} - \frac{\partial H}{\partial q_i} \frac{\partial \rho}{\partial p_i} \right) . \tag{2.73}$$

Any expression for the density which depends only on H solves the Liouville equation since $\frac{\partial \rho}{\partial t} = 0$, as well as $\{H, \rho\} = 0$ are fulfilled. Following Gibbs we know that for systems in a heat bath at temperature, T, and $\beta = 1/(k_B T)$, the best guess which leads to agreement with the data is the *canonical distribution*

$$\rho \sim e^{-\beta H} . \tag{2.74}$$

Let is discuss the corresponding quantum statistical canonical ensemble whose state is described by the wave function $\Psi\left(q_1 \ldots q_f, t\right)$. A physical system with the Hamilton-Operator, $\hat{H}$, is associated with a Schrödinger equation, which governs the time evolution of the wave function. Similar as in classical mechanics where (q, p) is a vector in phase space with components q_i, p_i, we consider $\Psi\left(q_1 \ldots q_f, t\right)$ as a vector in the Hilbert space. Wave functions can be represented by an orthonormal set of basis functions, ψ_k:

$$\Psi(q, t) = \sum_{k=0}^{\infty} a_k(t) \psi_k(q) . \tag{2.75}$$

The coefficients, a_k, correspond to projections to the axis of a cartesian system. In difference to the classical case, the Hilbert space has an infinite number of dimensions, that is, there are infinitely many basis functions spanning the Hilbert space. Further, the components, a_k, are complex numbers. In the simplest case, we use the energy eigenfunctions which are a special basis,

$$\hat{H} \psi_k = E_k \psi_k ; \qquad k = 1, 2, 3, \ldots \tag{2.76}$$

The mean value of the energy (the expectation value) in quantum mechanics is

$$\langle E \rangle = \int \Psi^* \hat{H} \Psi \, dq . \tag{2.77}$$

When measuring the energy of a system, we find always one of the eigenvalues, E_k, and after the measurement the system is for sure in the corresponding eigenstate $\Psi = \psi_k$. The expression $|a_k|^2 = a_k^* a_k$ is the probability to find the result E_k, that is, the probability that the system is in the state k. For the mean value we obtain

$$\langle E \rangle = \sum_k |a_k|^2 E_k .$$
(2.78)

The mean values of physical properties, $f(E)$, depending only on the energy are given by

$$\langle f(E) \rangle = \sum_k |a_k|^2 f(E_k) .$$
(2.79)

For the case that the system is described by a wave function we speak about a *pure ensemble*. The more general *mixed ensemble* corresponds to systems which are described by a mixture of pure k states. Instead of simple probabilities, $a_k^* a_k$, we have now mean values of $\overline{a_k^* a_k}$ and introduce new probabilities

$$w_k = \overline{a_k^* a_k} .$$
(2.80)

The *fundamental postulate of quantum statistics* due to Landau and Lifshitz (1976, 1990) and von Neumann (1932) states that a quantum statistical system cannot be described just by usual probabilities but by an ensemble of probabilities. Technically, this means that the classical density, $\rho(q, p)$, defined for the phase space is replaced by a density operator in Hilbert space, $\hat{\rho}$. The idea is related to the Gibbs ensemble picture. Instead of considering just one physical system we should consider a set of parallel quantum systems and consider an ensemble of probabilities, $a_k^* a_k$. As in the classical case, an ensemble is a set of virtual copies of the original system which are all in different microscopic states which are compatible with the macroscopic state.

In order to demonstrate this postulate, we first study a system with known energy eigenfunctions. In quantum statistics, averaging consists of two steps, the quantum mechanical and the ensemble average (expressed by the overline):

$$\langle E \rangle = \sum_k \overline{a_k^* a_k} E_k = \sum_k w_k E_k .$$
(2.81)

In analogy to the classical case, the canonical weights are

$$w_j = \frac{\exp\left(-\frac{E_j}{k_B T}\right)}{\sum_j \exp\left(-\frac{E_j}{k_B T}\right)} .$$
(2.82)

This is a simple example of a quantum statistical canonical ensemble in energy representation.

2.3 Standard Approximations for Many-Particle States

2.3.1 Hartree-Fock Approximation

The Hartree method is also based on the product Ansatz

$$\psi(1\ldots N) = \chi(\sigma_1 \ldots \sigma_N) \prod_{i=1}^{N} \phi_i(\mathbf{r}_i) .$$

(2.83)

We employ Ritz' variation to obtain the optimal test functions from the energy functional:

$$E = \int d\mathbf{r}_1 \ldots d\mathbf{r}_N \phi_1^*(\mathbf{r}_1) \ldots \phi_N^*(\mathbf{r}_N)$$

$$\times \left(\sum_{i=1}^{N} \hat{H}_i^{(1)} + \sum_{1 \le i < j \le N} \frac{e^2}{|\mathbf{r}_i - \mathbf{r}_j|} \right) \phi_1(\mathbf{r}_1) \ldots \phi_N(\mathbf{r}_N) .$$

(2.84)

Carrying out the integrations up to one with the index i, we obtain

$$E = \sum_{i=1}^{N} \int d\mathbf{r}_i \phi_i^* \hat{H}_i^{(1)} \phi_i + \sum_{1 \le i < j \le N} \int d\mathbf{r}_i d\mathbf{r}_j \frac{e^2}{|\mathbf{r}_i - \mathbf{r}_j|} \phi_i^* \phi_j^* \phi_i \phi_j .$$

(2.85)

We consider ϕ_k^* and ϕ_k as independent variational functions. By variation of ϕ_k^* at fixed ϕ_k with the extremal principle

$$\delta \left[E - \sum_{i=1}^{N} \varepsilon_i \left(\int d\mathbf{r}_i \phi_i^* \phi_i - 1 \right) \right] = 0$$

(2.86)

and the conditions

$$\int d\mathbf{r}_i \phi_i^*(\mathbf{r}_i) \phi_i(\mathbf{r}_i) = 1 ,$$

(2.87)

we obtain

$$\int d\mathbf{r}_k \delta \phi_k^* \left(\hat{H}^{(k)} \phi_k - \varepsilon_k \phi_k + \sum_{j=1,\ j\neq k}^{N} \int d\mathbf{r}_j \left| \phi_j \right|^2 \frac{e^2}{|\mathbf{r}_k - \mathbf{r}_j|} \right) = 0 \qquad (2.88)$$

$$-\frac{\hbar^2}{2m} \Delta_k \phi_k + \left(-Z \frac{e^2}{|\mathbf{r}_k|} + \sum_{j=1,\ j\neq k}^{N} \int d\mathbf{r}_j \left| \phi_j \right|^2 \frac{e^2}{|\mathbf{r}_k - \mathbf{r}_j|} \right) \phi_k = \varepsilon_k \phi_k. \qquad (2.89)$$

The variables ε_i are here the Lagrange multipliers which are used to satisfy the conditions (2.87). From this equation follows the Hartree equation which assumes the same form as a Schrödinger equation,

$$-\frac{\hbar^2}{2m} \Delta_k \phi_k + U_{\text{eff}} \phi_k = \varepsilon_k \phi_k , \qquad (2.90)$$

with an effective potential where the first term is the field of the nucleus and the second term is the mean field of the other electrons,

$$U_{\text{eff}} = -Z \frac{e^2}{|\mathbf{r}_k|} + \sum_{j=1,\ j\neq k}^{N} \int d\mathbf{r}_j \left| \phi_j \right|^2 \frac{e^2}{|\mathbf{r}_k - \mathbf{r}_j|} . \qquad (2.91)$$

The nonlinear partial differential equation (2.90) is called Hartree equation. It is the quantum mechanical counterpart of the classical *mean-field approximation*. Its solution can be found by iteration. Taking in addition the antisymmetry of the wave functions into account we arrive at the *Hartree-Fock method*. The Hartree-Fock method is self-consistent and therefore sometimes called *method of self-consistent fields*.

2.3.2 Born-Oppenheimer Approximation, Virial Theorem and Coulomb Stability

The Born-Oppenheimer approximation which we discussed already in brief is related to a system which consists of M heavy masses and N_e light electrons. The Born-Oppenheimer approximation is defined by the following procedure: first we solve the quantum mechanical problem for nuclei with infinite masses fixed at the locations $\mathbf{R}_1 \ldots \mathbf{R}_M$

$$H_\infty \psi = E (\mathbf{R}_1 \ldots \mathbf{R}_M) \psi . \qquad (2.92)$$

As a result, we obtain wave functions of the electrons where the positions of the nuclei are parameters. Due to the assumption, $m_k \gg m_e$, the wave-function, ψ, as well as the energy levels of the electrons have the positions of the nuclei as parameters. Subsequently, the electron energies are used as potentials for the dynamics of the nuclei itself. The negative ground state energy function, $E(\mathbf{R}_1 \ldots \mathbf{R}_M)$, decreases with the distance of the nuclei $|\mathbf{R}_k - \mathbf{R}_l|$, that is, in the ground state the electrons prefer the nuclei close by Thirring (1980). The collapse of the nuclei is, however, prevented by their Coulomb repulsion. In this way a stable configuration of the whole system is formed. We shall see later (Chap. 4) that in infinite systems the Fermi character of the electrons is necessary to prevent the collapse of the system, as shown first by Dyson (1967) and Lieb and Thirring (1975) in fundamental papers (Thirring 1980).

Let us consider now the ground state energy in Born-Oppenheimer approximation. From dimensional reasoning we expect

$$E_0(R_1 \ldots R_M) = \frac{m_e e^2}{\hbar^2} C(R_1 \ldots R_M) , \tag{2.93}$$

where the functional C is independent of the masses and the charges. By differentiating

$$E_0(R_1 \ldots R_M) = \langle H_\infty \rangle \tag{2.94}$$

with respect to e^2, we find

$$e^2 \frac{\partial E_0}{\partial e^2} = \left\langle \sum i < j \frac{e^2}{|r_i - r_j|} + \sum_{i,k} \frac{Z_k e^2}{|r_i - R_k|} \right\rangle = \langle U \rangle = \frac{2 m_e e^4 C}{\hbar^2} = 2E_0 . \tag{2.95}$$

This provides the important virial theorem for Coulomb systems,

$$E_0 = \langle T \rangle + \langle U \rangle = \frac{1}{2} \langle U \rangle = -\langle T \rangle . \tag{2.96}$$

The ground state of Coulomb systems is always negative (Thirring 1980). For the special case of the hydrogen atom, these relations are known already from the work by Niels Bohr. Indeed, the Bohr theory delivers for the energy, the kinetic, and for the potential energy in the ground state of the Bohr-atom the well-known relations

$$E_0 = -\frac{e^2}{2a_B} ; \qquad T_0 = -E_0 ; \qquad U_0 = -\frac{e^2}{a_B} . \tag{2.97}$$

The most essential feature of the Bohr theory was the finding that the atomic energy has a lower bound. The electron cannot fall into the Coulomb singularity, that is, atoms are asymptotically stable. The important question, whether this is still true for arbitrary finite systems of N nuclei and $N_e = ZN$, forming possibly N atoms could be answered only half a century later by Dyson and Lenard (1967, 1968) (see also Kraeft et al. 1986; Thirring 1980). Dyson and Lenard (1967, 1968) have shown that there exists indeed a lower bound proportional to N_e:

$$E_0 \leq -CN = -C_e N_e; \qquad 0 < C < \infty; \quad 0 < C_e < \infty. \tag{2.98}$$

The proof by Lenard and Dyson for the lower bound is a mathematical *tour de force* which shall not be presented here. Instead, we will provide some arguments making the theorem plausible. A different and more strict argumentation will be given in Chap. 4 in the context of density-functional theories. As we have shown above, the virial theorem states that for the ground state energy, it holds

$$E_0 = -\langle T \rangle \tag{2.99}$$

or, more explicitly,

$$E_0 = -\left\langle \sum_{k=1}^{N} \frac{\mathbf{p}_k^2}{2M_k} \right\rangle - \left\langle \sum_{i=1}^{N_e} \frac{\mathbf{p}_i^2}{2m_e} \right\rangle = -N \left\langle \frac{\mathbf{p}_1^2}{2M_1} \right\rangle - N_e \left\langle \frac{\mathbf{p}_1^2}{2M_1} \right\rangle. \tag{2.100}$$

We can consider this as a simplified proof of the theorem, provided the kinetic energy contributions of the (first) nucleus and the (first) electron do not increase with the number of the other particles. This is at least plausible, but needs, of course, a more rigorous proof (Thirring 1980).

2.3.3 Thomas-Fermi Theory of Multi-Electron Atoms

The Thomas-Fermi theory was developed independently (1927/1928) in papers by Thomas (1927) and Fermi (1928). Originally this was just an ingeniously simple theory of many-electron atoms. Its basic idea is to handle the electrons in a many-electron system like a dense gas (Fermi gas) whose momentum states are approximately the free states of a gas in a box, see Fig. 2.1. Further it is assumed that the energy states are occupied from below like in a sphere (Fermi sphere). Each cell corresponds to a volume h^3 or Vh^3 in phase space. The maximal momentum, p_M, which is the radius of the sphere is given by the condition

$$2\frac{4\pi}{3h^3}V = N_e; \qquad p_m = \left(\frac{3}{8\pi}\right)^{1/3} h n_e^{1/3}, \tag{2.101}$$

where $n_e = N_e/V$ is the electron density. The total kinetic energy can be calculated by summing the lowest energies forming a sphere in the momentum

Fig. 2.1 Thomas-Fermi theory: the momentum states occupied by electrons (Fermi sphere)

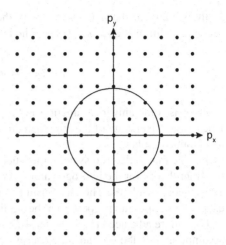

space. According to Fermi (1966), the sum of the maximal kinetic energy and the potential energy, $U(r) = -eV(r)$, plays the rôle of an electrochemical potential of the system,

$$\mu_e = \frac{1}{2m} p_m(r)^2 + U(r) . \tag{2.102}$$

Just as the electrochemical potentials by Helmholtz, Planck and Nernst, this quantity must be constant over the system, otherwise the charges would move. We obtain

$$\mu_e = \frac{h^2}{2^{4/3} m_e} \left(\frac{3}{4\pi} \right)^{2/3} n_e^{2/3} . \tag{2.103}$$

This gives finally the Thomas-Fermi distribution of electrons in the atom:

$$n_e(r) = \frac{2^{9/2} \pi \, m_e^{3/2}}{3h^2} [\mu_e + eV(r)]^{3/2} . \tag{2.104}$$

The value of the chemical potential follows from the normalization

$$\int n_e(r) d\mathbf{r} = N_e . \tag{2.105}$$

The remaining electronic part of the theory is quite similar to the Debye theory (Fermi 1966). Since the electric charge density is $\rho_e(r) = -en_e(r)$, following Debye's scheme, we get finally the Poisson equation

$$\Delta V = \frac{d^2 V}{dr^2} + \frac{2}{r} \frac{dV}{dr} = 4\pi e n_e(r) = \frac{2^{13/2} m_e^{3/2} e^{3/2}}{3h^3} [\mu_e + eV(r)] . \tag{2.106}$$

In the vicinity of the nucleus, $r \to 0$, the potential should approach Ze/r. The second condition is $Z = \int n_e(r)d\mathbf{r}$. The integration gives the solution

$$V(r) = \frac{Ze}{r}\phi\left(\frac{r}{r_{TF}}\right), \qquad (2.107)$$

where ϕ is some universal function and r_{TF} is a characteristic radius (Fermi 1966). Due to the universality of the approach this calculation is valid for any Z, that is, for any atomic number.

For the many-electron system, the Thomas-Fermi theory plays the same rôle as the Bohr theory for the hydrogen atom. As a remarkable fact, the Bohr theory is an exact theory, while the Thomas-Fermi theory has the property to be asymptotically exact. This important property will be studied later (see also Thirring 1980).

On a more advanced level, the term *Thomas-Fermi approximation* stands for the assumption, that the ground state energy of an electron-nucleus system is given by the functional (Thomas-Fermi functional)

$$E_{TF} = \frac{3}{5}\left(2\pi^2\right)^{2/3}\int d\mathbf{r}\, n^{5/3}(\mathbf{r}) - \sum_k \int d\mathbf{r}\,\frac{e_k^2 n(\mathbf{r})}{|\mathbf{r} - \mathbf{R}_k|}$$

$$+ \int d\mathbf{r}\int d\mathbf{r}'\,\frac{e^2}{|\mathbf{r} - \mathbf{r}'|} + U(\mathbf{R}_1 \ldots, \mathbf{R}_M). \qquad (2.108)$$

The last three terms represent respectively, the electron-nuclear, the electron-electron, and nuclear-electron energy. These terms are indeed exact, only the first term is an approximation. The Thirring-Lieb-Simon theorem (Thirring 1980) states that

$$E_{TF}(n(\mathbf{r})) = \inf\left(E\left[n(\mathbf{r})\right]\right); \qquad \int d\mathbf{r}\, n(\mathbf{r}) = N_e. \qquad (2.109)$$

The Thomas-Fermi theory provides us a lower limit and states that this limit is assumed for infinite electron numbers, $N_e \to \infty$. This statement is very important for the physics of Coulomb systems, since it guarantees the stability of matter against collapse. In Chap. 4 we will come back to this problem, referring to the stability of matter. But already here we understand that the existence of a lower bound of the ground state energy proportional to the particle number, which is called *H-stability of the system* is of fundamental importance for the structure of our world (Thirring 1980).

2.4 Quantum Statistical Ensemble Theory

2.4.1 Microcanonical and Canonical Ensembles

Let us study a system with given energy, E. With "given" we mean that the energy is fixed with respect to macroscopic scales. In analogy to the classical statistics we define the quantum statistical microcanonical ensemble as the set of states which are located in a narrow shell around $H = E$ with $\left\langle \hat{H} \right\rangle = E$. That is, a quantum state belongs to the microcanonical ensemble if its energy belongs to $(E - \delta E/2, E + \delta E/2)$, where δE is small and related to the uncertainty of the macroscopic energy, E. We require now

(1) *Postulate about equal probability of micro-states in the energy shell*
  The probability for finding the microstate E_k is

$$\rho_k = \begin{cases} \text{const} & \text{if } E - \frac{\delta E}{2} \leq E_k \leq E + \frac{\delta E}{2} \\ 0 & \text{else}. \end{cases} \tag{2.110}$$

(2) *Postulate about the independence of the micro-states*
(3) *Entropy postulate (Boltzmann-Planck postulate)*

$$S = k_B \ln \Gamma(E), \tag{2.111}$$

  where Γ and k_B are the number of states in the shell and Boltzmann's constant, respectively.

Note that there are several problems related to this entropy postulate which, strictly speaking, hold true only for "typical systems". For this and related problems, we refer to recent literature (Hänggi et al. 2016; Hilbert et al. 2014). Further, we refer to the maximum entropy principle by Gibbs and Jaynes discussed below in Chap. 4.

  In analogy to the classical case, we consider now *canonical ensembles*: Splitting a large microcanonical system into two parts which exchange energy, we can derive for the smaller system which is so to say imbedded into a heat bath the distribution given already above

$$w_j = \frac{e^{-\frac{E_j}{k_B T}}}{\sum_j e^{-\frac{E_j}{k_B T}}}. \tag{2.112}$$

Here we used the thermodynamic definition of the temperature of the imbedding system,

$$\frac{1}{T} = \left(\frac{\partial S}{\partial E} \right)_{N,V}. \tag{2.113}$$

The denominator follows from the normalization, $\sum_j w_j = 1$. We define the partition function

$$Q(T, V) = \sum_k e^{-\frac{E_k}{k_B T}} . \tag{2.114}$$

The summation runs over all microscopically allowed states. Note that we sum over states but *not* over energy levels. By differentiating with respect to T follows

$$\frac{\partial}{\partial T} \ln Q = \frac{\sum_k \frac{E_k}{k_B T^2} e^{-\frac{E_k}{k_B T}}}{Q} = \frac{1}{k_B T^2} \langle E \rangle$$

$$\langle E \rangle = k_B T^2 \frac{\partial}{\partial T} \ln Q = -k_B T \ln Q - T \frac{\partial}{\partial T} (-k_B T \ln Q) . \tag{2.115}$$

When we compare this relation with the Helmholtz equation,

$$F = U - TS = U + T \frac{\partial F}{\partial T} , \tag{2.116}$$

we obtain finally an expression for the free energy:

$$F(T, V) = -k_B T \ln Q . \tag{2.117}$$

2.4.2 Grand Canonical Ensembles

Let us now study a macroscopic open system which allows exchange of energy and particles with the surrounding world. This is, of course, a more realistic assumption about real systems which are always in contact with the external world—completely isolated systems do not exist in our world. Just for example, it is absolutely impossible to isolate a system completely from the influence of cosmic radiation. In analogy to classical statistics, we postulate:

The probability to find an open system in the microstate j_N which has the energy $E_{j,N}$, is given by the distribution

$$w_{jN} = C e^{\frac{\mu N - E_{jN}}{k_B T}} , \tag{2.118}$$

$$\frac{1}{C} = \varXi = \sum_{N=0}^{\infty} \sum_{j}^{\{\sum N_j = N\}} e^{\frac{\mu N - E_{jN}}{k_B T}} . \tag{2.119}$$

The number of particles is a fluctuating quantity, whose mean is given by

$$\bar{N} = \frac{\sum\limits_{N=0}^{\infty} \sum\limits_{j} N e^{\frac{\mu N - E_{jN}}{k_B T}}}{\sum\limits_{N=0}^{\infty} \sum\limits_{j} e^{\frac{\mu N - E_{jN}}{k_B T}}} = \frac{\partial}{\partial \mu}(k_B T \ln \varXi) . \tag{2.120}$$

We introduce the grand canonical potential

$$\varOmega = -k_B T \ln \varXi \tag{2.121}$$

by the formula

$$\varOmega = -k_B T \ln \sum_{N=0}^{\infty} \sum_{j} e^{\frac{\mu N}{k_B T} - \frac{E_{j,N}}{k_B T}} . \tag{2.122}$$

In the special case where the particle number shows only minor fluctuations, we may in some approximation neglect the fluctuations. It follows

$$\varOmega = -k_B T \ln \sum_{i} e^{\frac{\mu \tilde{N}}{k_B T} - \frac{E_{j,N}}{k_B T}} = -k_B T \ln \left(e^{\frac{\mu \tilde{N}}{k_B T}} Q_{\tilde{N}}(T, V) \right)$$

$$= -\mu \bar{N} + F(T, V, \tilde{N}) , \tag{2.123}$$

where $\tilde{N}$ is the average number of particles and

$$Q_N(T, V) = e^{-\frac{F(T,V,N)}{k_B T}} . \tag{2.124}$$

By using the Gibbs-Duhem relation,

$$G = \sum_{k} \mu_k N_k , \tag{2.125}$$

which holds for the case that G is strictly extensive, and the relations for thermodynamic potentials we find

$$-\varOmega = \mu N - F = G - F = F + pV - F = pV. \tag{2.126}$$

and the closure relation for the average number of particles, (2.120),

$$\bar{N} = \frac{\partial}{\partial \mu}(pV) . \tag{2.127}$$

Note that for large systems, the fluctuations around the mean are rather small (as we know from classical statistics). In the next section, we will discuss the fluctuations in more detail.

2.5   Theory of Fluctuations and Relaxation Processes

2.5.1   Einstein-Onsager Relaxation Theory

The theory of fluctuations goes mainly back to the work by Einstein (1905) and Onsager (1931a,b) and is a central part of the quantum statistical theory. In this section, we present the theory of near-equilibrium linear irreversible processes in a nutshell (for more details see, e.g., Ebeling and Sokolov 2005; Klimontovich 1982, 1986; Lifshitz and Pitaevskii 1981; Zubarev et al. 1996). For a recent critical review of fluctuation theory see, e.g., (Mishin 2015). Most important results of the statistical theory of near to equilibrium processes is based on the intimate relation of fluctuations and dissipation. On the first glance both phenomena are not related one to the other. The insight that there is a deep connection is due to the work of Einstein, Onsager and Kubo. We will study, first for classical variables and subsequently for quantum variables, the interaction of a specific degree of freedom with the surrounding. It will be shown that the concrete way of interaction with the particles of the surrounding is not relevant for the description of the variables under study. The essential aspect of the interaction is

(1) dissipation, that is, the distribution of energy to many degrees of freedom, and
(2) small deviations from the equilibrium states due to fluctuations.

According to Einstein, Onsager and Kubo, the correlation functions and spectra of the fluctuations stand in close relation to the response function on external forces, to the dissipative and transport coefficients as long as the linear approximation around the equilibrium state holds. Let us start with Einstein's ideas to apply the formalism of the Boltzmann-Planck entropy to fluctuations, worked out in several seminal papers published between 1902 and 1906. Suppose x is a classical fluctuating quantity of a thermodynamic system which is function of the microscopic variables, $x = x(q_1 \ldots p_f)$. Suppose further an isolated thermodynamic system, that is, E, N, V are (macroscopic) constants. We aim to find the probability distribution, $w(x)$, of the fluctuating quantity, x, under iso-energetic conditions. With the assumption of thermodynamic equilibrium at fixed E, V, N we introduce the conditional entropy, $S(x|E, V, N)$, for the microscopic states corresponding to a certain given value of x. Keeping the variable $x = $ const., corresponds to some subset of the hyper-surface $E = $ const. and the entropy determined by the number of microscopic states, $\Omega(x|E, V, N)$, of the states corresponding to the value of x, thus,

$$S(x|E, V, N) = k_B \ln \Omega(x|E, V, N). \qquad (2.128)$$

The probability distribution to find a certain value of x is then defined by

$$w(x|E, V, N) = \frac{\Omega(x|E, V, N)}{\Omega(E, V, N)} \qquad (2.129)$$

or, respectively.

$$w(x) = e^{-\frac{S(E,V,N)-S(x|E,V,N)}{k_B T}}. \qquad (2.130)$$

Following Einstein's approach we calculate now the entropy difference, δS, for two states based on thermodynamic relations. Since we use a normalized distribution, we do not need the full thermodynamic entropy $S(E, N, V)$, but only relative changes. Using the notation $S(x|E, V, N) = S(x)$, we find

$$w(x) = \frac{e^{\frac{S(x)}{k_B}}}{\int dx\, e^{\frac{S(x)}{k_B}}}. \qquad (2.131)$$

We study now the general properties of the probability distribution looking for some x around the equilibrium value, x_0. The thermodynamic potentials have extremal properties with respect to x. The second law requires $S \rightarrow$ max, that is, the maximum is assumed in equilibrium when $x \rightarrow x_0$. An expansion around a stable equilibrium state gives

$$S(x) = S(x_0) - \frac{1}{2}k_B \beta (x - x_0)^2. \qquad (2.132)$$

Correspondingly, the fluctuations around equilibrium states are Gaussian:

$$w(x) = \sqrt{\frac{\beta}{2\pi}} e^{-\frac{1}{2}\beta(x-x_0)^2}; \qquad \left\langle (x - x_0)^2 \right\rangle = \frac{1}{\beta}. \qquad (2.133)$$

We see that the deviations are determined by the second derivatives of the entropy. For other thermodynamic embeddings, we obtain in an analogous way Gaussian distributions where β is the positive second derivative of the corresponding thermodynamic potential.

The fluctuations of many classical variables, $x_1 \ldots x_s$, can be treated as a vector, $\mathbf{x}$, with components x_i, $(i = 1 \ldots n)$ (Klimontovich 1982). The resulting Gaussian distribution near the equilibrium state, $\mathbf{x}_0$, reads

$$w(\mathbf{x}) = \sqrt{\frac{\det \hat{\beta}}{(2\pi)^n}} \exp\left[-\frac{1}{2} \sum_{i,j} \beta_{ij}(x_i - x_{i0})(x_j - x_{j0}) \right], \qquad (2.134)$$

where $\hat{\beta}$ is defined by

$$\langle (x_i - x_{i0}) (x_j - x_{j0}) \rangle = \left(\hat{\beta}^{-1} \right)_{ij} .$$ (2.135)

As an application we apply these formulae to the fluctuations of the thermodynamic variables T, V, N etc. of a subvolume which is embedded into a thermal bath with temperature T (Klimontovich 1982, 1986). The fluctuations of quantum variables can be treated in a quite similar way (Landau and Lifshitz 1990).

Linear Relaxation Processes

The method by Einstein and Onsager is based on the idea that—in principle— any macroscopic quantity, x, can be considered as a fluctuating variable, which is determined by a certain probability distribution, $\omega(x)$. Its mean value is given as the first moment of the probability distribution,

$$x_0 = \langle x \rangle = \int x \, \omega(x) \, dx .$$ (2.136)

In a stationary state, we can shift the origin and assume $x_0 = 0$. Assuming that the stationary state is the state of thermodynamic equilibrium, $x_0 = 0$ corresponds to equilibrium and any value of $x(t)$ different from zero is strictly speaking a non-equilibrium state. According to the Second Law, the equilibrium state is an attractor of the dynamics since it corresponds to the maximum of entropy:

$$S(x = 0) = \max ; \quad \left(\frac{\partial S}{\partial x} \right)_{x=0} = 0 ; \quad \left(\frac{\partial^2 S}{\partial x^2} \right)_{x=0} \leq 0 .$$ (2.137)

Onsager postulates that the relaxation dynamics of the variable x is determined by the first derivative of the entropy, which is different from zero near equilibrium. Starting at a deviation from the equilibrium (corresponding to a value of entropy below the maximum), spontaneous irreversible processes should drive the entropy to increase,

$$\frac{d}{dt} S(x) = \frac{\partial S}{\partial x} \frac{dx}{dt} \geq 0 .$$ (2.138)

In this expression there appear two factors which Onsager interpreted as the driving force of the relaxation to equilibrium and the corresponding thermodynamic flux or thermodynamic flow:

$$X = -\frac{\partial S}{\partial x} ; \quad J = -\frac{dx}{dt} .$$ (2.139)

Further Onsager postulated a linear relation between the thermodynamic force and the corresponding flux,

$$J = LX. \tag{2.140}$$

The argument for this assumption is that the thermodynamic force causes the thermodynamic flow, and both should disappear at the same time, when equilibrium is reached. The coefficient L is called *Onsager's phenomenological coefficient*, or *Onsager's kinetic coefficient*. From the Second Law follows that Onsager-coefficients are strictly positive:

$$P = \frac{d}{dt} S(x) = J X = L X^2 \geq 0. \tag{2.141}$$

The assumption of a linear relation between thermodynamic forces and fluxes is the origin of the evolution of the thermodynamics of linear dissipative system, called *linear irreversible thermodynamics*. The linearity of the force-flow relation corresponds to bi-linearity of the entropy production,

$$P = J X. \tag{2.142}$$

For states near equilibrium, we obtain the relation

$$X = -\frac{\partial S}{\partial x} = k_B \beta x. \tag{2.143}$$

Using the previous equations, we find, finally, the following linear relaxation dynamics:

$$\dot{x} = -k_B L \beta x. \tag{2.144}$$

With the *relaxation coefficient* of the quantity x defined by

$$\lambda = L k_B \beta, \tag{2.145}$$

we, finally, arrive at

$$\dot{x} = -\lambda x; \qquad x(t) = x(0)e^{-\lambda t}. \tag{2.146}$$

This linear kinetic equation describes the relaxation of a thermodynamic system starting from a non-equilibrium state. We see that $t_0 = \lambda^{-1}$ plays the rôle of the decay time of the initial deviation from equilibrium.

On the other hand, this coefficient which is responsible for the relaxation to equilibrium is in close relation to the fluctuation properties of the considered system. Indeed, Eq. (2.145) relates a kinetic property, λ, to a fluctuation quantity, β. Relations of this type are called *fluctuation-dissipation relation*.

Historically, the first fluctuation-dissipation relation was found by Einstein (1905) as a relation between the mean square displacement of a Brownian particle and the (spatial) diffusion coefficient, D, in the form

$$\left\langle (x(t) - x(0))^2 \right\rangle = 2Dt . \tag{2.147}$$

Further early examples of fluctuation-dissipation relations are the relations between the diffusion coefficient and velocity-velocity correlations found by Taylor (1922) and extended by Kubo (1957) to a relation between the conductivity, σ, of a system of electrons with density n and the velocity-velocity correlation function,

$$D = \int_0^\infty \langle v(t)\, v(0) \rangle\, dt ; \qquad \sigma = \frac{n\, e^2}{k_B T} \int_0^\infty \langle v(t)\, v(0) \rangle\, dt . \tag{2.148}$$

Quite similar as in the case of one variable we might argue for the case of several thermodynamic values. Indeed, taking the entropy as a function of x_i, $(i = 1 \ldots N)$, the entropy production reads

$$S(x_1 \ldots x_n) = S_{\max} - \frac{1}{2}\beta_{ij}\, x_i\, x_j . \tag{2.149}$$

For notational brevity, here and in the following we apply the Einstein convention, that is, when an index variable appears twice in a single term and is not otherwise defined, summation of that term over all the values of the index is implied.

Following Onsager's idea described above we obtain the relations

$$X_i = -k_B\, \beta_{ij}\, x_j ; \qquad J_i = -\dot{x}_i \tag{2.150}$$

between the thermodynamic forces and fluxes. The generalized linear Onsager-ansatz and the conditions from the Second Law read

$$J_i = L_{ij} X_j ; \qquad L_{ij} X_i X_j \geq 0 . \tag{2.151}$$

The inequality holds for any value of X_i and the term disappears only for $X_i = 0$, $i = 1 \ldots n$. This corresponds to the requirement of positive definiteness of the matrix L_{ij}. By inserting Eq. (2.150) into Eq. (2.151) we obtain

$$\dot{x}_i = -k_B\, L_{ij}\, X_j . \tag{2.152}$$

When we, finally, introduce the matrix of the relaxation coefficients of the linear processes near equilibrium states we arrive at

$$\dot{x}_i = -\lambda_{ij} x_j ; \qquad \lambda_{ij} = k_B\, L_{ik}\, \beta_{kj} . \tag{2.153}$$

Since the matrix β_{ij} determines the dispersion of the stationary fluctuations, we have found again a close relation between fluctuations and dissipation, that is, a fluctuation-dissipation relation for a matrix of fluctuating and relaxing variables. The physical meaning of Onsager's assumption can be interpreted as follows: silently Onsager assumed that deviations from equilibrium and fluctuations around the equilibrium obey the same kinetics.

2.5.2 Correlations, Spectra, and Symmetry Relations

This section is devoted to the time correlation functions and their spectra (Ebeling and Sokolov 2005; Klimontovich 1982, 1986; Lifshitz and Pitaevskii 1981; Toda et al. 1983; Zubarev et al. 1996, 1997). The time correlation functions are defined here as averages over the stationary probability distribution, $\omega(x)$. We consider only stationary processes and the corresponding stationary probability distributions. As a consequence of stationarity all characteristic functions depending on two times, t, t', are functions of the time difference, $t - t'$, only. There are two equivalent ways to define correlation functions, namely,

(1) the time average over a long (infinite) time interval, and
(2) the ensemble average based on a certain probability distribution, $\omega(x)$.

For the case of many fluctuating variables, $x_i(t)$, $(i = 1 \ldots n)$, the time correlation function is defined as

$$C_{ij}(\tau) = \langle x_i(t) x_j(t + \tau) \rangle_t = a \int dx_1 \ldots dx_n \, \omega(x_1 \ldots x_n) \, x_i(t) \, x_j(t + \tau),$$

(2.154)

where $\tau > 0$ and $\omega(x_1 \ldots x_n)$ stand for the simultaneous probability distribution of all x_i variables. General properties of the time correlation functions are:

(1) The time correlation functions vanish for infinitely large time and are equal to the covariance coefficients for small time,

$$\lim_{t \to \infty} C_{ij}(\tau) = 0; \qquad C_{ij}(\tau = 0) = \langle x_i x_j \rangle. \qquad (2.155)$$

(2) Due to the assumed stationarity, $C_{ij}(\tau)$ does not depend on the actual time. Substituting $t \to t' - \tau$, we obtain

$$C_{ij}(\tau) = \langle x_i(t) x_j(t + \tau) \rangle = \langle x_i \left(t' - \tau \right) x_j \left(t' \right) \rangle = C_{ji}(-\tau). \qquad (2.156)$$

We introduce now the Fourier-component of the time correlation function,

$$S_{ij}(\omega) = \int_{-\infty}^{\infty} d\tau \, C_{ij}(\tau) \, e^{i\omega\tau} . \qquad (2.157)$$

This matrix function is called the spectrum of the fluctuating quantities. It can be calculated either directly from the time correlation function via Eq. (2.157), or by analysis of the dynamics of the fluctuating values. We introduce first the Fourier components of $x_i(t)$:

$$x_{i\omega} = \int_{-\infty}^{\infty} dt\, x_i(t) e^{i\,\omega t} . \qquad (2.158)$$

We multiply this expression by $x_{j\omega}$ and average over the stationary distribution,

$$\langle x_{i\omega} x_{j\omega'}\rangle = \int_{-\infty}^{\infty} dt\, dt' \, \langle x_i(t)\, x_j(t')\rangle \, e^{i\,(\omega+\omega')\,t + i\omega'\,(t'-t)} . \qquad (2.159)$$

Due to stationarity, the correlator $\langle x_i(t)\, x_j(t')\rangle$ depends only on the difference $t'-t$:

$$\langle x_{i\omega} x_{j\omega'}\rangle = 2\pi\delta\left(\omega+\omega'\right) S_{ij}(\omega) . \qquad (2.160)$$

This is a form of the *Wiener-Khintchin theorem* which relates the averaged product of the modes of a fluctuating stationary system to the spectrum of the fluctuating value.

Above, we derived equations for the linear relaxation dynamics of macroscopic variables. Onsager stated that these equations are valid for the relaxation of fluctuations too. Indeed, for the derivation of the relaxation dynamics we did not specify whether the initial non-equilibrium state was prepared as the result of an external force or it appeared as the result of a spontaneous fluctuation. In order to calculate the correlation functions, we start from the relaxation equations for the variables $x_i(t)$ which reads in the simplest case of one component $\dot{x}(t) = -\lambda x(t)$ (see previous section). We assume now that this relation is valid also for a deviation caused by a spontaneous fluctuation. We multiply the relaxation equation with the initial value $x(0)$ and find

$$\frac{d}{dt}[x(t)\, x(0)] = -\lambda[x(t)\, x(0)] . \qquad (2.161)$$

After averaging with respect to an ensemble of realizations, we find a kinetic equation for the time correlation function:

$$\frac{d}{d\tau}C(\tau) = -\lambda C(\tau) , \qquad (2.162)$$

with the initial conditions

$$C(\tau = 0) = \langle x^2\rangle = \int x\,\omega\,dx = \beta^{-1} ; \qquad C(\tau) = \frac{1}{\beta}e^{-\lambda|\tau|} . \qquad (2.163)$$

The generalization to several fluctuating variables is straightforward. The application of Onsager's regression hypothesis leads to the kinetic equations

$$\frac{d}{d\tau} C_{ij}(\tau) = -\lambda_{ik} C_{kj} \tag{2.164}$$

with the initial conditions

$$C_{ij}(0) = \langle x_i \, x_j \rangle . \tag{2.165}$$

For the most elegant way to solve these equations, we employ the one-sided Fourier-transforms (Klimontovich 1982, 1986). We represent the time correlation functions as

$$S_{ij}^+(\omega) = \int_0^\infty d\tau \, C_{ij}(\tau) e^{i\omega\tau} . \tag{2.166}$$

The negative part of the spectrum is just the complex conjugate

$$S_{ij}^-(\omega) = \int_{-\infty}^0 d\tau \, C_{ij}(\tau) \, e^{i\omega\tau} = [S_{ij}^+(\omega)]^* . \tag{2.167}$$

Taking into account the initial conditions, we find for the positive part

$$(-i\omega\delta_{ik} + \lambda_{ik}) \, S_{kj}^+(\omega) = \langle x_i \, x_j \rangle . \tag{2.168}$$

From here on we consider C_{ij}, λ_{ik}, and $S_{ij}^\pm$ as elements of matrices, $\hat{C}$, $\hat{\Lambda}$, and $\hat{S}^\pm$. Moreover, we introduce the matrix $\hat{B} = \hat{C}^{-1}(0)$, and use the notation $\hat{I}$ for the unit matrix with elements δ_{ij}. By adding positive and negative parts we find the full spectrum

$$\hat{S}(\omega) = (-i\omega\hat{I} + \hat{\Lambda})^{-1}\hat{B}^{-1} + \hat{B}^{-1}\left(-i\omega\hat{I} + \hat{\Lambda}\right)^{-1} . \tag{2.169}$$

Here we used the matrix inversion of Eq. (2.168) and the symmetry relation $C_{ij}(\tau) = C_{ji}(-\tau)$, valid for stationary processes. The correlation function $C_{ij}(\tau)$ follows then from the inverse Fourier transform.

More concrete illustrations of this procedure will be given in Chap. 8 where applications to Brownian motion will be studied. The approach presented so far is restricted to the dynamics of classical variables close to equilibrium. In Chap. 8 we will discuss extensions to quantum variables. More difficult is the extension to situations far from equilibrium. There exist many approaches to this problem, here we mention in particular the early work by Machlup and Onsager (1953). An advanced theory of nonlinear irreversible processes is given by Stratonovich (1984). We shall not go into further details of these methods as the field is still subject of intensive research.

References

Blochinzew, D.J. 1953. *Quantenmechanik*. Berlin: Deutscher Verlag der Wissenschaften.

Dirac, P.A.M. 1932. *Principles of Quantum Mechanics*. Oxford: Clarendon Press.

Dyson, F.J. 1967. Ground-State Energy of a Finite System of Charged Particles. *Journal of Mathematical Physics* 8: 1538–1545.

Dyson, F.J., and A. Lenard. 1967. Stability of Matter. I. *Journal of Mathematical Physics* 8: 423–434.

Dyson, F.J., and A. Lenard. 1968. Stability of Matter. II. *Journal of Mathematical Physics* 9: 698–711.

Ebeling, W., and D. Hoffmann. 2014. Eine Vorlage Einsteins in der Preußischen Akademie der Wissenschaften. *Leibniz Online*. http://www.leibnizsozietaet.de/wp-content/uploads/2014/12/EbelingHoffmann.pdf.

Ebeling, W., and I. Sokolov. 2005. *Statistical Thermodynamics and Stochastic Theory of Nonequilibrium Systems*. Singapore: World Scientific.

Einstein, A. 1905. Über die von der Molekularkinetischen Theorie der Wärme Geforderte Bewegung von in Ruhenden Flüssigkeiten Suspendierten Teilchen. *Annales de Physique* 17: 549–560.

Einstein, A. 1924. Quantentheorie des Einatomigen Idealen Gases. *Sitzungsber. Preuss. Akad. Wiss. Phys.-math. Kl.* 22: 261–267.

Einstein, A. 1925a. Quantentheorie des Einatomigen Idealen Gases. Zweite Abhandlung. *Sitzungsber. Preuss. Akad. Wiss. Phys.-math. Kl.* 23: 3–14.

Einstein, A. 1925b. Zur Quantentheorie des Idealen Gases. *Sitzungsber. Preuss. Akad. Wiss. Phys.-math. Kl.* 23: 18–25.

Eyring, H., J. Walter, and G.E. Kimball. 1944. *Quantum Chemistry*. New York: J. Wiley & Sons.

Fermi, E. 1928. Eine Statistische Methode zur Bestimmung Einiger Eigenschaften des Atoms und ihre Anwendung auf die Theorie des Periodischen Systems der Elemente. *Zeitschrift für Physik* 48 (1): 73–79.

Fermi, E. 1966. *Molecules, Crystals and Quantum Statistics*. New York: Benjamin.

Fetter, A.I., and J.D. Walecka. 1971. *Quantum Theory of Many Particle Systems*. New York: Mc Graw Hill.

Feynman, R.P. 1972. *Statistical Mechanics*. Reading Mass: Benjamin.

Fick, E. 1981. *Einführung in die Grundlagen der Quantentheorie*. Leipzig: Akademischer Verlag.

Hänggi, P., S. Hilbert, and J. Dunkel. 2016. Meaning of Temperature in Different Thermostatistical Ensembles. *Philosophical Transactions. Royal Society of London A* 374: 2064.

Hilbert, S., P. Hänggi, and J. Dunkel. 2014. Thermodynamic Laws in Isolated Systems. *Physical Review E* 90: 062116.

Kadanoff, L.P., and G. Baym. 1962. *Quantum Statistical Mechanics*. New York: Benjamin.

Klimontovich, Yu. L. 1982. *Statistical Physics*. Moscow: Nauka (in Russian).

Klimontovich, Yu. L. 1986. *Statistical Physics*. New York: Harwood.

Kraeft, W.D., D. Kremp, W. Ebeling, and G. Röpke. 1986. *Quantum Statistics of Charged Particle Systems*. Berlin: Akademie-Verlag.

Kremp, D., M. Schlanges, and W.D. Kraeft. 2005. *Quantum Statistics of Nonideal Plasmas*. Berlin: Springer.

Kubo, R. 1957. Statistical-Mechanical Theory of Irreversible Processes. I. General Theory and Simple Applications to Magnetic and Conduction Problems. *Journal of the Physical Society of Japan* 12: 570–586.

Landau, L.D., and E.M. Lifshitz. 1976. *Statistical Physics (part I)*. Moscow: Nauka.

Landau, L.D., and E.M. Lifshitz. 1990. *Statistical Physics*. New York: Pergamon.

Lieb, E.H., and W.E. Thirring. 1975. Bound for the Kinetic Energy of Fermions which proves the Stability of Matter. *Physical Review Letters* 35: 687–689.

Lifshitz, E.M., and L.P. Pitaevskii. 1981. *Physical Kinetics*. Vol. 10, Course of Theoretical Physics S. New York: Pergamon.

Machlup, S., and L. Onsager. 1953. Fluctuations and Irreversible Processes. *Physics Review* 91: 1512–1515.

Martin, P.C., and J. Schwinger. 1959. Theory of Many-Particle Systems I. *Physics Review* 115: 1342–1373.

Mishin, Y. 2015. Thermodynamic Theory of Equilibrium Fluctuations. *Annals of Physics* 363: 48–97.

Onsager, L. 1931a. Reciprocal Relations in Irreversible Processes. I. *Physics Review* 37: 405–426.

Onsager, L. 1931b. Reciprocal Relations in Irreversible Processes. II. *Physics Review* 38: 2265–2279.

Reichl, L.E. 1980. *A Modern Course in Statistical Physics*. Austin: University of Texas Press.

Slater, J.C. 1939. *Introduction to Chemical Physics*. New York, London: Mc Graw Hill.

Stratonovich, R.L. 1984. *Nonlinear Nonequlibrium Thermodynamics*. Berlin: Springer.

Taylor, G.I. 1922. Diffusion by Continuous Movements. *Proceedings of the London Mathematical Society Series 2* 20: 196–212.

ter Haar, D. 1995. *Elements of Statistical Mechanics*. Oxford: Butterworth-Heinemann.

Thirring, W.E. 1980. *Lehrbuch der Mathematischen Physik*. Vol. 3, Quantenmechanik von Atomen und Molekülen, Quantenmechanik großer Systeme. Berlin: Springer.

Thomas L.H. 1927. The Calculation of Atomic Fields. *Mathematical Proceedings of the Cambridge Philosophical Society* 23: 542–548.

Toda, M., R. Kubo, and N. Saito. 1983. *Statistical Physics*. Vols. I and II. Berlin: Springer.

Tolman, R. 1938. *The Principles of Statistical Physics*. Oxford: Oxford University Press.

von Neumann, J. 1932. *Mathematische Grundlagen der Quantenmechanik*. Berlin: Springer.

Zubarev, D.N., V. Morozov, and G. Röpke, eds. 1996. *Statistical Mechanics of Nonequilibrium Processes*. Vol. 1, Basic Concepts, Kinetic Theory. Weinheim: Wiley-VCH.

Zubarev, D.N., V. Morozov, and G. Röpke, eds. 1997. *Statistical Mechanics of Nonequilibrium Processes*. Vol. 2, Relaxation and Hydrodynamic Processes. Weinheim: Wiley-VCH.

Chapter 3
Ideal Quantum Gases

3.1 Quantum Statistics of Oscillator and Phonon Gases

3.1.1 Einstein Model of Oscillations in Crystals

Planck's theory of radiation which is the origin of quantum statistics, was semi-phenomenological based on concepts of electrodynamics, classical thermodynamics, and classical radiation theory. The essential new element of Planck's theory was the assumption of the quantization of energy and a quantum statistical theory of linear oscillators. According to Planck's hypothesis, linear oscillators with frequency ω have the quantized energies

$$E_n = \hbar \omega n ; \qquad n = 0, 1, 2, 3, \dots \tag{3.1}$$

Averaging this by means of Boltzmann weights we find the mean energy,

$$\langle E \rangle = \frac{\sum_n E_n e^{-\beta E_n}}{\sum e^{-\beta E_n}} . \tag{3.2}$$

Here, $\hbar$ is a new constant of nature related to Planck's constant, h, by

$$\hbar = \frac{h}{2\pi} \simeq 1.0545718 \cdot 10^{-34} \, \mathrm{J\,s} . \tag{3.3}$$

The new constant in Planck's law was determined by measurements of radiation. Using the sum Z over all states we obtain

$$Z = \sum_n e^{-\beta \hbar \omega n} ; \qquad \langle E \rangle = -\frac{\partial \ln Z}{\partial \beta} . \tag{3.4}$$

© Springer Nature Switzerland AG 2019
W. Ebeling, T. Pöschel, *Lectures on Quantum Statistics*,
Lecture Notes in Physics 953, https://doi.org/10.1007/978-3-030-05734-3_3

The sum is a geometric series with $q = e^{-\beta\hbar\omega}$ which can be performed to yield

$$Z_\omega = \frac{1}{1 + e^{-\beta\hbar\omega}} ; \qquad E_\omega = \frac{\hbar\omega}{e^{\beta\hbar\omega} - 1} . \qquad (3.5)$$

We note that in this calculation, Planck omitted the contributions of the zero point energy, $\hbar\omega/2$ which he found only in 1912. This leads to essential corrections, in particular at low temperature.

Einstein (1907) applied Planck's theory of radiation to a model of oscillations in a solid crystal. The problem posed by Nernst's heat law was to explain, why the observed specific heat of substances (see Fig. 3.1) agrees at low T with Nernst's law, $C_v \to 0$, but disagrees with Dulong-Petit's law, $c_v = 3k_B T$. From classical statistics it was expected that a solid crystal has about $6N$ degrees of freedom, 3 of them of kinetic and 3 of them of potential origin, and that each degree contributes $1/2\,k_B$ to the specific heat. In order to explain, why this is not true at low T, Einstein combined the simple model of $6N$ oscillators with Planck's quantum statistics of linear oscillators. Einstein's result for the energy distribution,

$$E = 6N \frac{\hbar\omega}{e^{\beta\hbar\omega} - 1} , \qquad (3.6)$$

leads to the prediction that indeed $c_v(T) \to 0$ with decreasing T. We note that the consideration of the zero-point energy yields the additional contribution $6N\,(\hbar\omega/2)$ in Eq. (3.6). Nernst was very positive about this result and considered it as an excellent confirmation of his third law of thermodynamics. There remained, however, one open problem: Einstein's theory did not agree quantitatively with the T^3-law observed in many experiments and predicted instead an increase with $T^{3/2}$. It was a young researcher from The Netherlands, Peter Debye (1884–1966), who solved the problem in 1912 (Debye 1912). Einstein's theory suffered from the assumption that all oscillators in a solid oscillate with the same frequency. Debye

Fig. 3.1 The curve of the specific heat C_v of a gas of oscillators and several experimental points for Al, Cu, Au and Pb (redrawn after Huang (2001))

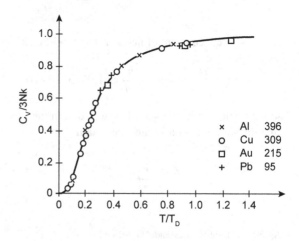

generalized this by assuming that in a crystal a whole spectrum of frequencies and waves is excited, later called phonons. The Debye model is a solid-state equivalent of Planck's law of black body radiation. Planck treated electromagnetic radiation as a gas of photons in a box. The Debye model treats the vibrations of atoms as a gas of phonons in a box.

3.1.2 Debye Theory of Phonon Excitations in Lattices

Einstein's theory of the specific heat developed more than 100 years ago found great interest among experimentalists, in particular we mentioned the Nernst group, which considered Einstein's result as a great success in explaining low temperature measurements. A drawback of Einstein's theory was the low temperature behavior of the specific heat which was qualitatively correct but quantitatively in disagreement with the data obtained by physicists and physico-chemists who found a T^3 law. To solve this problem, Peter Debye made two steps:

(1) the theory by Einstein was extended to three crystal dimensions,
(2) the spectrum of lattice oscillations was described more realistically and the finite number of excitations was taken into account.

The true spectrum of the vibrations in a lattice can be very difficult and differs from crystal to crystal. The fundamental idea by Peter Debye was that at low temperature, only the long waves corresponding to low frequencies are excited. These wave lengths are larger than the atomic spacing in the crystal, therefore at least for the long waves, the lattice may be approximated as a continuum. According to Debye, a reasonable approximation for the spectrum of frequencies below an upper limit, v_m, is (Huang 1963, 2001)

$$g(v) = 4\pi \left(\frac{1}{c_\ell^3} + \frac{2}{c_t^3} \right) v^2 . \tag{3.7}$$

Here, c_ℓ and c_t are the longitudinal and transversal sound velocities in the crystal.

Together with the Dulong-Petit-Einstein assumption that the total number of excitations is $3N$ and integrating up to the upper frequency limit, v_m, we have the normalization

$$3N = \int_0^{v_m} dv\, g(v) . \tag{3.8}$$

This way we find the energy distribution law

$$E = \int_0^{v_m} dv\, g(v) \frac{hv}{e^{\frac{hv}{k_B T}} - 1} \tag{3.9}$$

and correspondingly, after defining the so-called Debye function, the specific heat

$$c_v = 3Nk_B D\left(\frac{\Theta_D}{T}\right) \; ; \qquad D(x) = \frac{3}{x^3} \int_0^x d\xi \, \frac{\xi^3}{e^\xi - 1} \, . \tag{3.10}$$

An explicit calculation for low temperature shows that the Debye theory gives indeed the T^3 law (Huang 2001), as measured in experiments.

3.2 Statistics of Bose-Einstein and Fermi-Dirac Gases

3.2.1 Development of the Quantum Statistics of Gases

The first step towards quantum statistics of gases was done in 1924 by Bose who sent his manuscript *"Planck's law and the hypothesis on light quanta"* to Einstein, who translated it into German and published the article in Zeitschrift für Physik (Bose 1924a). Einstein added several remarks generalising Bose's approach. He wrote: *Anmerkung des Übersetzers. Boses Ableitung der Planckschen Formel bedeutet nach meiner Meinung einen wichtigen Fortschritt. Die hier benutzte Methode liefert auch die Quantentheorie des idealen Gases, wie ich an anderer Stelle ausführen will.*[1] Another article by Bose (1924b) on the "Wärmegleichgewicht im Strahlungsfeld bei Anwesenheit von Materie"[2] was commented more critically by Einstein: "I think that the hypothesis of Bose on the probability of elementary processes is not correct...". Nevertheless, he esteemed Bose's new method of counting the probabilities of the macro-states. In Boses's interpretation, the radiation field appears as a gas of photons. Since photons obey the Stefan-Boltzmann law, they appear as an entirely new kind of gases. We will show in the next section that the photon gas is the prototype of *relativistic plasmas*. Full of excitement about the new perspectives for the physics at low temperature, Einstein presented only 8 days after the official date of receipt of Bose's paper some new results in a session of the Prussian Academy held on July 10, 1924 in Berlin. In two articles *Quantentheorie des einatomigen idealen Gases* printed in the *Sitzungsberichte* of the Prussian Academy of Science in Berlin, Einstein (1924, 1925) applied Bose's new method to derive the quantum statistics of one-atomic ideal gases. In this work Einstein explained the gas degeneracy at low temperature by means of a new quantum statistics, which later was called *Bose-Einstein statistics* (Ebeling and Hoffmann 2014). Indeed, Einstein immediately realized the value of the new method and, moreover, he understood that the physical

---

[1]In English: "Translator's note. In my opinion, Bose's derivation of Planck's formula is an important step forward. The method used here gives also the quantum theory of the perfect gas, as I will discuss elsewhere."

[2]In English: "Thermal equilibrium in the radiation field in the presence of matter."

assumption underlying Bose's new method of counting is the indistinguishability of light quanta. Bose believed that his method is specific for light quanta but Einstein realized that only the identity of the constituents of the gas matters. This way, Einstein understood the essence of the new statistics and the great idea behind it. In particular, he understood, that the method could be applied also to identical atoms (see Kirsten and Körber 1975; Treder 1983). Let us note that the words *identity* or *indistinguishability* do not yet appear Bose's paper, but it was Einstein who understood first the rôle of identity and indistinguishability in this context and it was Einstein who defended this fundamental new principle in very critical discussions with Planck and Ehrenfest (Treder 1983).

The generalization of Bose's distribution of energy (3.11) to a gas of atoms with mass, m, momentum, p, and chemical potential, μ, reads

$$E = V \sum_p \frac{p^2}{2m} \left[\exp\left(\frac{\frac{p^2}{2m} - \mu}{k_B T} \right) - 1 \right]^{-1} . \tag{3.11}$$

We see that the main difference between gases of light quanta (photons) and atoms is the appearance of the chemical potential, $\mu < 0$, for a gas of atoms.

As mentioned above, initially the work by Bose and the first of Einstein's articles were heavily criticised by Ehrenfest, Planck and other colleagues. Since the authors did not provide the physical justification for the new way of counting. This aspect was discussed in detail in the second of Einstein's papers in the Proceedings of the Academy which he presented on January 8, 1925 (Einstein 1925). In this work, as a consequence of the theory Einstein predicted a novel phenomenon which by now we call *Bose-Einstein condensation*. Historically, it would be more correct to speak about the Bose-Einstein statistics, but about Einstein condensation, since the prediction of this phenomenon is clearly due only to Einstein. It should be mentioned that the derivation of a condensation phenomenon from the distribution (3.11) is by far not a trivial consequence but requires a series of complicated arguments given by Einstein. One of the authors of this book remembers that even in the fifties of the past century, the condensation predicted by Einstein was taught as a purely mathematical exercise without any physical meaning. This understanding of the Einstein condensation changed only later, when theoreticians like Bogolyubov and others included weak interactions into the theory, and finally experimentalists observed the Bose-Einstein condensation at extremely low temperature. The Einstein condition for condensation is

$$n \Lambda^3 \leq 2.612 ; \qquad \Lambda = \frac{h}{\sqrt{2\pi m k_B T}} , \tag{3.12}$$

where Λ is the thermal de Broglie wave length, that is, the wave length corresponding to the thermal momenta at temperature T. The young Erwin Schrödinger found this rôle of the De Broglie wavelength inspiring enough to start letter correspondence with Einstein. Schrödinger did not expect that one would ever be able to achieve such conditions in experiments.

The experimental verification of Einsteins prediction of a condensation of atomic gases at low temperature was confirmed only 70 years after its theoretical prediction: In 1995 Eric Cornell reported at the *International Conference of Laserspectroscopy* on the island Capri on experiments at the University of Boulder which confirmed Einsteins prediction (see Anderson et al. 1995; Petrich et al. 1995). Similar experiments were performed at the same time in the group of Wolfgang Ketterle at MIT (Davis et al. 1995). Presently, many physicists in the world work in this fruitful field, evidenced by several Nobel Prices. As we see, Einstein was right with his predictions and should be considered as the founder of one of the most fruitful fields of modern physics.

Evidently the first physicist who understood the general principle by Einstein was Pauli (1925) who formulated his exclusion principle. The first application of the new quantum statistics to electrons was given by Fermi (1926) and Dirac (1926). Fermi-Dirac statistics applies to gases of identical particles with half-integer spin in a system at thermodynamic equilibrium. Moreover, the system is assumed to be ideal, that is, particle-particle interaction is negligible. The many-particle system is described in terms of single-particle energy states as we will explain below. Note that extremely fruitful applications of the new statistics were given by Fowler and Sommerfeld who applied the new Fermi-Dirac statistics to electron plasmas.

Today, particles that obey the exclusion principle, such as particles of spin $s = 1/2$, are called *fermions*. Like Bose-Einstein gases, also dense Fermi gases behave at low temperature in a way different from classical ideal gas. Defining the energy density as the energy per unit of volume, $\rho_E = E/V$, the energy density and the pressure of Boltzmann-like gases or plasmas follow the relations

$$\rho_E = \frac{3}{2}nk_BT \; ; \qquad p = nk_BT \; , \tag{3.13}$$

that is, energy density increases linearly with density, n, and temperature, T. In contrast, for Fermi gases the dependence on temperature is only weak but the energy density and the pressure increase more strongly with density,

$$\rho_E \sim n^{4/3} \; ; \qquad p \sim n^{4/3} \; . \tag{3.14}$$

As a consequence of large pressure and energy density in dense systems, all bound states are suppressed. Thus, very dense fermion systems behave like ideal quantum gases (Fig. 3.2).

Fig. 3.2 Density-temperature
plane on log-scale. The line
$n\Lambda^3 = 1$ separates
degenerate from
non-degenerate gases

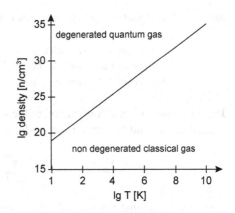

3.2.2   Gases as Particle Systems with Additive Hamiltonian

As the above examples demonstrate, the simplest object of quantum statistics are
systems with additive Hamiltonian (Bogolyubov 1991; Feynman 1972; Fick 1981;
Landau and Lifshitz 1976, 1980, 1990),

$$\hat{H} = \sum_{i=1}^{N} \hat{H}^{(i)} . \tag{3.15}$$

We have seen that there are many interesting physical systems which belong to this
class. Examples are:

Ideal gases	- molecules
Photons in black body	- eigenmodes of the electrical field
Phonons in lattices (ideal solids)	- eigenmodes of lattice oscillations

In spite of the fact that Eq. (3.15) is valid for a larger class of physical systems, we
consider here gases, which are the most important representatives of this class of
physical systems. The Schrödinger equation for the constituents can be written as

$$\hat{H}^{(i)}\psi_k^{(i)} = \varepsilon_k^{(i)}\psi_k^{(i)} . \tag{3.16}$$

Assuming that we have a discrete spectrum of one-particle eigenvalues, ε_0, ε_1,
$\varepsilon_2, \ldots$, we ask the question, how many particles, $N_1, N_2, N_3, \ldots$, of the gas are in
these states. We find this, in the tradition of Boltzmann, Planck etc., just by counting

particles after dividing the system into subsystems:

Subsystem 0 of all particles in state 0 with energy $E_0 = N_0 \varepsilon_0$,
Subsystem 1 of all particles in state 1 with energy $E_1 = N_1 \varepsilon_1$,
Subsystem 2 of all particles in state 2 with energy $E_2 = N_2 \varepsilon_2$, etc.

The total system is the conjunction of the subsystems. Energy, E, the total number of particles, N and pressure, p, are additive:

$$E = \sum_k E_k ; \qquad N = \sum_k N_k ; \qquad p = \sum_k p_k . \tag{3.17}$$

Note that the summation index, k, runs over all microscopic states but not only over the different energy states. Here the spin states are to be counted as extra quantum states. The partial pressures, p_k, of subsystems are like partial pressures in a mixture of ideal gases and are, therefore, additive too, according to Eq. (3.17). We apply the laws of grand canonical ensembles to the subsystem k, where N_k is the number of particles in state k:

$$p_k V = k_B T \ln \Xi_k ; \qquad \Xi_k = \sum_{N_k} e^{\frac{\mu N_k - E_k}{k_B T}} \tag{3.18}$$

$$\bar{N}_k = \frac{\partial}{\partial \mu} (p_k V) = \frac{\partial}{\partial \mu} k_B T \ln \Xi_k . \tag{3.19}$$

Applying Eq. (3.16), we obtain

$$\Xi_k = \sum_{N_k} e^{\frac{\mu N_k - \varepsilon_k N_k}{k_B T}} = \sum_{N_k} \left(e^{\frac{\mu - \varepsilon_k}{k_B T}} \right)^{N_k} = \sum_{N_k} q^{N_k} \tag{3.20}$$

and this way follow the partial pressures

$$p_k = \frac{k_B T}{V} \ln \sum_{N_k} q^{N_K} \quad \text{with} \quad q = e^{\frac{\mu - \varepsilon_k}{k_B T}} . \tag{3.21}$$

3.3   Fermi- and Bose Distributions

3.3.1   Bose-Einstein Gases

We study a gas consisting of particles having integer values of spin, $s = 0, 1, 2, 3, \ldots$, which we call *Bose-Einstein gases*. Such systems have wave functions which are symmetrical against the exchange of two particles as shown in the previous chapter. In this case, arbitrarily many particles can be in the same one-particle state, k, that is, N_k may assume any value. The grand canonical partition function follows by summing up. As a necessary condition, the geometric series must converge, which is the case for $|q| < 1$. We introduce the summation index $N_k \to n$ and perform the summation to obtain the geometric series

$$\Xi_k = \sum_{n=0}^{\infty} q^n = \frac{1}{1-q} . \tag{3.22}$$

This way we find the partial pressure, p_k, and the mean particle number, $\bar{N}_k$, of the subsystem in state k:

$$p_k = -\frac{k_B T}{V} \ln(1-q) = -\frac{k_B T}{V} \ln\left(1 - e^{\frac{\mu - \varepsilon_k}{k_B T}}\right) \tag{3.23}$$

$$\bar{N}_k = \frac{\partial}{\partial \mu}(p_k V) = \frac{e^{\frac{\mu - \varepsilon_k}{k_B T}}}{1 - e^{\frac{\mu - \varepsilon_k}{k_B T}}} = \frac{1}{e^{\frac{\varepsilon_k - \mu}{k_B T}} - 1} . \tag{3.24}$$

This is the Bose-Einstein distribution for the mean occupation number of the one-particle state k. The condition

$$N = \sum_k \bar{N}_k \tag{3.25}$$

provides the chemical potential, μ. The internal energy follows by summing up the energies over all states, k,

$$U = \sum_k \varepsilon_k \bar{N}_k . \tag{3.26}$$

In order to specify this formula, we have to define the concrete species, for instance phonons.

3.3.2 Fermi-Dirac Gases

Let us consider now *ideal Fermi-Dirac gases*. Fermions have antisymmetric wave functions and the one-particle states cannot be occupied by more than one fermion, that is, $N_k \in \{0, 1\}$. We obtain the grand canonical distribution

$$\Xi_k = 1 + \exp\left[\frac{\mu - \varepsilon_k}{k_B T}\right] \tag{3.27}$$

Fig. 3.3 Mean occupation of
a quantum state k with energy
ε_k for fermions

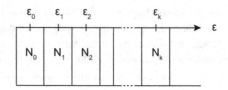

and, therefore,

$$p_k = \frac{k_B T}{V} \ln\left[1 + \exp\left(\frac{\mu - \varepsilon_k}{k_B T}\right)\right] \tag{3.28}$$

$$\bar{N}_k = \frac{\partial}{\partial \mu}(p_k V) = \frac{\exp\left(\frac{\mu - \varepsilon_k}{k_B T}\right)}{1 + \exp\left(\frac{\mu - \varepsilon_k}{k_B T}\right)} = \frac{1}{\exp\left(\frac{\varepsilon_k - \mu}{k_B T}\right) + 1}. \tag{3.29}$$

The last equation is the *Fermi-Dirac distribution* for the mean occupation numbers
of the one-particle states, k.

Problem 4 Draw the function $N(\varepsilon)$ and show for increasing values of the parameter μ that Eq. (3.29) converges to the Boltzmann distribution, $\bar{N}_k = \exp\left(\frac{\mu - \varepsilon}{k_B T}\right)$ for the limit $\exp\left(\frac{\varepsilon_k - \mu}{k_B T}\right) \gg 1$.

Figure 3.3 shows the mean occupation of the states, k, as a function of the energy,
ε_k. The chemical potential follows again from the normalization,

$$N = \sum_k \bar{N}_k, \tag{3.30}$$

and the internal energy is

$$U = \sum_k \varepsilon_k \bar{N}_k. \tag{3.31}$$

The thermodynamic properties of fermion gases will be studied later in more detail.

3.4   Thermodynamics Properties of Bose-Einstein Gases

In this Section, we will consider gases of bosons (Bose-Einstein gases) and perform
a similar calculation as for Fermi-Dirac gases. We will focus on the problem of
phase transitions and Einstein condensation. Previously, we obtained the mean

occupation of a state k,

$$\bar{N}_k = \left(e^{\frac{\varepsilon_k - \mu}{k_B T}} - 1 \right)^{-1} . \tag{3.32}$$

The total number of particles is

$$N = \sum_k \bar{N}_k . \tag{3.33}$$

Note that for bosons each state may be occupied by many particles. In particular this is true for the energetically lowest state, the ground state of the gas, $p_x = p_y = p_z = 0$. We denote the mean occupation of the ground state by $\bar{N}_0$. Following Einstein, a macroscopic occupation of this state is possible. Therefore, it is reasonable to separate this state and write

$$N = \sum_k \frac{1}{e^{\frac{\varepsilon_k - \mu}{k_B T}} - 1} = N_0 + \sum_{k > 0, \varepsilon_k > 0} \frac{2s + 1}{e^{\frac{\varepsilon_k - \mu}{k_B T}} - 1} , \tag{3.34}$$

with

$$\bar{N}_0 = \frac{2s + 1}{e^{-\frac{\mu}{k_B T}} - 1} = \frac{(2s + 1)z}{z - 1} . \tag{3.35}$$

Here, $z = \exp(\beta \mu)$ is the fugacity with $\beta = 1/(k_B T)$. By integration in spherical coordinates follows

$$N = \bar{N}_0 + \frac{2s + 1}{h^3} V \int_{0^+}^{\infty} dp \, 4\pi p^2 \frac{1}{\frac{1}{z} \exp\left(-\frac{p^2}{2k_B T m} \right) - 1}$$

$$= \bar{N}_0 + \frac{2s + 1}{\Lambda^3} V \frac{4}{\sqrt{\pi}} \int_{0^+}^{\infty} dx \, x^2 \frac{z}{e^{x^2} - z}$$

$$= \bar{N}_0 + \frac{2s + 1}{\Lambda^3} V g_{3/2}(z) \tag{3.36}$$

with the thermal wave length

$$\Lambda = \frac{h}{\sqrt{2\pi m k_B T}} . \tag{3.37}$$

Fig. 3.4 The Bose-Einstein
function $g_{3/2}(z)$

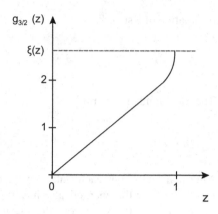

We used the substitution $p = x\sqrt{2mk_{B}T}$. Further we introduced in Eq. (3.36) the
Bose-Einstein function (Fig. 3.4)

$$g_{3/2}(z) \equiv \frac{4}{\sqrt{\pi}} \int_{0}^{\infty} dx\, x^{2} \frac{z}{e^{x^{2}} - z}\,, \tag{3.38}$$

which can be expanded into the series

$$g_{3/2}(z) = \sum_{l=1}^{\infty} \frac{z^{l}}{l^{3/2}}\,. \tag{3.39}$$

The convergence of this series is restricted to the interval $0 \leq z \leq 1$, therefore, the
chemical potential of Bose gases must be negative. For the case $z = 1$, we express
Eq. (3.39) through the zeta function, introduced by Riemann (1859)

$$\zeta(z) = \sum_{\ell=1}^{\infty} \frac{1}{\ell^{z}}\,; \,, \tag{3.40}$$

and obtain

$$g_{3/2}(1) = \sum_{\ell=1}^{\infty} \frac{1}{\ell^{3/2}} = \zeta\left(\frac{3}{2}\right) \approx 2.612\,. \tag{3.41}$$

In a similar way the pressure can be calculated:

$$pV = -k_{B}T(2s + 1) \sum_{k} \ln\left(1 - e^{\frac{\mu - \varepsilon_{k}}{k_{B}T}}\right)$$

$$= -k_{B}T(2s + 1) \ln\left(1 - e^{\frac{\mu}{k_{B}T}}\right) - k_{B}T(2s + 1) \sum_{k,\varepsilon_{k}>0} \ln\left(1 - e^{\frac{\mu - \varepsilon_{k}}{k_{B}T}}\right).$$

$$\tag{3.42}$$

We find

$$pV = -k_B T (2s + 1) \ln (1 - z) - k_B T \frac{2s + 1}{h^3} V \int_{0+}^{\infty} dp \, 4\pi \, p^2 \ln \left(1 - z e^{-\frac{p^2}{2\pi k_B T}} \right)$$

$$= -T k_B (2s + 1) \ln (1 - z) + T k_B \frac{2s + 1}{\Lambda^3} \frac{4}{\sqrt{\pi}} V \int_{0}^{\infty} dx \, x^2 \ln \left(1 - z e^{-x^2} \right)$$

$$= -T k_B (2s + 1) \ln (1 - z) + \frac{2s + 1}{\Lambda^3} g_{5/2}(z) \, T \, k_B V \,, \tag{3.43}$$

with

$$g_{5/2}(z) = \int_{0}^{\infty} dx \, x^2 \ln \left(1 - z e^{-x^2} \right) = \sum_{l=1}^{\infty} \frac{z^l}{l^{5/2}} \,. \tag{3.44}$$

The series (3.44) converges for $0 \le z \le 1$. Consequently, since $z = \exp(\mu/k_B T)$ we have $\mu < 0$. Again we see that Bose-Einstein gases cannot have positive chemical potential. A limiting case are photons for which $\mu = 0$, thus, $z = 1$.

Studying Eq. (3.36), we find expressions for the occupation of the ground state. The function $g_{3/2}(z)$ is shown in Fig. 3.4. We see that this function has a vertical tangent at $z = 1$ but remains finite.

Problem 5 Study the function $g_{3/2}(z)$ in more detail. Show that $g_{3/2}(1) \approx 2.612$. Use the relations $g_{3/2}(z) = z \frac{\partial}{\partial z} g_{5/2}(z)$ and take into account that $g_{5/2}(z)$ is a harmonic series.

From Eq. (3.36) we obtain

$$\frac{\Lambda^3}{V} \frac{z}{1 - z} = \frac{n \Lambda^3}{2s + 1} - g_{3/2}(z) \tag{3.45}$$

and

$$n = n_0 + \frac{2s + 1}{\Lambda^3} g_{3/2}(z); \quad \text{with} \quad n_0 = \frac{\bar{N}_0}{V} = \frac{(2s + 1)}{V} \left(\frac{z}{1 - z} \right). \tag{3.46}$$

We see that if $n \Lambda^3/(2s + 1) > g_{3/2}(1)$ holds, the occupation $\frac{\Lambda^3}{V} \frac{z}{1-z}$ must be of finite positive value, see Fig. 3.5. This means that one value of momentum $\mathbf{p} = 0$ is occupied by a finite macroscopic fraction of the bosons which are in the ground state. This effect was first detected by Einstein (1925) after extending Bose's statistics to gases. We call this effect *Bose-Einstein condensation*. For temperature and density beyond the region of Bose-Einstein condensation, the ground state is

Fig. 3.5 The fugacity z as
function of the inverse
density $1/n$

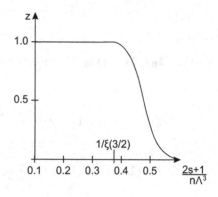

occupied just by an infinitesimal amount of the particles and we find

$$n = 0 + \frac{2s+1}{\Lambda^3} g_{3/2}(z).$$ (3.47)

From these relations follows the specific heat as shown in Fig. 3.6.

How can we observe a Bose-Einstein condensation in experiments? Increasing the density of a Bose-Einstein gas, we expect the phase transition at the critical density

$$n_{cr} = \frac{(2s+1)}{\Lambda^3} g_{3/2}(1),$$ (3.48)

where $\Lambda = \Lambda(T)$. For further increasing density, the ground state will be occupied by a finite number of gas particles since n_0 must assume a finite positive value. At this density also the equation of state changes. Below the phase transition we have

$$p = k_B T \frac{2s+1}{\Lambda^3} g_{5/2}(z),$$ (3.49)

Fig. 3.6 Specific heat of a
Bose gas as function of
temperature showing a phase
transition

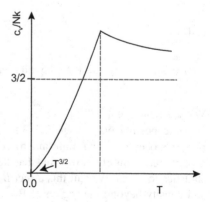

and after the start of the condensation

$$pV = -k_B T (2s + 1) \ln (1 - z) . \tag{3.50}$$

Let us calculate now the relative occupation of the ground state, $\bar{N}_0/N$, as a function of temperature. With Eq. (3.48), we find for the critical temperature

$$T_{cr} = \frac{h^2 n^{2/3}}{2\pi\, m\, k_B \left[(2s + 1)\, g_{3/2}(1)\right]^{2/3}} \tag{3.51}$$

and

$$\frac{\bar{N}_0}{N} = 1 - \left(\frac{T}{T_{cr}}\right)^{3/2} . \tag{3.52}$$

The specific heat as a function of temperature is shown in Fig. 3.6. Note the peak at $T = T_{cr}$ is not a pole but an edge. Since the shape reminds the Greek letter λ the transition is sometimes called λ-transition. In experiments with He_4 one observes a phase transition at the transition temperature $T_c \simeq 2.17$ K. Assuming for liquid Helium the density $\rho \sim 0.15$ g/cm^3, the theory predicts about 3.2 K which is sufficiently close to the experimental value. A closer inspection shows, however, that the experimentally observed transition in He_4 cannot be interpreted as an Einstein condensation. In particular, theory does not reproduce the characteristic asymmetrical λ-shape seen in the experiment. Instead, theory predicts an edge as shown in Fig. 3.6. Since He_4 evidently does not show an Einstein condensation, let us discuss the question, which physical systems are the right candidates for the condensation which Einstein (1925) predicted. The Einstein condition,

$$n \frac{h^3}{(2\pi m k_B T)^{3/2}} \simeq 2.612 \,(2s + 1) , \tag{3.53}$$

shows that we need low temperature, high density, small masses and weak interactions. These conditions are extremely difficult to fulfil simultaneously as one needs temperature in the range of milli- or even micro-Kelvin and relatively high density. Still Schrödinger could not believe, that such conditions can ever be reached for real gases in experiments. Because of the enormous experimental difficulties, a convincing experimental verification of Einsteins prediction of a condensation of atomic gases at low temperature was achieved only 70 years later. At present time, this field of research is intensively studied and already several Nobel Prices have been won for work in this field. In particular, the following systems were intensively studied:

⋆ Gases of He_4 atoms since He_4 atoms are relatively light bosons. Further He_4 gases show superfluidity at low temperature while fermionic He_3 gases do not show superfluid phases. The expectation that the experimentally observed λ-transition is a kind of Einstein condensation could not be clearly confirmed.

⋆ Gases of atoms of alkali and other elements in magnetic traps. In spite of intensive search only in 1995 Petrich et al. (group of Eric Cornell) reported at the *International Conference of* LASER *spectroscopy* on the Island Capri on experiments at the University of Boulder which confirmed Einsteins prediction. Similar experiments were performed at the same time in the group of Wolfgang Ketterle at MIT (Davis et al. 1995).

⋆ Excitonic systems: In highly excited semiconductor systems, a bosonic gas of electron-hole atoms can be created. These so-called excitons have a very small mass and are ideal candidates for Einstein condensation. A problem is, however, the existence of strong Coulomb interactions which can be responsible for other Coulomb induced phase transitions (see Chap. 6). Because of these problems, by now the ultimate evidence of the existence of Bose-Einstein condensation in excitonic systems seems to be still missing. Recent work suggests that electrons and holes in bilayers are promising candidates.

3.5 Black Body Radiation and Relativistic Gases

3.5.1 Planck's Law of Radiation

Let us start with a brief discussion of the original ideas by Planck which led him to the first formulation of a quantum theory. At the time when Planck developed his theory around 1897 the following experimental and theoretical knowledge was available::

⋆ Kirchhoff's law providing a relation between emission and energy density of black body radiation,

⋆ Stefan-Boltzmann's law

$$E = \int u_\nu \, d\nu = \sigma T^4 \,, \tag{3.54}$$

⋆ Wien's law relating the wavelength at the maximum of the distribution with temperature,

$$\lambda_m T = \text{const.} \tag{3.55}$$

The exact radiation law was still unknown but there existed already Wien's proposition for the general functional dependence of the spectral distribution,

$$u_\nu = \nu^3 F\left(\frac{\nu}{T}\right) . \tag{3.56}$$

The general law by Wien was compatible with the special Wien distribution,

$$F\left(\frac{\nu}{T}\right) = \alpha e^{-\beta \frac{\nu}{T}} \tag{3.57}$$

and with the Rayleigh-Jeans distribution

$$u_\nu = \frac{8\pi \nu^2}{c^3} k_B T . \tag{3.58}$$

When Planck started his studies in 1897, he tried to provide a physically sound foundation to Wien's exponential distribution which evidently was related to Boltzmann's distributions. Planck assumed that the interior of a black body is filled with electromagnetic radiation which is in equilibrium with oscillators at the boundaries. Being convinced that the clue of the solution is the concept of entropy, Planck studied first the thermodynamic relations between energy and entropy,

$$\frac{\partial S}{\partial U} = \frac{1}{T} . \tag{3.59}$$

Using Wien's law, we obtain the relations for the first and second derivatives of the entropy,

$$\frac{\partial S}{\partial U} \propto -\log U, \qquad \frac{\partial^2 S}{\partial U^2} \propto -\frac{1}{U} . \tag{3.60}$$

The Rayleigh-Jeans law, however, leads to different relations:

$$\frac{\partial S}{\partial U} \propto -\frac{1}{U}, \qquad \frac{\partial^2 S}{\partial U^2} \propto -\frac{1}{U^2} . \tag{3.61}$$

A possible interpretation is that at small energy, the Wien law determines the second derivative of the entropy and the decay at large energy is determined by the Rayleigh-Jeans law. In order to combine both branches, Planck guessed an interpolation formula:

$$\frac{\partial^2 S}{\partial U^2} \propto -\frac{1}{U(U+b)} . \tag{3.62}$$

From this interpolation formula follows by integration

$$U = \frac{b}{e^{\text{const.}/T} - 1} . \tag{3.63}$$

In combination with Wien's general distribution law, we obtain finally Planck's distribution,

$$u_\nu = A \frac{\nu^3}{e^{\frac{c\nu}{T}} - 1} \,, \tag{3.64}$$

where the constants A and c follow from comparison with Wien's law:

$$A = \frac{8\pi \nu^3 h}{c^3} \,; \qquad c = \frac{h}{k_B} \,. \tag{3.65}$$

Here, h is a new constant, nowadays called Planck's constant which follows from the experiments on black body radiation. Using the best available at that time data, Planck estimated values of h and of k_B, which were already rather close to the numbers we are know by today, $h \simeq 6.626070 \times 10^{-34}$ J s, $k_B = 1.38054 \times 10^{-23}$ J/K. Let us derive now Planck's formula from a more modern point of view making profit of our knowledge about Bose gases (Huang 2001): We consider the radiation of a black body as a box filled with photons treated as bosons with spin zero ($s = 0$). The number of photons in the black box is not constant, since photons are absorbed and emitted from the boundaries. We have, therefore, a condition for the free energy,

$$\left(\frac{\partial F}{\partial N} \right)_{V,T} = 0 \,, \tag{3.66}$$

to express the idea that in thermal equilibrium, for given T and V, the free energy assumes a minimum, $F(T, V, N) \to$ min. Since the number of photons should correspond to this minimum, the partial derivative of F with respect to N should disappear. Since the chemical potential is defined just as this derivative, we obtain the general result that the chemical potential of photons is zero,

$$\mu = 0 \,. \tag{3.67}$$

Again, we see that photons are quite special bosons. According to Planck, the energy is $\varepsilon_k = \hbar \omega_k$. Corresponding to the bosonic character, the mean occupation number of state k is

$$\bar{N}_k = \frac{1}{e^{\frac{\hbar\omega}{k_B T}} - 1} \,. \tag{3.68}$$

The quantization rule for the wave numbers in a rectangular box is

$$k_x a_x = 2\pi \, n_x \,; \quad k_y a_y = 2\pi \, n_y \,; \quad k_z a_z = 2\pi \, n_z \,. \tag{3.69}$$

The number of states in an element of volume is, therefore,

$$dn_x \, dn_y \, dn_z = \frac{a_x a_y a_z}{(2\pi)^3} \, dk_x \, dk_y \, dk_z = \frac{V}{(2\pi)^3} \, dk_x \, dk_y \, dk_z \,, \tag{3.70}$$

and the number of oscillator states in $(k, k + dk)$ is

$$2 \, \frac{V}{(2\pi)^3} \, 4\pi \, k^2 \, dk \,. \tag{3.71}$$

The factor 2 in Eq. (3.71) comes from the two polarization states of the photon. Inserting in Eq. (3.71) $k = \frac{\omega}{c}, c = \lambda\nu, k = \frac{2\pi}{\lambda}$, and $\omega = 2\pi\nu$, we obtain the number of eigen oscillations in $d\omega$:

$$\frac{V\omega^2 d\omega}{\pi^2 c^3} \,. \tag{3.72}$$

Each of these states can be occupied or empty. The mean number of photons in the interval $(\omega, \omega + d\omega)$ is

$$dN_\omega = \frac{V\omega^2 d\omega}{\pi^2 c^3} \, \frac{1}{e^{\frac{\hbar\omega}{k_B T}} - 1} \,. \tag{3.73}$$

The mean energy of these photons is

$$dE_\omega = \hbar\omega \, dN_\omega = \frac{\hbar V}{\pi^2 c^3} f(\omega) \,, \tag{3.74}$$

where

$$f(\omega) = \frac{\omega^3 d\omega}{e^{\beta\hbar\omega} - 1} \tag{3.75}$$

is the spectral energy density of black body radiation (see Fig. 3.7). Expanding Eq. (3.74) for small frequencies ($\beta\hbar\omega \ll 1$) or for high temperature, respectively, and using the approximation $e^{\beta\hbar\omega} \approx 1 + \beta\hbar\omega$, we obtain the Rayleigh-Jeans radiation law,

$$dE_\omega = \frac{V k_B T \omega^2}{\pi^2 c^3} d\omega \,. \tag{3.76}$$

This radiation law is of purely classical character, evidenced by the absence of Planck's constant. In the limit of high frequency ($\beta\hbar\omega \gg 1$) we recover Wien's radiation law

$$dE_\omega = \frac{\hbar V \omega^3}{\pi^2 c^3} e^{-\beta\hbar\omega} d\omega \,. \tag{3.77}$$

Fig. 3.7 Planck's law for the
spectral energy density of
radiation inside a black body

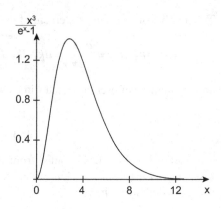

The maximum of the spectrum is given by Eq. (3.74); this way we find Wien's law
of spectral shift

$$\omega_{\max} = 2.822 \frac{k_B T}{\hbar} \,. \tag{3.78}$$

As follows from this discussion, on the basis of Planck's radiation law, a great
variety of experimental facts could be interpreted theoretically.

3.5.2   Radiation as a Relativistic Gas

The statistics of radiation is the statistics of photons which is completely different
from the statistics of particles with mass m at velocities far below the speed of light.
This is the main reason why photon gases behave differently from normal gases and
plasmas. As a first sign of this specific behavior, we remember the Stefan-Boltzmann
law which states that the relation between energy density, pressure, and temperature
of a photon gas is given by

$$\rho_E(T) = A T^4 \,; \qquad p = \frac{1}{3}\rho_E(T) = \frac{A}{3} T^4 \,. \tag{3.79}$$

This means that pressure and energy density increase very fast, with the fourth
power of the temperature. Another fundamental equation relates temperature and
mass density, ρ_M, for adiabatic processes of photon gases,

$$\rho_M(T) \propto T^3 \,. \tag{3.80}$$

The class of systems which follow these or similar laws is quite large and includes
besides the well-known photon gas also a large part of the plasmas in our Universe,
called relativistic plasmas (Landau and Lifshitz 1990).

3.6 Thermodynamic Functions of Fermi Gases

3.6.1 The Ideal Fermi Gas

We consider a Fermi gas in a rectangular box of volume $V = a_x a_y a_z$ with periodic boundary conditions filled by N fermions with spin $S = 1/2, 3/2, \ldots$. The solutions of the corresponding stationary Schrödinger equation are the wave functions

$$\psi(\mathbf{r}) = \frac{1}{\sqrt{V}} e^{\frac{i}{\hbar}(p_x x + p_y y + p_z z)} . \tag{3.81}$$

Periodic boundary condition, $\psi(x + a_x, y, z) = \psi(x, y, z)$, imply $\frac{a_x p_x}{\hbar} = 2\pi n_x$, thus, in general

$$p_x = n_x \frac{h}{a_x} ; \qquad p_y = n_y \frac{h}{a_y} ; \qquad p_z = n_z \frac{h}{a_z} . \tag{3.82}$$

For the case $a_x = a_y = a_z = L$ we find $p_k = \frac{nh}{L}$ with $n = 0, \pm 1, \pm 2, \ldots$ Per lattice point in the quantized momentum space, we have $2s + 1$ different possible spin states. The energy is quantized according to

$$\varepsilon_k = \frac{1}{2m} \left(p_{x_k}^2 + p_{y_k}^2 + p_{z_k}^2 \right) . \tag{3.83}$$

Each cell in the momentum space has the volume (see Fig. 3.8)

$$\Delta p_x \, \Delta p_y \, \Delta p_z = \frac{h}{a_x} \frac{h}{a_y} \frac{h}{a_z} = \frac{h^3}{V} , \tag{3.84}$$

Fig. 3.8 Fermi-sphere in momentum space and discrete quantum states

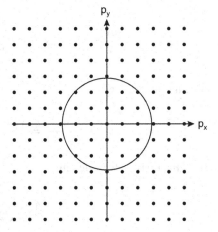

and in phase space, $\frac{h^3}{V} V = h^3$. For ideal gases without interactions, the elementary cell in phase space is a column as long as the macroscopic container, L, and as short as h/L in the direction of momentum. We replace now the summations by integrals over the phase space (note that in the case of infinite limits we often omit the integration intervals):

$$\sum_{\text{states}} \quad \rightarrow \quad (2s+1) \int \frac{dp_x \, dp_y \, dp_z}{h^3} \, dx \, dy \, dz \,, \tag{3.85}$$

where the integral covers the available phase space. Therefore,

$$N = \sum_k N_k = \sum_k \frac{1}{e^{\frac{\epsilon_k - \mu}{k_B T}} + 1} \rightarrow (2s+1)V \int \frac{dp_x dp_y dp_z}{h^3} \frac{1}{e^{\frac{p^2}{2m} - \mu}{k_B T} + 1}. \tag{3.86}$$

Introducing spherical coordinates, we obtain

$$\frac{N}{V} = n = (2s+1) \int_0^\infty dp \frac{4\pi p^2}{h^3} \frac{1}{e^{\frac{p^2}{2mk_B T} - \frac{\mu}{k_B T}} + 1} \,, \tag{3.87}$$

which is a formula for the chemical potential, $\mu = \mu(n, T)$. In the following we investigate several limits.

3.6.2 Fermi Gases in the High- and Low-Temperature Limits

For the case of high temperature (classical limit), we may neglect unity in the denominator of Eq. (3.87) and obtain

$$\frac{N}{V} = n = (2s+1) \frac{4\pi}{h^3} \int_0^\infty dp \, p^2 e^{-\frac{p^2}{2mk_B T}} \, e^{\frac{\mu}{k_B T}} \tag{3.88}$$

and after integration

$$\frac{n}{2s+1} = \frac{(2\pi m k_B T)^{3/2}}{h^3} \, e^{\frac{\mu}{k_B T}} \,. \tag{3.89}$$

Using the thermal wave length

$$\Lambda = \frac{h}{\sqrt{2\pi m k_B T}} \tag{3.90}$$

Fig. 3.9 Electron gas: sketch of the situation with respect to degeneracy in the density-temperature plane. The classical and slightly degenerate region is located at high temperature and low density below the line $\frac{n\Lambda^3}{2s+1} = 1$. The degenerate low-temperature regions is above this line

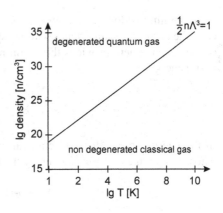

we obtain

$$\frac{n\Lambda^3}{2s+1} = e^{\frac{\mu}{k_BT}} \tag{3.91}$$

and eventually the chemical potential

$$\mu(n, T) = k_B T \ln\left(\frac{n\Lambda^3}{2s+1}\right). \tag{3.92}$$

Up to the spin term this is the same expression as known from Boltzmann statistics. In Fig. 3.9 we show the limit of the classical approximation. According to Eq. (3.87) unity can be neglected only if

$$\frac{n\Lambda^3}{2s+1} \ll 1. \tag{3.93}$$

The limit of the classical region corresponds to equality. In log-scale we find a line

$$\lg n = \frac{3}{2} \lg T + \text{const.} \tag{3.94}$$

For energy and pressure, we find in the non-degenerate region the classical expressions

$$U = \sum_k \frac{p_k^2}{2m}\bar{N}_k = \frac{3}{2}Nk_BT \tag{3.95}$$

$$p = \sum_k p_k = n\,k_BT. \tag{3.96}$$

In the limit of low temperature we speak about degenerate Fermi gases. In this case, the particles assume states within a sphere in the momentum space. The radius of the Fermi sphere provides the Fermi momentum, p_F, and corresponding to the Fermi energy

$$E_F = \frac{p_F^2}{2m}.$$

(3.97)

In the energy space, the low-temperature distribution is approximated by a box whose maximal extension is given by the Fermi energy

$$\bar{N}_k = \Theta\left(\varepsilon_F - \varepsilon_k\right),$$

(3.98)

with $\varepsilon_F = \mu(T = 0)$. For ε_F we have

$$\varepsilon_F = \lim_{T \to 0} \mu(n, T).$$

(3.99)

It follows

$$N = \frac{2s + 1}{h^3} V \int\limits_{|p| \le p_F} dp_x \, dp_y \, dp_z$$

(3.100)

and

$$\frac{N}{V} = (2s + 1)\frac{4\pi}{h^3} \int\limits_0^{p_F} dp \, p^2 = (2s + 1)\frac{1}{h^3}\frac{4\pi}{3} p_F^3 = (2s + 1)\frac{1}{\hbar}\frac{p_F^3}{6\pi^2}.$$

(3.101)

We obtain Fermi momentum and Fermi energy

$$p_F = \hbar \left(\frac{6\pi^2 n}{2s + 1}\right)^{1/3} ; \qquad \varepsilon_F = \frac{p_F^2}{2m} = \frac{\hbar^2}{2m}\left(\frac{6\pi^2 n}{2s + 1}\right)^{2/3}.$$

(3.102)

For the internal energy follows

$$U = \sum_k \frac{p^2}{2m} \bar{N}_k = \frac{(2s + 1)V}{h^3} \int\limits_0^{p_F} dp \, 4\pi \, p^2 \frac{p^2}{2m} = \frac{p_F^2}{2m}\frac{2s + 1}{h^3} V \frac{4\pi}{5} p_F^3$$

$$= \frac{3}{5} N\varepsilon_F.$$

(3.103)

In Figs. 3.10 and 3.11 we draw the internal energy, U/N, and the specific heat, C_v, as function of temperature.

Fig. 3.10 Internal energy per particle U/N as a function of the temperature

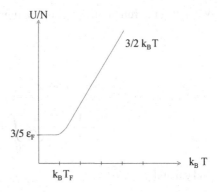

Fig. 3.11 Specific heat of a Fermi gas, C_V, as a function of temperature

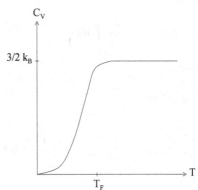

3.7 Density-Dependent Fermi Gas Functions

3.7.1 Expansions for Weakly Degenerated Fermi Gases

After studying the limiting cases, $T \to \infty$ and $T \to 0$, we consider now the region of weak degeneration and consider the Fermi integral (3.87) in more detail. With the abbreviations

$$p = x\sqrt{2m\,k_B T}\,; \quad z = e^\alpha\,; \quad \alpha = \frac{\mu}{k_B T} \tag{3.104}$$

we write Eq. (3.87) in the form

$$n = (2s + 1)\left(\frac{2mk_B T}{h^2}\right)^{3/2} 4\pi \int_0^\infty dx\, x^2 \frac{1}{e^{x^2-\alpha}+1}\,. \tag{3.105}$$

We define the function $f_{3/2}(z)$ of the fugacity,

$$z = e^{\alpha} = e^{\beta \mu},$$
(3.106)

by

$$\frac{n \Lambda^3}{2s+1} = \frac{4}{\sqrt{\pi}} \int\limits_0^{\infty} dx \, \frac{x^2 z}{e^{-x^2} + z} \equiv f_{3/2}(z)$$
(3.107)

and expand $f_{3/2}(z)$,

$$f_{3/2}(z) = \frac{4}{\sqrt{\pi}} \int\limits_0^{\infty} dx \, \frac{x^2 z e^{-x^2}}{1 + z e^{-x^2}} = \frac{4}{\sqrt{\pi}} z \int\limits_0^{\infty} dx \, x^2 e^{-x^2} \left(1 - z e^{-x^2} + \dots \right).$$
(3.108)

Exploiting

$$\int\limits_0^{\infty} dx \, x^2 e^{-ax^2} = \frac{1}{4} \sqrt{\frac{1}{a^3}},$$
(3.109)

we find the expansion

$$f_{3/2}(z) = z - \frac{z^2}{2^{3/2}} + \frac{z^3}{3^{3/2}} - \frac{z^4}{4^{3/2}} + \dots = \sum_{l=1}^{\infty} (-1)^{l+1} \frac{z^l}{l^{3/2}}.$$
(3.110)

In the general case, we define the *Fermi-functions* in the region of convergence by the expansion

$$f_{\lambda}(z) = \sum_{l=1}^{\infty} (-1)^{l+1} \frac{z^l}{l^{\lambda}}.$$
(3.111)

Instead of the present definition of Fermi functions we can use another definition which is in some respect more convenient (De Witt 1961, 1962, 1966; Kraeft et al. 1986):

$$I_{\nu}(y) = \frac{1}{\Gamma(\nu+1)} \int_0^{\infty} dx \, \frac{x^{\nu}}{e^{x-y} + 1}.$$
(3.112)

There exists an extended mathematical literature about the properties of this class of functions (see e.g. Fukushima 2015; Lether 2000). In first Taylor approximation, we find the corrections to the Boltzmann expressions for fugacity and chemical potential,

$$z = \frac{n\Lambda^3}{2s+1} + \frac{1}{2\sqrt{2}}\left(\frac{n\Lambda^3}{2s+1}\right)^2 \tag{3.113}$$

$$\frac{\mu}{k_B T} = \ln z = \ln \frac{n\Lambda^3}{2s+1} + \frac{1}{2\sqrt{2}}\frac{n\Lambda^3}{2s+1} + \ldots \tag{3.114}$$

Similar expressions follow for the pressure,

$$p = k_B T \frac{4\pi}{h^3}(2s+1)\int_0^\infty dp\, p^2 \ln\left(1 + z e^{-\frac{p^2}{2mk_B T}}\right). \tag{3.115}$$

With $p = x\sqrt{2mk_B T}$, we find

$$p = k_B T \frac{2s+1}{\Lambda^3} f_{5/2}(z) = \frac{k_B T(2s+1)}{\Lambda^3}\left(z - \frac{z^2}{2^{2/5}} + \cdots\right)$$

$$\simeq n k_B T\left(1 + \frac{1}{4\sqrt{2}}\frac{n\Lambda^3}{2s+1}\right). \tag{3.116}$$

We see that the pressure of a Fermi gas is larger than the pressure of a Boltzmann gas and increases with density (see also Fig. 3.12).

3.7.2 Thermodynamics in Full Density Range

So far, we followed the standard route (see e.g. Huang 1963, 2001; Kremp et al. 2005) and worked in the grand canonical ensemble, considering the fugacities as independent input parameters. Only in the last section, for low density we expressed the pressure in terms of density. For practical applications, however, the primary input parameters are almost always the densities. Fugacities (chemical potentials) are, thus, dependent quantities which are difficult to control in experiments. In order to compare theory with experiment, we need a rule to represent the thermodynamics functions as an explicit functions of the density, covering a wide range. The implicit representation of the equation of state by two equations for p and n dependent on fugacity is inconvenient and in most representations it does not converge. Therefore, we have to solve the so-called inversion problem. In principle it should be enough to get one thermodynamic potential, for instance, the free energy, $F(T, V, N)$, as a function of density. All other potentials can be found by differentiation using the Maxwell relations.

Fig. 3.12 Free energy per electron and chemical potential of an ideal electron gas in Fermi-Dirac approximation. The equation of state is found from the difference between both curves. The density is given in the dimensionless units $y = n_e \Lambda_e^3/2$. The deviations from the thermodynamics of a classical gas strongly increase with the density (degeneracy)

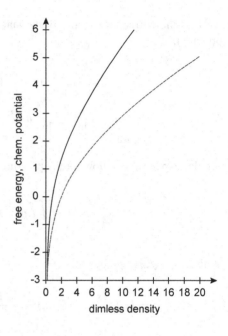

In spite of the fact, that one thermodynamic potential is enough, all others could be derived, we find it more convenient, to start from two or three explicit thermodynamic functions. This procedure allows to obtain all other functions just by arithmetic operations as addition or subtraction. Note that differentiation is frequently problematic with regards to numerical precision. We study the Gibbs potential, G_e, free energy, F_e, and the internal energy, U_e and derive the pressure, p_e, the entropy, S_e, and further quantities by simple arithmetic relations, such as

$$p_e V = F_e - G_e ; \qquad T S_e = U_e - F_e . \qquad (3.117)$$

In the previous sections, we expressed these functions in dependence of activity (chemical potential). The problem to replace the dependence on activity by a dependence on density is known as *inversion problem*. We find it most convenient to start with the chemical potential of electrons (which is the Gibbs potential per electron) and with the free energy per electron, both measured in $k_B T$ (or possibly in Rydberg) as primary quantities (Ebeling 1985; Ebeling et al. 1986),

$$\alpha_e = \frac{\beta G_e}{N_e} ; \qquad \varphi_e = \frac{\beta F_e}{N_e} . \qquad (3.118)$$

The main contribution to these functions comes from the ideal terms which were considered already in detail on a rather elementary basis. The extension to the region of weak correlations requires a more advanced analysis of the series with respect to the degeneracy parameter. There is a large body of mathematical literature on Fermi integrals (see e.g. Fukushima 2015; Lether 2000). The problem of inversion, that is, replacing the variable α to n is certainly not trivial (Zimmermann 1988). Note that the known limiting cases have to be represented:

$$\alpha_e \rightarrow \begin{cases} \ln y & \text{for } y \ll 1 \\ \left(\frac{3\sqrt{\pi}}{4}\right)^{2/3} y^{2/3} & \text{for } y \gg 1 . \end{cases} \tag{3.119}$$

We will use a convenient formula proposed by Zimmermann (1988) consisting of two different analytical expressions for low and high density respectively, given in dimensionless units, $y = n_e \Lambda_e^3 / 2$, which are smoothly interpolated

$$\varphi_e^{FD}(y) = w_1(y - 5.5) \left[\ln(y) - 1 + 0.1768\, y - 0.0165\, y^2 + 0.000031\, y^3 \right]$$

$$+ w_2(y - 5.5) \left[0.7254\, y^{2/3} - 2.040\, y^{-2/3} - 0.45\, y^{-2} \right] \tag{3.120}$$

$$\alpha_e^{FD} = w_1(y - 2.5) \left[\ln(y) + .3536\, y - 0.00198\, y^2 + 0.000124\, y^3 \right]$$

$$+ w_2(y - 2.5) \left[1.209\, y^{2/3} - 0.6803\, y^{-2/3} + 0.45\, y^{-2} \right] , \tag{3.121}$$

where the interpolating weight factors are defined by

$$w_1(x) = \frac{1}{1 + e^{bx}} ; \qquad w_2(x) = \frac{1}{1 + e^{-bx}} . \tag{3.122}$$

(with $b = 10$). These expressions are very convenient for practical calculations of the thermodynamic functions including the equation of state

$$\pi = \beta n\, p^{FD} = \varphi_e^{FD}(y) - \alpha_e^{FD} . \tag{3.123}$$

3.8 Hartree-Fock Theory of Weakly Interacting Electron Gases

Let us first specify the conditions under which the Hartree-Fock approximation is justified (Fig. 3.13). As we already know, there is a large region in the density-temperature plane, where plasmas behave like ideal Fermi gases. The Hartree-Fock approximation covers a region near to the boundary of the ideal region, where correlations due to interactions are still very weak but noticeable. Here, the correlations are mainly due to exchange effects between the plasma particles. One denotes these effects often as Hartree-Fock effects. In this section we consider only the simple case of electron plasmas imbedded in a uniform positive background.

Fig. 3.13 Different regions
of non-ideality in plasmas; we
consider here the corridor of
weak interactions $1 \lesssim r_s \lesssim 5$
(horizontal lines) and
$1 \lesssim \xi \lesssim 5$ (vertical lines)

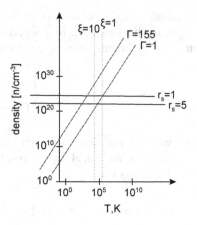

We assume the model by Joseph Mayer (1950) regularising the interaction potential
between two electrons and the corresponding Fourier transform:

$$V_{ee}(r) = \frac{e^2}{r} ; \quad \tilde{V}_{ee}(\mathbf{t}) = \int_V d(\mathbf{r})\, V_{ab}(r) e^{i\mathbf{t}\cdot\mathbf{r}} = 4\pi \frac{e^2}{t^2} . \tag{3.124}$$

Note that the integrals extend over the entire volume which is assumed infinite,
$V \rightarrow \infty$. In case the volume is finite, we use instead of the Mayer model
the regularization by Pines (1963) given as a representation through a Fourier
series,

$$V_{ee}(r) = \frac{1}{V} \sum_{\mathbf{q}} e^{-i\mathbf{q}\cdot\mathbf{r}}\, \tilde{V}_{ee}(\mathbf{q}) , \tag{3.125}$$

where the sum is to be extended over a discrete spectrum of wave modes, $\mathbf{q}$, in
the finite volume. The Fourier transform is then defined as (Bohm and Pines 1951;
Pines 1963; Pines and Nozieres 1966)

$$\tilde{V}_{ee}(\mathbf{q}) = \int_V d(\mathbf{r})\, V_{ab}(r) e^{i\mathbf{q}\cdot\mathbf{r}} = 4\pi \frac{e^2}{q^2} [1 - \delta(\mathbf{q}, 0)] . \tag{3.126}$$

As these formulae show, the case $\mathbf{q} = 0$ plays a special rôle for Coulomb systems
since the corresponding contribution is missing in the Bohm-Pines-Nozieres series
representation, indicated by the Kronecker symbol. In a recent investigation by
Bobrov et al. (2015) based on relativistic electrodynamics, it is shown that the
case $\mathbf{q} = 0$ does not contribute, since in parallel moving charges do not exchange
photons.

We introduce now a dimensionless interaction parameter, ξ, defined as the
relation of the Coulomb energy at a distance equal to the thermal wave length,

λ, to the mean kinetic energy of electrons, Θ_e, as it results due to Fermi statistics:

$$\xi = \frac{e^2}{\Theta_e \lambda}; \qquad \lambda = \frac{\hbar}{\sqrt{mk_B T}}; \qquad \Lambda_e = \frac{h}{\sqrt{2\pi m_e k_B T}} \qquad (3.127)$$

$$\Theta_e = \frac{2k_B T}{n_e \Lambda_e^3} I_{3/2}(\alpha_e); \qquad \frac{n_e \Lambda_e^3}{2s_k + 1} = I_{1/2}(\alpha_e). \qquad (3.128)$$

This is an implicit equation for the chemical potential (according to Fermi statistics) as a function of the density.

In order to study the influence of weak correlations to the properties of ideal gases, we start with an electron plasma on a smeared out positive background. So far we neglected interactions which are now taken into account to lowest order, which is the Hartree-Fock approximation. This approximation refers to certain regions in the density-temperature plane shown in Fig. 3.13. Outside these regions the laws of (ideal) Fermi-Dirac gases are valid.

There are different ways to calculate the Hartree-Fock corrections (Ichimaru 1992; Klimontovich 1982; Toda et al. 1983). A first simple way is to compute the corrections through the mean Coulomb energy of the electrons expressed in coordinate space

$$\langle V_c \rangle = \frac{1}{2} V n_e^2 \langle V_{ee}(1,2) \, F_{ee}(1,2) \rangle, \qquad (3.129)$$

where $F_{ee}(1,2) = 1 + g_{ee}(1,2)$ is the quantum statistical pair distribution function of an ideal system and $g_{ee}(1,2)$ the corresponding correlation function of two non-interacting electrons. In the limit of low density, we have

$$g_{ee}(1,2) = -\frac{1}{2} e^{-r^2/\lambda_{ee}^2}; \qquad \lambda^2 = \lambda_{ee}^2 = \frac{\hbar^2}{m_e k_B T}. \qquad (3.130)$$

The first term in the integral is formally infinite, but this term cancels with the contribution of the background (Bohm and Pines 1951; Kremp et al. 2005). Note that this term does not appear in several regularized models (Bobrov et al. 2015; Pines 1963; Pines and Nozieres 1966). The second term is convergent and delivers after elementary integration the low density limit for the Hartree-Fock contribution to the Coulomb energy and the free energy (Ebeling et al. 1976):

$$E^{HF} = F^{HF} = -V n_e e^2 \pi \int dr \, r e^{-r^2/\lambda^2} = -\frac{V}{2} \pi n_e e^2 \lambda^2. \qquad (3.131)$$

In the opposite case of high density (strong degeneracy), the correlation function of noninteracting electrons is given by (Pines and Nozieres 1966)

$$g_{ee}(r) = -\frac{9}{2}\left(\frac{\sin(p_F\,r) - p_F\,r\cos(p_F\,r)}{p_F^3\,r^3}\right)^2 ; \qquad p_F^1 = 0.521\,a_B\,r_s .$$

(3.132)

By integration follows the ground state energy,

$$E^{HF} = -N_e\,\frac{3}{5}\,\frac{p_F^2}{2m} .$$

(3.133)

In order to find the general expression of the Hartree-Fock contributions valid for any degree of degeneracy, one chooses a more convenient representation in the momentum space for the Hartree-Fock expression (Ebeling et al. 1976; Kraeft et al. 1986; Kremp et al. 2005):

$$\langle V_c \rangle = V \int \frac{d\mathbf{p}_1\,d\mathbf{p}_2}{(2\pi)^6}\,V_{ee}\,(\mathbf{p}_1\,\mathbf{p}_2)\,f_e\,(\mathbf{p}_1)\,f_e\,(\mathbf{p}_2) .$$

(3.134)

These analytic expressions depend on the full Fermi functions, f_e, taken as functions of the momenta. By some transformations of the integral we obtain (De Witt 1961, 1962, 1966; Kraeft et al. 1986)

$$\langle V_c \rangle = -2V\,\frac{e^2}{\Lambda_e^4}\int_{-\infty}^{\alpha} dy\,I_{-1/2}^2(y) .$$

(3.135)

The Hartree-Fock contribution to the free energy follows then by charging

$$\varphi_e^{HF}(\alpha_e) = -\frac{2e^2}{2\,n_e\,k_B T\,\Lambda_e^4}\int_{-\infty}^{\alpha} dy\,I_{-1/2}^2(y) .$$

(3.136)

By differentiation with respect to α we find a particularly simple expression for the chemical potential at any degree of degeneracy,

$$\alpha_e^{HF} = -\frac{e^2}{k_B T\,\Lambda_e}\,I_{-1/2}(\alpha_e) .$$

(3.137)

Note that density and ideal chemical potential are related by the implicit equations

$$\frac{1}{2}n_e\Lambda_e^3 = I_{1/2}(\alpha_e) ; \qquad \alpha_e = \beta\mu_e^{FD} .$$

(3.138)

The Hartree-Fock approximations given here were derived as the first corrections to ideality in the grand-canonical ensemble. If we are interested in the corrections for an ensemble with given density and temperature, we have to introduce the canonical ideal chemical potentials from the Fermi expressions, Eq. (3.138), which have to be inverted (Kraeft et al. 1986). The problem to replace the dependence on the activities (chemical potential) by a dependence on density is a similar inversion problem as for the ideal Fermi gas. Again, as pointed out above, for the ideal case we find it most convenient to start with the Gibbs potential per electron which is the chemical potential of electrons, and with the free energy per electron measured in $k_B T$ as primary quantities (Ebeling 1985; Ebeling et al. 1986)

$$\alpha_e = \frac{\beta \, G_e}{N_e} \; ; \qquad \varphi_e = \frac{\beta F_e}{N_e} \, . \tag{3.139}$$

After obtaining these two thermodynamic functions in Hartree-Fock approximation, all other quantities such as pressure follow just by algebraic operations. The first step is to find $\alpha = \alpha_e$ as a function of the dimensionless electron density, $y = n_e \Lambda_e^3 / 2$. The limiting cases of small or large density,

$$\alpha^{HF} \to -\frac{e^2}{k_B T \, \Lambda_e} \, y \; ; \qquad \alpha^{HF} \to \left(\frac{3\sqrt{\pi}}{4} \right)^{2/3} y^{2/3} \, , \tag{3.140}$$

respectively, must be compatible. For concrete calculations, we consider most useful a (modified) method of piecewise representations and interpolations developed for the ideal Fermi contributions by Zimmermann (1988). We find

$$\varphi_e^{HF}(y) = -\frac{e^2}{k_B T \, \Lambda_e} \left(w_1(y - 1.8) \left[0.5 \, y - 0.3540 \, y^2 + 0.1825 \, y^3 \right] \right.$$
$$\left. + w_2(y - 1.8) \left[0.9307 \, y^{1/3} - 0.6976 \, y^{-1} + 0.7635 \, y^{-7/3} \right] \right) \tag{3.141}$$

$$\alpha_e^{HF} = -\frac{e^2}{k_B T \, \Lambda_e} \left(w_1(y - 0.8) \left[y - 1.0612 \, y^2 + 0.7299 \, y^3 \right] \right.$$
$$\left. + w_2(y - 0.8) \left[1.241 \, y^{1/3} - 0.6981 \, y^{-1} + 1.018 \, y^{-7/3} \right] \right) . \tag{3.142}$$

Here the interpolating weight factors were again defined by

$$w_1(x) = \frac{1}{1 + e^{bx}} \; ; \qquad w_2(x) = \frac{1}{1 + e^{-bx}} \; ; \qquad b = 10 \, . \tag{3.143}$$

These expressions are compatible with the approximations for low density and for high degeneracy, Eq. (3.140), and are useful for explicit representations of thermodynamic functions as, e.g., the equation of state which reads

$$\beta p^{HF}(y) = \alpha_e^{HF}(y) - \varphi_e^{FD}(y) \, . \tag{3.144}$$

Fig. 3.14 Free energy per electron (upper curve) and chemical potential (lower curve) of a weakly interacting electron gas in Hartree-Fock approximation given in units $(e^2/k_B T \Lambda_e)$. Note that the difference of the curves provides the Hartree-Fock pressure. The Hartree-Fock contribution increases first linearly with density and changes then at the transition to strong degeneracy its analytical shape to a $n^{1/3}$-law

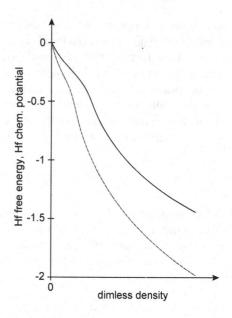

We see that the Hartree-Fock contributions increase with density and change their analytical form at certain degeneracy parameter, where the non-degenerate behavior transits sharply into the degenerate behavior, see Fig. 3.14. In certain regions of the density and temperature (in Fig. 3.13 this is the region with $r_s > 1$ and $\Gamma > 1$), the Hartree-Fock contribution is not small and delivers a substantial correction to the thermodynamic properties of the ideal Fermi-gas.

References

Anderson, M.H., J.R. Ensher, M.R. Matthews, C.E. Wieman, and E.A. Cornell. 1995. Observation of Bose-Einstein Condensation in a Dilute Atomic Vapor. *Science* 269: 198–201.

Bobrov, V.B., A.G. Zagorodny, and S.A. Trigger. 2015. Coulomb Potential of Interaction and Bose-Einstein Condensate. *Low Temperature Physics* 41: 1154–1163 (in Russian).

Bogolyubov, N.N. 1991. *Selected Works*. Vol. II. Quantum and Classical Statistical Mechanics. New York: Gordon and Breach.

Bohm, D., and D. Pines. 1951. A Collective Description of Electron Interactions I. Magnetic Interactions. *Physics Review* 82: 625–634.

Bose, S.N. 1924a. Plancks Gesetz und Lichtquantenhypothese. *Zeitschrift für Physik* 26: 178–181.

Bose, S.N. 1924b. Wärmegleichgewicht im Strahlungsfeld bei Anwesenheit von Materie. *Zeitschrift für Physik* 27: 178–181.

Davis, K.B., M.O. Mewes, M.R. Andrews, N.J. van Druten, D.S. Durfee, D.M. Kurn, and W. Ketterle. 1995. Bose-Einstein Condensation in a Gas of Sodium Atoms. *Physical Review Letters* 75: 3969–3973.

DeWitt, H.E. 1961. Thermodynamic Functions of a Partially Degenerate, Fully Ionized Gas. *Journal of Nuclear Energy, Part C Plasma Physics* 2: 27–45.

DeWitt, H.E. 1962. Evaluation of the Quantum-Mechanical Ring Sum with Boltzmann Statistics. *Journal of Mathematical Physics* 3: 1216–1228.

DeWitt, H.E. 1966. Statistical Mechanics of High-Temperature Quantum Plasmas Beyond the Ring Approximation. *Journal of Mathematical Physics* 7: 616–626.

Debye, P. 1912. Zur Theorie der Spezifischen Wärmen. *Annalen der Physik* 39: 789–839.

Dirac, P.A.M. 1926. On the Theory of Quantum Mechanics. *Proceedings of Royal Society London A: Mathematical, Physical and Engineering Sciences* 112: 661–677.

Ebeling, W. 1985. Statistical Thermodynamics of Fluid Hydrogen at High Energy Density. *Physica A* 130: 587–596.

Ebeling, W., and D. Hoffmann. 2014. Eine Vorlage Einsteins in der Preußischen Akademie der Wissenschaften. *Leibniz Online*. http://www.leibnizsozietaet.de/wp-content/uploads/2014/12/EbelingHoffmann.pdf.

Ebeling, W., W.D. Kraeft, and D. Kremp. 1976. *Theory of Bound States and Ionisation Equilibrium in Plasmas and Solids*. Berlin: Akademie-Verlag.

Ebeling, W., W.D. Kraeft, D. Kremp, and G. Röpke. 1986. Quantum Statistics of Coulomb Systems: Thermodynamic Functions and Phase Transitions. *Physica* 140a: 160–168.

Einstein, A. 1907. Die Plancksche Theorie der Strahlung und die Theorie der Spezifischen Wärme. *Annals of Physics* 22: 180–190.

Einstein, A. 1924. Quantentheorie des Einatomigen Idealen Gases. *Sitzungsber. Preuss. Akad. Wiss. Phys.-math. Kl* 22: 261–267.

Einstein, A. 1925. Quantentheorie des Einatomigen Idealen Gases. Zweite Abhandlung. *Sitzungsber. Preuss. Akad. Wiss. Phys.-math. Kl* 23: 3–14.

Fermi, E. 1926. Über die Wahrscheinlichkeit der Quantenzustände. *Zeitschrift für Physik* 26: 54–56.

Feynman, R.P. 1972. *Statistical Mechanics*. Reading, MA: Benjamin.

Fick, E. 1981. *Einführung in die Grundlagen der Quantentheorie*. Leipzig: Akademischer Verlag.

Fukushima, T. 2015. Precise and Fast Computation of Fermi-Dirac Integral of Integer and Half Integer Order by Piecewise Minimax Rational Approximation. *Applied Mathematics and Computation* 259: 708–729.

Huang, K. 1963. *Statistical Mechanics*. New York: Wiley.

Huang, K. 2001. *Introduction to Statistical Physics*. London: Taylor & Francis.

Ichimaru, S. 1992. *Statistical Plasma Physics*. Redwood: Addison-Wesley.

Kirsten, C., and H.G. Körber, eds. 1975. *Physiker über Physiker*. Berlin: Akademie-Verlag.

Klimontovich, Yu. L. 1982. *Statistical Physics*. Moscow: Nauka (in Russian).

Kraeft, W.D., D. Kremp, W. Ebeling, and G. Röpke. 1986. *Quantum Statistics of Charged Particle Systems*. Berlin: Akademie-Verlag.

Kremp, D., M. Schlanges, and W.D. Kraeft. 2005. *Quantum Statistics of Nonideal Plasmas*. Berlin: Springer.

Landau, L.D., and E.M. Lifshitz. 1976. *Statistical Physics (Part I)*. Moscow: Nauka.

Landau, L.D., and E.M. Lifshitz. 1980. *Statistical Physics*. Oxford: Butterworth-Heinemann.

Landau, L.D., and E.M. Lifshitz. 1990. *Statistical Physics*. New York: Pergamon.

Lether, F.G. 2000. Analytical Expansions and Numerical Approximations of Fermi-Dirac Integrals $Fj(x)$ of Order $j = -1 = 2$ and $j = 1/2$. *Journal of Science Communication* 15: 479–497.

Mayer, J.E. 1950. The Theory of Ionic Solutions. *The Journal of Chemical Physics* 18: 1426–1436.

Pauli, W. 1925. Über den Zusammenhang des Abschlusses der Elektronengruppen im Atom mit der Komplexstruktur der Spektren. *Zeitschrift für Physik* 31: 765–783.

Petrich, W., M.H. Anderson, J.R. Ensher, and E.A. Cornell. 1995. Stable, Tightly Confining Magnetic Trap for Evaporative Cooling of Neutral Atoms. *Physical Review Letters* 74: 3352–3355.

Pines, D. 1963. *Elementary Excitations in Solids*. Lecture Notes and Supplements in Physics. New York: Benjamin.

Pines, D., and P. Nozieres. 1966. *The Theory of Quantum Liquids*. New York: Benjamin.

Riemann, B. 1859. Ueber die Anzahl der Primzahlen unter einer gegebenen Grösse. *Monatsberichte der Berliner Akademie*, 671–680.

Toda, M., R. Kubo, and N. Saito. 1983. *Statistical Physics*. Vols. I and II. Berlin: Springer.
Treder, H.J. 1983. *Große Physiker und ihre Probleme – Studien zur Geschichte der Physik*. Berlin: Akademie Verlag.
Zimmermann, R. 1988. *Many Particle Theory of Highly Excited Semiconductors*. Leipzig: Teubner.

Chapter 4
Density Operators and Other Tools of Quantum Statistics

In this chapter, we will introduce useful tools of Quantum Statistics. Most of them will be used in later chapters of this book to solve concrete problems. Our survey covers, of course, the most prominent methods such as density operators introduced by von Neumann and Landau, Wigner's phase-space functions method, and Bogolyubov's method of reduced density operators. Matsubara's thermodynamical Green's functions and real-time Green's functions are important methods in the field of quantum plasmas but are discussed here only rather briefly. We consider this method as very relevant, evidenced by many important results (Ebeling et al. 1976; Kraeft et al. 1986), however, here we wish to refer to the excellent and still available book by Kremp et al. (2005), based on these particular methods where "an introduction is given into the quantum statistics of of equilibrium and non-equilibrium systems on the basis of the methods of real-time Green's functions" (quotation from the introduction, l.c.). The authors state that the Green's function method is the "unified point of view of quantum field theoretical methods" (Kremp et al. 2005). We will show in this book that there are many further beautiful and fruitful methods, such as Wigner's phase space description, Bogolyubov's hierarchy, Bohm-Pines' collective variables, Kelbg's method of effective potentials, Feynman's method of path integrals, and Klimontovich's method of random density operators. We share the point of view that in quantum statistics the diversity of methods including the elegant methods developed by Wigner, Bohm-Pines, Feynman, Bogolyubov, Kelbg, Klimontovich, and many others are part of a beautiful culture of quantum statistics and our students should have at hand at least an overview about the most important theoretical methods in order to make their personal choice.

For an overview of other methods, the reader is referred to the extensive literature (e.g. Ashcroft and Mermin 1976; Feynman and Hibbs 1965; Huang 2001; Ichimaru 1992; Pines 1963; Pines and Nozieres 1966; ter Haar 1995; Thirring 1980; Toda et al. 1983; Tolman 1938).

© Springer Nature Switzerland AG 2019
W. Ebeling, T. Pöschel, *Lectures on Quantum Statistics*,
Lecture Notes in Physics 953, https://doi.org/10.1007/978-3-030-05734-3_4

4.1   Density Matrices and Operators

4.1.1   Density Matrices

Quantum mechanics can be formulated in different representations, among them the representation by energy eigenfunctions plays a special rôle. As shown in Chap. 2, this is true also for mixed ensembles. For example, for a mixed ensemble the mean value of energy can be represented by

$$\langle E \rangle = \sum_k \omega_k E_k ; \qquad \langle f(E) \rangle = \sum_k \omega_k f(E_k) . \tag{4.1}$$

In the special case of a pure ensemble which is in the eigenstate k, we have $w_k = 1$ and in the canonical ensemble the weight factor

$$\omega_k = Q^{-1} e^{-\beta E_k} . \tag{4.2}$$

We shall generalize these formulae and start again with the case of normal quantum mechanics, that is, with pure ensembles. Following von Neumann and Landau, first we formulate quantum mechanics by means of a matrix representation. To this end, we introduce density matrices and corresponding density operators. For example, in energy representation, alternatively to the usual notation we write

$$\langle E \rangle = \sum a_k^* a_k E_k ; \qquad E_k = \int \psi_k^* \hat{H} \psi_k dq . \tag{4.3}$$

Transforming this to an arbitrary orthonormal system,

$$\psi = \sum_k a_k \psi_k , \tag{4.4}$$

we find

$$\langle E \rangle = \sum_k \sum_l a_l a_k^* E_{kl} ; \qquad E_{kl} = \int \Psi_k^* \hat{H} \Psi_l dq . \tag{4.5}$$

In energy representation, the matrix E_{kl} is diagonal, $E_{kl} = E_k \delta_{kl}$. Following von Neumann and Landau, we define a density matrix by

$$\rho_{lk} = a_l a_k^* . \tag{4.6}$$

Using this notation, we write all mean values as a trace over a product of matrices:

$$\langle E \rangle = \mathrm{Tr} \left(\hat{\rho} \hat{H} \right) . \tag{4.7}$$

A corresponding formula can be written for the mean value of any observable, R, with the operator $\hat{R}$. The corresponding matrix of the observable is then

$$R_{lk} = \int \psi_l^* \hat{R} \psi_k dq \,, \tag{4.8}$$

where ψ_l and ψ_k are orthonormal eigenfunctions. This way we find again for the mean value a representation by a trace,

$$\langle R \rangle = \mathrm{Tr}\left(\hat{\rho} \hat{R} \right) . \tag{4.9}$$

In Dirac's notation (Dirac 1932), the density operator reads

$$\hat{\rho} = |\psi\rangle\langle\psi| . \tag{4.10}$$

The formula for the mean value assumes the form

$$\left\langle \hat{R} \right\rangle = \left\langle \psi \left| \hat{R} \right| \psi \right\rangle . \tag{4.11}$$

So far, this is nothing more than an alternative way to formulate the standard quantum mechanics. Following, however, the fundamental work by von Neumann (1932) and Landau and Lifshitz (1976, 1980), this altenative form is indeed more general as this representation allows a direct extension to quantum statistics. Following von Neumann and Landau, we postulate that for mixed ensembles, we can define a matrix,

$$\rho_{lk} = \overline{a_l a_k^*} \,, \tag{4.12}$$

where the average is performed—in the spirit of Gibbs—over many copies of the original system which are under the same conditions. This way, the simple quantum mechanical average is replaced by an average over an ensemble of quantum mechanical systems. We imagine an ensemble of quantum mechanical systems, where the individual systems are possibly in different states (phases) but all consistent with the assumed quantum statistical macro state. This way and using the symbol "Tr" for the operation "trace", we arrive for mixed ensembles at the formula for the mean of energy:

$$\langle E \rangle = \sum_k \sum_l \overline{a_l^* a_k} E_{kl} = \sum_k \sum_l \rho_{lk} E_{kl} = \mathrm{Tr}\left(\hat{\rho} \hat{E} \right) . \tag{4.13}$$

For physical quantities which depend on the energy we obtain the corresponding formula

$$\langle f(E) \rangle = \sum_k \sum_l \rho_{lk} f(E_{kl}) = \text{Tr}\left[\hat{\rho} f\left(\hat{E} \right) \right]. \qquad (4.14)$$

For an arbitrary physical quantities with the operators $\hat{R}$, we write for the mean value

$$\langle R \rangle = \text{Tr}(\hat{\rho} \hat{R}). \qquad (4.15)$$

The density matrix, ρ_{kl}, is just the matrix of the density operator in a concrete representation,

$$\rho_{kl} = \begin{pmatrix} \rho_{11} & \rho_{12} & \cdots \\ \rho_{21} & \rho_{22} & \cdots \\ \cdots & \cdots & \cdots \end{pmatrix}. \qquad (4.16)$$

The advantage of this representation is that for calculations any systems of orthonormal sets can be used (Landau and Lifshitz 1980). Dirac proposed to write the density operator for mixed ensembles in the form (Dirac 1932)

$$\hat{\rho} = \sum_n w_n |\psi_n\rangle\langle\psi_n|, \qquad (4.17)$$

where w_n are agains the normalized weights of the set of states.

The density matrix is a Hermitian $n \times n$ matrix with (in general) infinite number of lines and columns, $n \to \infty$. Since matrices can be understood as representations of operators, we will assume that the density operator is just the Hermitian operator which is associated to our density matrices. The density operator, $\hat{\rho}$, is defined in the Hilbert space of the quantum mechanical system. the operator $\hat{\rho}$ is also called the statistical operator. It is defined by its actions in the Hilbert space, $\mathcal{H}$: $\hat{\rho}\,\psi \in \mathcal{H}$.

4.1.2 von Neumann's Density Operators and Time Evolution

Density operators are new quantities extending the usual quantum theory, therefore, strictly speaking it is not possible to define them in a satisfying way within quantum mechanics as we tried for educational purpose above. From the logical point of view, according to von Neumann, it is more satisfying to introduce density operators axiomatically. The axiomatic foundation of density operators was developed by von Neumann in his famous book *Mathematische Grundlagen der Quantenmechanik* (von Neumann 1932). The postulates are the quantum mechanical assumptions that each physical quantity, R, is associated with an operator, $\hat{R}$, whose spectrum provides the possible results of a measurement and whose mean value is the expectation value of many measurements,

$$E\left(\hat{R}\right) = \left\langle \phi \middle| \hat{R}\phi \right\rangle = \int \phi^* \hat{R} \phi \, dq \,. \tag{4.18}$$

For macroscopic systems, in general the exact state, ϕ, is not known. Measuring the quantity R we know, however, that it should be one of the eigenstates, ϕ_i, of the operator R, that is,

$$\hat{R}\phi_i = R_i \phi_i \,. \tag{4.19}$$

Consequently, for the expectation value we have

$$E\left(\hat{R}\right) = \sum_n w_n \left\langle \phi_n \middle| \hat{R}\phi_n \right\rangle = \sum_n w_n E\left(R_n\right) \,, \tag{4.20}$$

with the probabilities $w_n = a_n^* a_n$. We define a projection operator, $\hat{P}_{\phi_i}$, which projects a state on the eigenstate ϕ_i (von Neumann 1932). Then the relation

$$\left\langle \phi_i \middle| \hat{R}\phi_i \right\rangle = \mathrm{Tr}\left(\hat{P}_{\phi_i} \hat{R}\right) \tag{4.21}$$

holds and

$$\mathrm{Tr}\left(\hat{P}_{\phi_i} \hat{R}\right) = \sum_m \sum_n \left\langle \phi_m \middle| \hat{P}_{\phi_i} \phi_n \right\rangle \left\langle \phi_n \middle| \hat{R}\phi_m \right\rangle$$

$$= \sum_m \left\langle \phi_m \middle| \phi_i \right\rangle \left\langle \phi_i \middle| \hat{R}\phi_n \right\rangle = \left\langle \phi_i \middle| \hat{R}\phi_i \right\rangle = E\left(\hat{R}\right)_{\phi_i} \,. \tag{4.22}$$

Here we used the representation of the identity operator

$$\hat{1} = \sum_n |\phi_n\rangle \langle \phi_n| \,. \tag{4.23}$$

With Eq. (4.22) we write

$$E\left(\hat{R}\right) = \sum_n w_n \mathrm{Tr}\left(\hat{P}_{\phi_n} \hat{R}\right) = \mathrm{Tr}\left(\sum_n w_n \hat{P}_{\phi_n} \hat{R}\right) = \mathrm{Tr}\left(\hat{\rho}\hat{R}\right) \,. \tag{4.24}$$

The expression (4.24) defines then the density operator,

$$\hat{\rho} = \sum_n w_n \hat{P}_{\phi_n} \,. \tag{4.25}$$

In the calculations we used the linearity of the trace operation and invariance of the trace with respect to a commutation of the members,

$$\text{Tr}(AB) = \text{Tr}(BA). \tag{4.26}$$

Important mathematical properties of projection operators, $\hat{P}_{\phi_i}$, are

* $\hat{P}_{\phi_i}$ projects Ψ to ϕ_i,
* per definitionem, $\hat{P}_{\phi_i} \Psi = a_i \phi_i$
* $\hat{P}_{\phi_i}^2 = \hat{P}_{\phi_i}$
* $\hat{P}_{\phi_i} \phi_k = \delta_{ik} \phi_k$
* $\text{Tr}\left(\hat{P}_{\phi_i} \hat{R}\right) = \langle \phi_i | R | \phi_i \rangle$.
* $w_n = \overline{a_n^* a_n} \geq 0$.

Further we note

$$\text{Tr}\left(\hat{P}_{\phi_n}\right) = \sum_m \left\langle \phi_m \left| \hat{P}_{\phi_n} \phi_m \right\rangle = \sum_m \langle \phi_m | \phi_n \rangle = \sum_m \delta_{mn} = 1 \tag{4.27}$$

and taking into account the normalization $\sum_n w_n = 1$ we have the relation

$$\text{Tr}\left(\hat{\rho}\right) = 1. \tag{4.28}$$

For the case of pure ensembles, that is, for the case of usual quantum mechanics, $\hat{\rho} = \hat{P}_\phi$, the general description reduces to the standard quantum mechanical form

$$E\left(\hat{R}\right) = \text{Tr}\left(\hat{\rho}\hat{R}\right) \tag{4.29}$$

which is equivalent to

$$E\left(\hat{R}\right) = \text{Tr}\left(\hat{P}_\phi \hat{R}\right) = \left\langle \phi \left| \hat{R} \phi \right\rangle. \tag{4.30}$$

von Neumann Equation

The von Neumann equation describes the time evolution of the density operator. It is a generalization of the Schrödinger equation to mixed ensembles and can be considered as the quantum mechanical counterpart of the classical Liouville equation. Let us start with a reformulation of the Schrödinger-equation in operator form. The standard form in quantum mechanics is

$$i\hbar \frac{\partial}{\partial t} \phi(t) = \hat{H} \phi(t), \tag{4.31}$$

with the formal solution

$$\phi(t) = e^{-\frac{i}{\hbar}\hat{H}t}\phi(0) \, . \tag{4.32}$$

In quantum statistics, the wave function ϕ corresponds to a projection operator, $\hat{P}_\phi$, or possibly a superposition of projection operators. Note that in the Dirac-formalism, projection operators assume the form

$$\hat{P}_\phi = |\phi\rangle\langle\phi| \, . \tag{4.33}$$

The action of a projection operator on an arbitrary function in the Hilbert space, $f \in \mathcal{H}$, is defined by

$$\hat{P}_\phi |f\rangle = \langle\phi|f\rangle |\phi\rangle \, . \tag{4.34}$$

For time-dependent ϕ we have

$$\hat{P}_{\phi(t)} |f\rangle = \langle\phi(t)|f\rangle |\phi(t)\rangle \, . \tag{4.35}$$

Assuming that f is a function which does not depend on time, we find

$$
\begin{aligned}
\frac{\partial}{\partial t}\hat{P}_{\phi(t)} |f\rangle &= \frac{\partial}{\partial t} \langle\phi(t)|f\rangle |\phi(t)\rangle \\
&= \frac{i}{\hbar}\left\langle\phi(t)\hat{H}\middle| f\right\rangle |\phi(t)\rangle - \frac{i}{\hbar} \langle\phi(t)|f\rangle \hat{H} |\phi(t)\rangle \\
&= \frac{i}{\hbar} |\phi(t)\rangle \left\langle\phi(t)\hat{H}\middle| f\right\rangle - \frac{i}{\hbar}\hat{H} |\phi(t)\rangle \langle\phi(t)|f\rangle \\
&= \frac{i}{\hbar}\left(\hat{P}_{\phi(t)}\hat{H} - \hat{H}\hat{P}_{\phi(t)}\right) |f\rangle \tag{4.36}
\end{aligned}
$$

$$\frac{\partial}{\partial t}\hat{P}_{\phi(t)} = \frac{i}{\hbar}\left(\hat{P}_{\phi(t)}\hat{H} - \hat{H}\hat{P}_{\phi(t)}\right) \, . \tag{4.37}$$

With

$$\hat{\rho}(t) = \sum_n w_n \hat{P}_{\phi_n(t)} \tag{4.38}$$

we find an analogue of the time-dependent Schrödinger equation which we call *von Neumann equation*,

$$i\hbar\frac{\partial}{\partial t}\hat{\rho}(t) = \hat{H}\hat{\rho}(t) - \hat{\rho}(t)\hat{H} = \left[\hat{H}, \hat{\rho}(t)\right] \, . \tag{4.39}$$

The description of systems by density operators is very general. It includes the quantum mechanics as a special case. According to von Neumann, quantum mechanics and quantum statistics can be defined by the same axiomatic system:

1. Observables are described by Hermitian operators in Hilbert space.
2. The state of the system is given by a density operator, $\hat{\rho}$, with the normalization $\mathrm{Tr}\left(\hat{\rho}\right) = 1$. The density operators have the representation

$$\hat{\rho} = \sum_n w_n \hat{P}_{\phi_n} \tag{4.40}$$

with the properties:

* $w_n \geq 0$
* $\sum_n w_n = 1$
* $\hat{\rho}\, \phi_k = w_k \phi_k$
* for pure ensembles, that is, if the system is in a pure state, ϕ, the density operator is given by $\hat{\rho} = \hat{P}_\phi$.

3. The time-dependence of a quantum statistical systems is given by the von Neumann equation

$$i\hbar \frac{\partial}{\partial t}\hat{\rho}(t) = \left[\hat{H}, \hat{\rho}(t)\right] . \tag{4.41}$$

4. For the mean values of a physical quantity with the operator $\hat{R}$ we have:

$$E\left(\hat{R}\right) = \mathrm{Tr}\left(\hat{\rho}\hat{R}\right) = \mathrm{Tr}\left(\hat{R}\hat{\rho}\right) . \tag{4.42}$$

This way the foundation of quantum statistics and quantum mechanics can be reduced to four postulates.

In classical statistics, the evolution of the probability density, $\rho(t)$, is governed by

$$\frac{d\rho}{dt} = \frac{\partial\rho}{\partial t} + \sum_{i=1}^{f}\left(\frac{\partial\rho}{\partial q_i}\dot{q}_i + \frac{\partial\rho}{\partial p_i}\dot{p}_i\right)$$

$$= \frac{\partial\rho}{\partial t} + \sum_{i=1}^{f}\left(\frac{\partial H}{\partial p_i}\frac{\partial\rho}{\partial q_i} - \frac{\partial\rho}{\partial p_i}\frac{\partial H}{\partial q_i}\right) = \frac{\partial\rho}{\partial t} + \{H, \rho\} , \tag{4.43}$$

and with the condition of conservation of probability, $d\rho/dt = 0$, we obtain the Liouville-equation

$$\frac{\partial\rho}{\partial t} + \{H, \rho\} = 0 . \tag{4.44}$$

The von Neumann equation can be written in a similar form,

$$\frac{\partial \hat{\rho}}{\partial t} = \frac{1}{i\hbar} \left(\hat{H}\hat{\rho} - \hat{\rho}\hat{H} \right) \tag{4.45}$$

$$i\hbar \frac{\partial \hat{\rho}}{\partial t} + \left[\hat{H}\hat{\rho} \right] = 0; \qquad \left[\hat{H}\hat{\rho} \right] = \left(\hat{H}\hat{\rho} - \hat{\rho}\hat{H} \right). \tag{4.46}$$

Comparing the von Neumann equation with the Liouville equation we see that the transition from classical to quantum statistics corresponds to the transition from Poisson brackets to commutators. Let us define the unitary operator

$$\hat{U}(t) = e^{-\frac{i}{\hbar}\hat{H}t}. \tag{4.47}$$

For the case that $\hat{H}$ does not explicitly depend on time, the solution of the Schrödinger equation can be written as

$$\phi(t) = \hat{U}(t)\phi(0). \tag{4.48}$$

Further the evolution of the density operator can be expressed by

$$\hat{\rho}(t) = \hat{U}(t)\hat{\rho}(0)\hat{U}^{\dagger}(t). \tag{4.49}$$

To prove this statement, we differentiate Eq. (4.49) with respect to time and obtain

$$\frac{\partial}{\partial t}\hat{\rho}(t) = \frac{1}{i\hbar}\hat{H}\hat{U}\hat{\rho}(0)\hat{U}^{\dagger}(t) - \hat{U}\hat{\rho}(0)\frac{1}{i\hbar}\hat{H}\hat{U}^{\dagger}(t) = \frac{1}{i\hbar}\left[\hat{H}, \hat{\rho}\right], \tag{4.50}$$

which is nothing else but the von Neumann equation. If $\hat{R}$ is a time-independent operator, we can at fixed mean value move the time-dependence of $\hat{\rho}(t)$ to the time-dependence of $\hat{R}(t)$. From the cyclic invariance of the trace follows

$$E\left(\hat{R}\right) = \left\langle \hat{R} \right\rangle = \mathrm{Tr}\left[\hat{R}\,\hat{\rho}(t)\right] = \mathrm{Tr}\left[\hat{R}\hat{U}(t)\hat{\rho}(0)\hat{U}^{\dagger}(t)\right] = \mathrm{Tr}\left[\hat{R}(t)\,\hat{\rho}(0)\right]. \tag{4.51}$$

By the transformation

$$\hat{R}(t) := \hat{U}^{\dagger}(t)\,\hat{R}\hat{U}(t) \tag{4.52}$$

we obtain this way a Heisenberg picture for the operators. As in standard quantum mechanics, in Heisenberg representation the density operators are time-independent and the time-dependence is shifted to the operators of physical quantities, $\hat{R}(t)$:

$$\frac{d\hat{R}(t)}{dt} = \frac{i}{\hbar}\hat{H}\hat{U}^{\dagger}\hat{R}\hat{U} - \frac{i}{\hbar}\hat{U}^{\dagger}\hat{R}\hat{H}\hat{U} = \frac{1}{i\hbar}\left[\hat{R}(t), \hat{H}\right]. \tag{4.53}$$

These two representations exist on equal rights. Usually we will stay within the Schrödinger-von Neumann representation and have time-dependencies in the density operator

$$\hat{\rho}(t) = \hat{U}(t)\rho(0)\hat{U}^{\dagger}(t), \tag{4.54}$$

while the operators corresponding to physical observables are time-independent.

4.1.3   Maximum Entropy Principle and Thermodynamic Functions

Based on the relations found for the simplest ensembles of quantum statistics, we derive now relations of equilibrium thermodynamics expressed through density operators. The expression for the probability in the canonical ensemble as found in Chap. 2, Sect. 2.4.1:

$$\omega_k = Q^{-1}e^{-\beta E_k}, \tag{4.55}$$

corresponds to a canonical density operator

$$\hat{\rho}_{\text{can}} = Q^{-1}e^{-\beta \hat{H}}. \tag{4.56}$$

The grand canonical operator assumes the form

$$\hat{\rho}_{\text{grc}} = \varXi^{-1}e^{\beta \hat{N}\mu - \beta \hat{H}_N}. \tag{4.57}$$

As an alternative derivation of the canonical (as well as of the grand-canonical) ensemble we can use the *maximum entropy principle*: (Ebeling and Sokolov 2005; Jaynes 1985; Subarew 1976; Zubarev et al. 1996, 1997). This fundamental principle of statistical mechanics was first formulated by J. W. Gibbs still in the nineteenth century. For the quantum statistical case we find it in a simplified form in the monograph by von Neumann (1932). It was further developed by Stratonovich (1975), Subarew (1976) and Zubarev (1974) and formulated as a fundamental new principle by the great follower of Gibbs' work, Jaynes (1985). In our view, this principle is of fundamental importance for statistical physics (see Ebeling and Sokolov 2005; Zubarev et al. 1996, 1997). The method of maximum entropy for the derivation of equilibrium and non-equilibrium density operators was developed by the great pioneer of statistical physics Dmitri N. Zubarev (1917–1992). Zubarev, however, considered the method "not as a strict derivation of the statistical distribution functions", but only "as a useful heuristic method". We tend to share, however, the opinion of Stratonovich (1975) and Jaynes (1985) who stated that it is, indeed, a new principle.

We can show that all Gibbs ensembles can be derived in a unique way just from this principle, called the *Gibbs-Jaynes maximum entropy principle*.

Jaynes criticized the traditional approach to derive the canonical and grand canonical ensembles with the following remarks (Jaynes 1985): *A moment's thought makes it clear how useless for this purpose is our conventional textbook statistical mechanics, where the basic connections between micro and macro are sought in ergodic theorems. These suppose that the microstate will eventually pass near every one compatible with the total macroscopic energy; if so, then the then the long-time behavior of a system must be determined by its energy. What we see about us does not suggest this.* In order to find an alternative, and possibly more elegant procedure, let us turn the question around, starting now from Gibbs' original work as Jaynes (1985) did: *Why is that knowledge of micro-phenomena does not seem sufficient to understand macro-phenomena? Is there an extra general principle needed for this? Our message is that such a general principle is indeed needed and already exists, having been given by J. Willard Gibbs 110 years ago. . . . A macrostate has a crucially important further property - entropy - that is not determined by the microstate.*

As Gibbs and Jaynes, we start directly from the variational principle and formulate first the following abstract problem: Consider a macroscopic system, given incomplete information A. Then we postulate (Gibbs-Jaynes principle): If an incomplete information, A, holds true for a macroscopic system, the best prediction we can make about other quantities are those obtained from the ensemble, ρ, that yields maximum information entropy, $H = S/k_B$, while agreeing with A. By "agreeing with A" we mean that the average $\langle A \rangle$ calculated with ρ corresponds to the given information, A.

We apply now this principle to von Neumann's entropy with the supplementary condition of agreeing with the energy E:

$$S = -k_B \text{Tr} \left(\hat{\rho} \ln \hat{\rho} \right) \rightarrow \max \tag{4.58}$$

with the conditions of normalization $\text{Tr} \left(\hat{\rho} \right) = 1$ and $\text{Tr} \left(\hat{H} \hat{\rho} \right) = E$. By variation of the entropy, S, we obtain

$$\delta S = -k_B \text{Tr} \left(\hat{\rho} \frac{1}{\hat{\rho}} \delta \hat{\rho} + \delta \hat{\rho} \ln \hat{\rho} \right) = -k_B \text{Tr} \left[\delta \hat{\rho} \left(1 + \ln \hat{\rho} \right) \right] . \tag{4.59}$$

The supplementary conditions are satisfied by Lagrangian multipiers,

$$k_B \beta \left(\text{Tr} \left(\hat{H} \hat{\rho} \right) - E \right) = 0 ; \qquad k_B \gamma \left(\text{Tr} \left(\hat{\rho} \right) - 1 \right) = 0 . \tag{4.60}$$

By adding the variations of entropy and the Lagrangian terms we get

$$\delta \left[S + k_B \beta \left(\text{Tr} \left(\hat{H} \hat{\rho} \right) - E \right) + k_B \gamma \left(\text{Tr} \left(\hat{\rho} \right) - 1 \right) \right] = 0 \qquad (4.61)$$

$$-k_B \text{Tr} \left(\delta \hat{\rho} \ln \hat{\rho} + \delta \hat{\rho} + \delta \hat{\rho} \beta \hat{H} + \gamma \delta \hat{\rho} \right) = 0 \qquad (4.62)$$

$$\hat{\rho} = e^{-(1+\gamma)} e^{-\beta \hat{H}} . \qquad (4.63)$$

The normalization of the density operator leads to the canonical density operator

$$\hat{\rho} = Q^{-1} e^{-\beta \hat{H}} ; \qquad Q(T, V, N) = \text{Tr} \left[e^{-\beta \hat{H}} \right] . \qquad (4.64)$$

We can easily show as in Sect. 2.4.1 that the expression $-k_B T \ln Q(T, V, N)$ satisfies standard thermodynamic relations and can be identified with the free energy of the system,

$$F = -k_B T \ln Q(T, V, N) . \qquad (4.65)$$

This relation was given already in Sect. 2.4.1 in the special energy representation.

So far we considered the number of particles fixed and given. We extend this by assuming that the number of particles is given only on average, N. By this assumption, we postulate an operator corresponding to this observable, $\hat{N}$. Its mean represents the number of particles, N, through

$$\langle N \rangle = \text{Tr} \left(\hat{N} \hat{\rho} \right) . \qquad (4.66)$$

The maximum-entropy condition for this ensemble leads to

$$\delta S = -k_B \text{Tr} \left[\delta \hat{\rho} \left(\ln \hat{\rho} + \beta \hat{H}_N - \beta \mu \hat{N} + (1 + \gamma) \right) \right] = 0 \qquad (4.67)$$

$$\hat{\rho}_N = e^{-(\gamma + 1) - \beta \hat{H}_N + \beta \hat{N} \mu} = \varXi^{-1} e^{\beta \hat{N} \mu - \beta \hat{H}_N} . \qquad (4.68)$$

With the normalization

$$\sum_{N=0}^{\infty} \text{Tr} \hat{\rho}_N = 1 , \qquad (4.69)$$

we obtain eventually the grand canonical partition function

$$\varXi = \sum_{N=0}^{\infty} e^{\beta \hat{N} \mu} Q(T, V, N) . \qquad (4.70)$$

We can generalize the definition of the trace by including the summations over the number of particles, denoting the new trace as "Trace". This leads to a more general expression,

$$\langle R \rangle = \text{Trace} \left(\hat{R} \hat{\rho} \right) ; \quad \Xi = \text{Trace} \left[e^{\beta \mu \hat{N} - \beta \hat{H}} \right] . \tag{4.71}$$

Similar as in Sect. 2.4.1, from the thermodynamical relations we find that the grand-canonical potential is related to the pressure of the system,

$$pV = -\Omega = k_B T \ln \Xi (T, V, \mu) . \tag{4.72}$$

We can easily check that the canonical as well as the grand canonical density operators fulfil the von Neumann equation and that the partition functions satisfy thermodynamic conditions.

Problem 6 Show that the canonical and the grand canonical density operators

$$\hat{\rho}_{can} = Q^{-1} e^{-\beta \hat{H}} ; \quad \hat{\rho}_{gcan} = \Xi^{-1} e^{\beta \left(\mu \hat{N} - \hat{H} \right)} \tag{4.73}$$

fulfil the von Neumann equation (4.39) and that the normalization functions, $Q(T, V, N)$ and $\Xi(T, V, \mu)$, are in agreement with thermodynamic differential relations.

4.2   Representations in Coordinate Space and Two-Time Functions

4.2.1   Coordinate Representations and Bloch Equations

The quantum mechanical density operator of an N particle system,

$$\hat{\rho}(t) = \sum_{n=1}^{N} w_n \hat{P}_{\phi_n} , \tag{4.74}$$

acts in the space of N coordinates. This can be written in the form

$$\hat{\rho}(t) = \sum_n w_n \hat{U}_t^{\phi_n} \quad \text{von Neumann notation} \tag{4.75}$$

$$\hat{\rho}(t) = \sum_n w_n |\phi_n\rangle \langle \phi_n| \quad \text{Dirac notation} , \tag{4.76}$$

where the sum is performed over the spectrum. In a basis, ψ_n, the elements of the density matrix are given by

$$\rho_{kl} = \sum_n w_n \langle \psi_k | \phi_n \rangle \langle \phi_n | \psi_l \rangle \quad \text{with respect to} \quad |\psi_n\rangle . \tag{4.77}$$

In coordinate space representation, we write

$$\hat{\rho}\left(x, x'\right) = \sum_n w_n \langle x | \phi_n \rangle \langle \phi_n | x' \rangle = \sum_n w_n \phi_n(x) \phi_n^*(x') . \tag{4.78}$$

For N particles with coordinates (including the spins) $q_i = \left(\hat{r}_i, \sigma_i\right)$ we write

$$\hat{\rho}\left(q_1 \ldots q_N \, q_1' \ldots q_N'\right) = \sum_{n=1}^{N} w_n \phi_n(q_1 \ldots q_N) \phi_n^*(q_1' \ldots q_N') . \tag{4.79}$$

The expectation value (mean) of an operator, $\hat{R}$, is

$$\left\langle \hat{R} \right\rangle = \sum_n w_n \int \phi_n^*(q_1 \ldots q_N) \hat{R} \, \phi_n(q_1 \ldots q_N) \, dq_1 \ldots dq_N \tag{4.80}$$

$$= \int \hat{R} \hat{\rho} \left(q_1 \ldots q_N \, q_1' \ldots q_N'\right)_{q_i' = q_i} dq_1 \ldots dq_N . \tag{4.81}$$

The canonical ensemble

$$\hat{\rho}\left(x, x'\right) = \frac{1}{Q} \sum_n e^{-\beta E_n} \, \psi_n(x) \psi_n^*(x') \tag{4.82}$$

with energy eigenfunctions, ψ_n, gives the expression

$$\hat{\rho}\left(x, x'\right) = \left\langle x \left| \frac{1}{Q} e^{-\beta \hat{H}} \right| x' \right\rangle . \tag{4.83}$$

Let us define a non-normalized canonical operator by[1]

$$\check{\rho}(\beta) = e^{-\beta \hat{H}} \tag{4.84}$$

with the initial condition $\check{\rho}(0) = \mathbf{1}$. By differentiation

$$\frac{\partial \check{\rho}}{\partial \beta} = -\hat{H} e^{-\beta \hat{H}} \tag{4.85}$$

---

[1] The symbol $\check{x}$ indicates that the operator is not normalized. The corresponding normalized operator would be denoted by $\hat{x}$.

follows the *Bloch equation*

$$\frac{\partial \check{\rho}}{\partial \beta} + \hat{H}\check{\rho} = 0.$$

(4.86)

In energy representation, we have $\rho_{ij} = e^{-\beta E_i}\delta_{ij}$ and the Bloch equation reads

$$\frac{\partial \rho_{nm}}{\partial \beta} + E_n \rho_{nm} = 0.$$

(4.87)

Correspondingly, in coordinate representation we obtain

$$-\frac{\partial}{\partial \beta}\check{\rho}\left(x, x', \beta\right) = \hat{H}_x \check{\rho}\left(x, x', \beta\right),$$

(4.88)

with the initial condition

$$\check{\rho}\left(x, x', 0\right) = \delta\left(x - x'\right).$$

(4.89)

For a free one-dimensional particle, thus, we find

$$\frac{\partial}{\partial \beta}\check{\rho}\left(x, x', \beta\right) - \frac{\hbar^2}{2m}\frac{\partial^2}{\partial x^2}\check{\rho}\left(x, x', \beta\right) = 0$$

(4.90)

$$\check{\rho}\left(x, x', 0\right) = \delta\left(x - x'\right).$$

(4.91)

The solution of this partial differential equation of diffusion type is

$$\check{\rho}\left(x, x', \beta\right) = \sqrt{\frac{m}{2\pi \hbar^2 \beta}}\, e^{-\left(\frac{m}{2\hbar^2 \beta}\right)(x-x')^2}.$$

(4.92)

In three dimensions, we obtain, correspondingly,

$$\check{\rho}\left(\mathbf{r}, \mathbf{r}', \beta\right) = \left(\frac{m}{2\pi \hbar^2 \beta}\right)^{3/2} e^{-\left(\frac{m}{2\hbar^2 \beta}\right)(\mathbf{r}-\mathbf{r}')^2}.$$

(4.93)

As a second example we study a harmonic oscillator (Feynman 1972) with

$$\hat{H} = \frac{\hat{p}^2}{2m} + \frac{1}{2}m\omega^2 x^2.$$

(4.94)

The corresponding Bloch equation,

$$\frac{\partial}{\partial \beta}\check{\rho}\left(x, x', \beta\right) - \frac{\hbar^2}{2m}\frac{\partial^2}{\partial x^2}\check{\rho}\left(x, x', \beta\right) + \frac{1}{2}m\omega^2 x^2 \check{\rho}\left(x, x', \beta\right) = 0,$$

(4.95)

has the solution

$$\check{\rho}(x, x, \beta) = \sqrt{\frac{m\omega}{2\pi \hbar \, \sinh(\hbar \, \omega \, \beta)}} \, e^{-\frac{m\omega}{\hbar} \tanh\left(\frac{\beta\hbar\omega}{2}\right)x^2} . \tag{4.96}$$

In the limit of high temperature $(T \to \infty$ or $\beta \to 0)$ follows with $\tanh(x) \approx x$ the classical limit

$$\check{\rho}(x, x, \beta) \to e^{-\frac{1}{2}\beta m\omega^2 x^2} = e^{-\beta U(x)} , \tag{4.97}$$

where $U(x) = \frac{1}{2}m\omega^2 x^2$. With

$$\int_{-\infty}^{\infty} dx\, e^{-\alpha x^2} = \sqrt{\frac{\pi}{\alpha}}; \quad \sinh(2\phi) = 2 \sinh(\phi) \cosh(\phi) \tag{4.98}$$

follow for the partition function

$$Q(T, V) = \int \check{\rho}(x, x, \beta)\, dx = \sqrt{\frac{m\omega}{2\pi \hbar \, \sinh(\hbar\omega\beta)}} \int dx\, e^{-\frac{m\omega}{\hbar} \tanh\left(\frac{1}{2}\beta\hbar\omega\right)x^2}$$

$$= \sqrt{\frac{m\omega}{2\pi\hbar \sinh(\hbar\omega\beta)}} \sqrt{\frac{\pi\hbar}{m\omega \tanh\left(\frac{1}{2}\beta\hbar\omega\right)}} = \frac{1}{2 \sinh\left(\frac{1}{2}\beta\hbar\omega\right)} \tag{4.99}$$

and the free energy

$$F = k_B T \ln\left[2 \sinh\left(\frac{1}{2}\beta\hbar\omega\right)\right] = k_B T \ln\left(e^{\frac{\hbar\omega}{2k_B T}} - e^{-\frac{\hbar\omega}{2k_B T}}\right)$$

$$= \frac{1}{2}\hbar\omega + k_B T \ln\left(1 - e^{-\frac{\hbar\omega}{k_B T}}\right) . \tag{4.100}$$

Problem 7 Derive for the internal energy

$$U = \frac{\partial}{\partial\beta}(\beta F) = \frac{1}{2}\hbar\omega + \frac{\hbar\omega e^{-\beta\hbar\omega}}{1 - e^{-\beta\hbar\omega}} . \tag{4.101}$$

Problem 8 Derive for the mean square deviation

$$\langle x^2 \rangle = \frac{\int x^2 \check{\rho}(x, x, \beta)}{\int \check{\rho}(x, x, \beta)} = \frac{\hbar}{2m\omega} \coth\left(\frac{1}{2}\beta\hbar\omega\right) . \tag{4.102}$$

For the case of low temperature ($T \to 0$ or $\beta \to \infty$), we find from Eq. (4.96) the density

$$\tilde{\rho}(x, x, \beta) \to e^{-\frac{1}{2}\beta\hbar\omega} \, |\phi_0|^2 \, , \tag{4.103}$$

with

$$\phi_0 = \left(\frac{m\omega}{\pi\hbar}\right)^{1/4} e^{-\frac{m\omega}{2\hbar}x^2} \, . \tag{4.104}$$

Note that for the normalized density we find the asymptotics $\hat{\rho}(x, x, \beta \to \infty) \to |\phi_0|^2$.

We have shown in this section that the Bloch equation is a rather useful tool for solving concrete problems (Feynman 1972). The Bloch equation and most of the results can be generalized to N-particle systems:

$$\hat{\rho}\left(q_1 \ldots q_N \, q_1' \ldots q_N', \beta\right) = \sum_{n=1}^{N} w_n \psi_n \left(q_1 \ldots q_N\right) \psi_n^* \left(q_1' \ldots q_N'\right) \, , \tag{4.105}$$

where ψ_n are the symmetric or antisymmetric eigenfunctions. Neglecting symmetry and spin, we represent $\hat{\rho}$ as a product of one-particle densities

$$\tilde{\rho}_D \left(\mathbf{r}_1 \ldots \mathbf{r}_N \, \mathbf{r}_1' \ldots \mathbf{r}_N', \beta\right) = \left(\frac{m}{2\pi\hbar^2\beta}\right)^{\frac{3N}{2}} \exp\left[-\frac{m}{2\hbar^2\beta} \sum_{k=1}^{N} (\mathbf{r}_k - \mathbf{r}_k')^2\right] \, . \tag{4.106}$$

In the case of symmetrical wave functions, we obtain the sum

$$\hat{\rho}_S \left(q_1 \ldots q_N, q_1' \ldots q_N', \beta\right) = \frac{1}{N!} \sum_{P} \tilde{\rho}_D \left(q_1 \ldots q_N \, P(q_1') \ldots P(q_N'), \beta\right) \tag{4.107}$$

while for antisymmetric wave functions,

$$\hat{\rho}_A \left(q_1 \ldots q_N \, q_1' \ldots q_N', \beta\right)$$

$$= \frac{1}{N!} \sum_{P} (-1)^{\#P} \tilde{\rho}_D \left[q \ldots q_N \, P(q_1') \ldots P(q_N'), \beta\right] \, . \tag{4.108}$$

The sum runs over all permutations, P, of particles and $\#P$ is the number of pair exchanges in a permutation.

4.2.2   Two-Time Density Operators

So far, density operators in coordinate representation are defined by

$$\hat{\rho}\left(x, x', t\right) = \sum_n w_n \phi_n(x)\, \phi_N^*(x') . \tag{4.109}$$

We call this kind of operators *one-time density operator*. The time-dependency is contained either in the coefficients (Heisenberg representation) or in the wave functions (Schrödinger representation). Here, we describe the Schrödinger picture with $\phi_n(x, t)$ and define the *two-time density matrix* for ($N = 1$) by

$$\hat{\rho}\left(x, x', t, t'\right) = \sum_n w_n \phi_n(x, t)\, \phi_n^*\left(x', t'\right) . \tag{4.110}$$

In the analog way, for many particles, $N > 1$, we define the density operator (matrix) by

$$\hat{\rho}\left(x_1 \ldots x_N\, x_1' \ldots x_N', t, t'\right) = \sum_{n=1}^{N} w_n\, \phi_n\left(x_1 \ldots x_N, t\right) \phi_n^*\left(x_1' \ldots x_N', t'\right) . \tag{4.111}$$

The limit $t \rightarrow t'$ leads us back to the usual one-time operators (matrices). An advantage of two-time operators is that the partial differential equations are even simpler.

From the Schrödinger equation

$$i\hbar \frac{\partial \phi_n(x, t)}{\partial t} = \hat{H}_x \phi_n(x, t) \tag{4.112}$$

follows for the two-time case instead of the von Neumann equation a simpler equation:

$$i\hbar \frac{\partial}{\partial t} \hat{\rho}\left(x, x', t, t'\right) = \hat{H}_x \hat{\rho}\left(x, x', t, t'\right) . \tag{4.113}$$

In the case of uniform time we find with $\tau = t - t'$ again the known equation

$$i\hbar \frac{\partial}{\partial \tau} \hat{\rho}\left(x, x', \tau\right) = \hat{H}_x \hat{\rho}\left(x, x', \tau\right) . \tag{4.114}$$

On the other hand, in thermodynamic equilibrium, also the Bloch-equation (4.86) holds true,

$$\frac{\partial}{\partial \beta} \check{\rho}\left(x, x', \beta\right) = \hat{H}_x \check{\rho}\left(x, x', \beta\right) . \tag{4.115}$$

Introducing a complex time $\tilde{\tau}$ with $\mathrm{Re}\,(\tilde{\tau}) = t - t'$ and $\mathrm{Im}\,(\tilde{\tau}) = \hbar\beta$, we extend the density operator, $\hat{\rho}\,(x, x', \tilde{\tau})$, to a complex time plane (Fetter and Walecka 1971; Kadanoff and Baym 1962; Kraeft et al. 1986; Kremp et al. 2005; Martin and Schwinger 1959; Toda et al. 1983). In this complex plane one finds for systems in thermodynamic equilibrium certain conditions of periodicity along the imaginary axis as

$$\hat{\rho}\,(\tilde{\tau}) = \pm e^{-\beta\mu}\,\hat{\rho}\,(\tilde{\tau} + i\hbar\beta)\,. \tag{4.116}$$

Relations of this type are called Kubo-Martin-Schwinger (KMS) relations.

4.3 Bogolyubov's Reduced Density Operators

For many applications we do not need the full information contained in the density operator. For example, in order to calculate the mean kinetic energy we need the average

$$\langle T \rangle = \left\langle \sum_{k=1}^{N} \frac{\hat{p}_k^2}{2m} \right\rangle = N \left\langle \frac{\hat{p}_1^2}{2m} \right\rangle \tag{4.117}$$

and for the case of N identical particles it suffices to know $\hat{\rho}(q_1, q_1')$ since

$$
\begin{aligned}
\left\langle \frac{\hat{p}_1^2}{2m} \right\rangle &= \mathrm{Tr}_{(N)} \left(\hat{\rho}\,\frac{\hat{p}_1^2}{2m} \right) \\
&= \int \left(-\frac{\hbar^2}{2m} \right) \nabla_1^2 \hat{\rho}\,(q_1 \ldots q_N\, q_1' \ldots q_N') \bigg|_{q_i'=q_i} dq_1 \ldots dq_N \\
&= \int \left(-\frac{\hbar^2}{2m} \nabla_1^2 \right) \hat{\rho}_1\,(q_1, q_1') \bigg|_{q_1'=q_1} dq_1\,.
\end{aligned}
\tag{4.118}
$$

Here we introduced the definition of $\hat{\rho}_1\,(q_1, q_1')$ as an integral of the density matrix over $(N - 1)$ coordinates $q_2 \ldots q_N$:

$$\hat{\rho}_1\,(q_1, q_1') = \int \hat{\rho}\,(q_1\,q_2 \ldots q_N\, q_1'\,q_2 \ldots q_N)\,dq_2 \ldots dq_N\,. \tag{4.119}$$

In the same way, we calculate the mean potential energy for the cases that the potential energy is a sum of pair interaction terms,

$$\langle U \rangle = \left\langle \sum_{1 \le i < j \le N}^{N} U(q_i, q_j) \right\rangle = \frac{N(N-1)}{2} \langle U(q_1, q_2) \rangle$$

$$= \frac{N(N-1)}{2} \int U(q_1, q_2) \hat{\rho}_2(q_1, q_2) dq_1 dq_2 , \qquad (4.120)$$

with the two-particle density operator

$$\hat{\rho}_2 \left(q_1, q_2, q_1', q_2' \right) = \int \hat{\rho} \left(q_1 q_2 q_3 \ldots q_N q_1' q_2' q_3 \ldots q_N \right) dq_3 \ldots dq_N .$$

$$(4.121)$$

More general we define s-particle functions by[2] (Bogolyubov 2005-2009)

$$\hat{\rho}_s (1 \ldots s) = \text{Tr}_{(s+1 \ldots N)} \left(\hat{\rho}(1 \ldots N) \right) \qquad (4.122)$$

or with a different normalization,

$$\hat{F}_s (1 \ldots s) = V^s \hat{\rho}_s (1 \ldots s) = \text{Tr}_{(s+1 \ldots N)} \left(V^s \hat{\rho}(1 \ldots N) \right) . \qquad (4.123)$$

With these definitions follows

$$\left\langle \hat{T} \right\rangle = \frac{N}{V} \text{Tr}_{(1)} \left(\hat{F}_1 \frac{\hat{p}_1^2}{2m} \right) = n \, \text{Tr}_{(1)} \left(\hat{F}_1 \frac{\hat{p}_1^2}{2m} \right) \qquad (4.124)$$

and

$$\langle U \rangle = \frac{N(N-1)}{2V^2} \text{Tr}_{(1,2)} \left[\hat{F}_2 \, U(1,2) \right] \approx \frac{n^2}{2} \text{Tr}_{(1,2)} \left[\hat{F}_2 \, U(1,2) \right] . \qquad (4.125)$$

The corresponding energy is

$$E = \left\langle \hat{H} \right\rangle = \left\langle \hat{T} \right\rangle + \langle U \rangle . \qquad (4.126)$$

In order to derive equations of motion for the reduced density operators, we apply the trace, $\text{Tr}_{(s+1 \ldots N)}$, to the von Neumann equation:

$$i\hbar \frac{\partial}{\partial t} \hat{\rho}_s = \text{Tr}_{(s+1 \ldots N)} \left[\hat{H}, \hat{\rho}_N \right] . \qquad (4.127)$$

---

[2]For brevity of the notation we will use frequently the short-hand notation of the arguments of the operators as introduced on page 25.

The Hamilton operator

$$\hat{H} = \sum_{k=1}^{N} \hat{T}_k \left(\hat{p}_k^2 \right) + \sum_{i<j}^{N} U_{(ij)} \left(\mathbf{r}_i, \mathbf{r}_j \right) \tag{4.128}$$

can be treated by considering the sub-groups $UG_1 = 1 \ldots s$ and $UG_s = s+1 \ldots N$. It follows

$$i\hbar \frac{\partial}{\partial t} \hat{\rho}_s = \text{Tr}_{(s+1 \ldots N)} \left(\left[\hat{H}_s, \hat{\rho}_N \right] \right) + \text{Tr}_{(s+1 \ldots N)} \left(\left[\hat{H}_{N-s}, \hat{\rho}_N \right] \right)$$

$$+ \text{Tr}_{(s+1 \ldots N)} \left(\left[\sum_{i=1}^{s} \sum_{j=s+1}^{N} U_{ij}, \hat{\rho}_N \right] \right). \tag{4.129}$$

We can simplify this expression taking into account that $\hat{H}_s$ acts only in the subspace of the s-particle system:

$$\text{Tr}_{(s+1 \ldots N)} \left(\left[\hat{H}_s, \hat{\rho}_N \right] \right) = \left[\hat{H}_s, \hat{\rho}_s \right]. \tag{4.130}$$

Since the second contribution disappears, we find

$$i\hbar \frac{\partial}{\partial t} \hat{\rho}_s = \left[\hat{H}_s, \hat{\rho}_s \right] + \sum_{i=1}^{s} \sum_{j=s+1}^{N} \text{Tr}_{(s+1 \ldots N)} \left(\left[U_{ij}, \hat{\rho}_N \right] \right)$$

$$= \left[\hat{H}_s, \hat{\rho}_s \right] + \sum_{i=1}^{s} (N-s) \text{Tr}_{s+1} \left(\left[U_{i,s+1}, \hat{\rho}_{s+1} \right] \right). \tag{4.131}$$

The interpretation of this equation is that we need $\hat{\rho}_{s+1}$ in order to find $\hat{\rho}_s$. This leads us to the Bogolyubov hierarchy which provides a prescription how to compute $\hat{\rho}_s$ through iteration.

4.4 Slater-, Wigner- and Klimontovich Representations

For the sake of transparency and better comparability with classical statistics, sometimes it is useful to supplement the operators in Hilbert space by representations in the phase space, that is, in the space of coordinates and momenta. Such representations of quantum statistics were introduced in the thirties of last century by Wigner, Kirkwood, Slater and since the fifties by Klimontovich.

4.4.1 Slater Representations

We start from a representation of the density operator in the coordinate space,

$$\left\langle \mathbf{x}_1 \dots \mathbf{x}_N \left| \hat{\rho} \right| \mathbf{x}_1' \dots \mathbf{x}_N' \right\rangle = \hat{\rho} \left(\mathbf{x}_1 \dots \mathbf{x}_N \, \mathbf{x}_1' \dots \mathbf{x}_N' \right) , \tag{4.132}$$

where $\mathbf{x}_i$ is the vector of coordinates (including spin) of the ith particle, $\mathbf{x}_i = \{\mathbf{r}_i, \sigma_i\}$. The diagonal elements of the density matrix, $\hat{\rho}\,(\mathbf{x}_1 \dots \mathbf{x}_N, \mathbf{x}_1 \dots \mathbf{x}_N)$ are the probability density to find in a measurement particle 1 at the coordinate $\mathbf{x}_1$, particle 2 at the coordinate $\mathbf{x}_2$, ..., etc., and particle N at the coordinate $\mathbf{x}_N$. This way, the diagonal elements of the density matrix are a kind of generalization of the quantum mechanical density, $|\psi\,(\mathbf{x}_1 \dots \mathbf{x}_N)|^2$. In thermodynamic equilibrium, we write

$$\hat{\rho}^{\text{eq}}\,(\mathbf{x}_1 \dots \mathbf{x}_N \, \mathbf{x}_1 \dots \mathbf{x}_N) = \left\langle \mathbf{x}_1 \dots \mathbf{x}_N \left| Q_N^{-1} e^{-\beta \hat{H}_N} \right| \mathbf{x}_N \dots \mathbf{x}_1 \right\rangle . \tag{4.133}$$

The free energy is a thermodynamic function given by

$$F(T, V, N) = -k_{\mathrm{B}} T \ln Q_N(T, V) , \tag{4.134}$$

with the partition function

$$Q_N(T, V) = \mathrm{Tr}\!\left(e^{-\beta \hat{H}_N} \right) = \sum_{\sigma_i} \int d\mathbf{r}_1 \dots d\mathbf{r}_N \left\langle \mathbf{x}_1 \dots \mathbf{x}_N \left| e^{-\beta \hat{H}_N} \right| \mathbf{x}_N \dots \mathbf{x}_1 \right\rangle . \tag{4.135}$$

We introduce now the *Slater sum* or *Slater function* of N particles by

$$S^{(N)}\,(\mathbf{r}_1 \dots \mathbf{r}_N)$$

$$= \frac{N! \Lambda^{3N}}{(2s+1)^N} \sum_{\sigma_i} \left\langle \mathbf{r}_1, \sigma_1 \dots \mathbf{r}_N, \sigma_N \left| e^{-\beta \hat{H}} \right| \mathbf{r}_N, \sigma_N \dots \mathbf{r}_1, \sigma_1 \right\rangle , \tag{4.136}$$

where the thermal wave length is as earlier defined

$$\Lambda = \frac{h}{\sqrt{2\pi \, m \, k_{\mathrm{B}} \, T}} . \tag{4.137}$$

Then the free energy takes the form

$$F = k_{\mathrm{B}} T N \left[\ln\!\left(\frac{n \Lambda^3}{2s+1} \right) - 1 \right] - k_{\mathrm{B}} T \ln \int \frac{d\mathbf{r}_1 \dots d\mathbf{r}_N}{V^N} S^{(N)}\,(\mathbf{r}_1 \dots \mathbf{r}_N) . \tag{4.138}$$

This is analog to the classical expression for the configuration integral (the configuration part of the partition function)

$$Q_N^{\text{conf}}(T, V) = \frac{1}{V^N} \int d\mathbf{r}_1 \ldots d\mathbf{r}_N e^{-\beta \phi_N (\mathbf{r}_1 \ldots \mathbf{r}_N)}. \tag{4.139}$$

Defining the quantum statistical configuration integral by using Eq. (4.138),

$$Q_N^{\text{conf}} = \frac{1}{V^N} \int d\mathbf{r}_1 \ldots d\mathbf{r}_N S^{(N)} (\mathbf{r}_1 \ldots \mathbf{r}_N), \tag{4.140}$$

we represent the quantum statistical free energy in the same form as the classical expression,

$$F = k_{\text{B}} T N \left(\ln n \Lambda^3 - 1 \right) - k_{\text{B}} T \ln Q_N^{\text{conf}}. \tag{4.141}$$

The analogy to the classical expression suggests also (following Morita) to introduce per definitionem effective potentials which are temperature-dependent by

$$S^{(N)} (\mathbf{r}_1 \ldots \mathbf{r}_N) = e^{-\beta U_N (\mathbf{r}_1 \ldots \mathbf{r}_N)}. \tag{4.142}$$

According to Morita, Kelbg and others, this formulation allows to transfer many methods of classical statistics, such as cluster expansions, to quantum statistics.

4.4.2 Wigner-Representation

We consider first a one-dimensional case and start with the Hamilton operator, the density operator, and the von Neumann equation in the usual form,

$$i\hbar \frac{\partial \hat{\rho}}{\partial t} = \left[\hat{H}, \hat{\rho} \right]; \qquad \hat{H} = \frac{\hat{p}^2}{2m} + U(x). \tag{4.143}$$

In matrix representation we get

$$i\hbar \frac{\partial \hat{\rho}(x, x', t)}{\partial t} = \left\langle x \left| \hat{H} \hat{\rho} x' \right\rangle - \left\langle x \left| \hat{\rho} \hat{H} x' \right\rangle \right.$$

$$= -\frac{\hbar^2}{2m} \left(\frac{d^2}{dx^2} - \frac{d^2}{dx'^2} \right) \hat{\rho}(x, x', t)$$

$$+ \left[U(x) - U(x') \right] \hat{\rho}(x, x', t). \tag{4.144}$$

Following Wigner, we define a function depending on coordinate and momentum,

$$W(x, p, t) = \int \frac{dx'}{h} \hat{\rho}(x, x', t) e^{-\frac{i}{\hbar} p(x-x')} . \tag{4.145}$$

This is an analogue of the classical probability on phase space, however, the Wigner function is only a quasi-probability since $W(x, p)$ is not necessarily real and non-negative.

Problem 9 Show that by integration of Wigner functions over the momentum follows the diagonal element of the density matrix:

$$W(x, t) = \int dp W(x, p, t) = \hat{\rho}(x, x' = x, t) . \tag{4.146}$$

Problem 10 Show that all operators with the structure

$$\hat{A} \rightarrow \hat{A}(x, \hat{p}) ; \qquad \left\langle \hat{A}(x, \hat{p}) \right\rangle = A(x, p) \tag{4.147}$$

can be calculated in classical way by means of $W(x, p, t)$:

$$\left\langle \hat{A} \right\rangle = \frac{2\pi}{\hbar} \int dx \int dp W(x, p, t) A(x, p) . \tag{4.148}$$

Problem 11 Show that the Wigner function satisfies the classical Liouville equation

$$\frac{\partial W(x, p, t)}{\partial t} + \frac{p}{m} \frac{\partial W(x, p, t)}{\partial x} - \frac{\partial W(x, p, t)}{\partial p} \frac{\partial U(x)}{\partial x} + \mathcal{O}(\hbar) = 0 \tag{4.149}$$

up to terms of order $\mathcal{O}(h)$.

The relations given here can be generalized to the three-dimensional case using the definition

$$W(\mathbf{x}, \mathbf{p}, t) = \int \frac{d\mathbf{x}'}{h^3} \rho(\mathbf{x}, \mathbf{p}, t) \, e^{-\frac{i}{\hbar} \mathbf{p} \cdot (\mathbf{x}-\mathbf{x}')} . \tag{4.150}$$

4.4.3 Klimontovich's Microscopic Density

Yuri Klimontovich was a student of Bogolyubov and developed since the 1950th a formalism of classical and quantum statistics which is very fruitful for many applications, in particular in plasma physics (Klimontovich 1982, 1986). In some sense, Klimontovich followed the idea by Einstein and Onsager that the fluctuations of macroscopic quantities obey the same dynamics as the microscopic quantities themself (see Sect. 2.5.1 for discussion). We start from the classical case of a system of N particles with coordinates and momenta, $x = \mathbf{r}_i, \mathbf{p}_i$, and the Hamilton equations

$$\frac{d\mathbf{r}_i}{dt} = \mathbf{v}_i = \frac{\mathbf{p}_i}{m_i} \; ; \qquad \frac{d\mathbf{p}_i}{dt} = -\sum_j \frac{dV_{ij}}{\partial \mathbf{r}_i} \; . \tag{4.151}$$

Here, $x = \mathbf{r}_i, \mathbf{p}_i$ plays the rôle of fluctuating microscopic quantities. Klimontovich defines now a microscopic phase density,

$$N(x, t) = N(\mathbf{r}_i, \mathbf{p}_i, t) = \sum_i \delta(r - \mathbf{r}_i(t))\delta(p - \mathbf{p}_i(t)), \tag{4.152}$$

$$N = \int N(x, t)dx \; . \tag{4.153}$$

The microscopic density satisfies a partial differential equation which is similar to the Liouville equation:

$$\frac{\partial N(x, t)}{\partial t} + \frac{\mathbf{p}}{m} \cdot \frac{\partial N(x, t)}{\partial \mathbf{r}} + (\mathbf{F}_0 + \mathbf{F}_M) \cdot \frac{\partial N(x, t)}{\partial \mathbf{p}} = 0 \; , \tag{4.154}$$

where $\mathbf{F}_0$ is the external force. The microscopic force due to interactions is given by

$$\mathbf{F}_M = -\frac{\partial}{\partial \mathbf{r}} \int dx' \Phi \left(\mathbf{r} - \mathbf{r}'\right) N \left(x', t\right) \; . \tag{4.155}$$

The (one-particle) distribution function in phase space follows from the microscopic density by averaging over a Gibbs ensemble,

$$f(\mathbf{r}, \mathbf{p}, t) = \frac{1}{N} \overline{N(\mathbf{r}_i, \mathbf{p}_i, t)} \; . \tag{4.156}$$

We will come back to the Klimontovich kinetic equation in Chap. 8. In order to extend his formalism to quantum phenomena, Klimontovich uses two methods:

- Wigner functions and density operators,
- second quantization.

We introduce here only the first of these methods. According to Klimontovich (1982, 1986), the N-particles Wigner function can be expressed by the N-particles density matrix in coordinate representation,

$$f_N(\mathbf{q}, \mathbf{p}, t) = \frac{\hbar^{3N}}{V^N} \int d\mathbf{g} \, \rho_N \left(\mathbf{q} + \frac{\hbar}{2}\mathbf{g}, \mathbf{p} - \frac{\hbar}{2}\mathbf{g}, t\right) e^{-i\mathbf{g}\cdot\mathbf{p}} \; . \tag{4.157}$$

By using this representation, many quantum statistical relations assume a similar form as known from classical physics.

4.5   Density Functionals, Virial Theorems and Stability

4.5.1   Kohn-Sham and Thomas-Fermi Functionals

As already pointed out in Chap. 2, the Thomas-Fermi theory and the density functional theory (DFT) which is its modern form is of fundamental importance for many electron systems, and plays for the many-electron system the same rôle as the Bohr theory for the hydrogen atom (Karasiev et al. 2012; Parr and Yang 1989). Although density functional theory has its conceptual roots in the Thomas-Fermi model, DFT was put on a firm theoretical foundation by the two Hohenberg-Kohn theorems. Following the original approach by Fermi, we explained already several applications of this theory to the problem of many-electron atoms. The remarkable fact is, that not only the Bohr theory is an exact theory, but also the Thomas-Fermi theory has the property to be exact, at least in asymptotic sense, $N_e \to \infty$ (Lieb and Simon 1973; Thirring 1980). This important property considered already in Chap. 2, will be discussed here again in the context of density functional theory. The Thomas-Fermi theory is just a special case of a class of more general theories named now density functional theories, a direction of the plasma theory whose importance is still increasing. The general density functional approach is based on an important theorem by Kohn, Sham and Hohenberg saying that the energy can be expressed by some functional of the electron density,

$$E = E\,[n_e\,(\mathbf{r})] \; ; \quad n_e\,(\mathbf{r}) = N_e \sum_{\sigma_i} \int d\mathbf{r}_2 \ldots d\mathbf{r}_{N_e}\, |\psi\,(\mathbf{r}\,\mathbf{r}_2 \ldots \mathbf{r}_N)|^2 , \quad (4.158)$$

where ψ is the N-electron wave function. The theorem due to Kohn, Sham and Hohenberg (Kohn and Sham 1965) states that the E-functional is unique if the exact energy is considered. The exact form of the functional is not known, however, we are interested in finding a lower bound only and, therefore, it will be enough to construct an approximate functional. Let us consider first the kinetic energy where

$$T_\psi = N_e \sum_{\sigma_i} \int d\mathbf{r}_2 \ldots d\mathbf{r}_{N_e}\, |\nabla_i \psi\,(\mathbf{r}\,\mathbf{r}_2 \ldots \mathbf{r}_N)|^2 . \quad (4.159)$$

A lower bound to T_ψ can be found by means of the Lieb-Thirring theorem and the Hölder inequality (see Kraeft et al. 1986)

$$T_\psi \geq T_{\mathrm{TF}} = K \int d\mathbf{r}\, n_e^{5/2}\,(\mathbf{r}) \; ; \quad K = \frac{3}{5}\left(2\pi^2\right)^{2/3} \simeq 4.382 \quad (4.160)$$

with the Hölder relation we find

$$\int d\mathbf{r}\, n_e^{5/2}(\mathbf{r}) \geq \frac{N_e^{5/3}}{V^{2/3}}. \tag{4.161}$$

Problem 12 Prove the Lieb-Thirring inequality, Eq. (4.161), using the Hölder inequality.

This shows, finally, that the kinetic energy is larger than the corresponding kinetic energy of an ideal Fermi gas:

$$T_\psi \geq T_{\mathrm{id}} \geq K \frac{N_e^{5/3}}{V^{2/3}}. \tag{4.162}$$

The relation $T_{\mathrm{TF}} \leq T_\psi$ means that the Thomas-Ferm kinetic energy is a lower bound to the exact kinetic energy of the electron systems. The complete Thomas-Fermi energy of an electron-nucleus system consists of the kinetic and the potential energy contributions. It is given by the Thomas-Fermi functional

$$E_{\mathrm{TF}} = \frac{3}{5}\left(2\pi^2\right)^{2/3} \int d\mathbf{r}\, n_e^{5/3}(\mathbf{r}) - \sum_k \int d\mathbf{r} \int d\mathbf{r}_k \frac{e_k^2\, n_e(\mathbf{r})}{|\mathbf{r} - \mathbf{R}_k|}$$
$$+ \int d\mathbf{r} \int d\mathbf{r}' \frac{e^2}{|\mathbf{r} - \mathbf{r}'|} + U(\mathbf{R}_1 \ldots \mathbf{R}_M). \tag{4.163}$$

The last three terms represent respectively, the electron-nuclear, the electron-electron, and the nuclear energy. Here only the first term is an approximation, the other terms are indeed exact. The Thirring-Lieb-Simon theorem for the kinetic energy leads now to the conclusion that the functional for the total energy gives us also a lower limit of the exact energy, which is assumed for $N_e \rightarrow \infty$ (Lieb and Simon 1973; Thirring 1980),

$$E_{\mathrm{TF}}[n(\mathbf{r})] = \inf\left[E(n(\mathbf{r})), \int d\mathbf{r}\, n(\mathbf{r}) = N_e\right]. \tag{4.164}$$

We have shown this way that the ground state energy can be expressed by the local electron density. This approach plays a central rôle in plasma physics in spite of the fact that, strictly speaking, it is originally only a large-Z theory. Being initially only a collection of theorems and theoretical energy bounds, the density functional approach is developing now also to a tool of great practical importance for simulations of Coulomb systems (Parr and Yang 1989; Redmer et al. 2010).

4.5.2 Coulomb Stability and Virial Theorem

The Born-Oppenheimer approximation (see also Sect. 2.3.2) considers a system of M heavy masses and N_e light electrons. As shown in Sect. 2.3.2, dimensional reasoning leads to the remarkable virial theorem for Coulomb systems,

$$E_0 = \langle T \rangle + \langle U \rangle = \frac{1}{2}\langle U \rangle = -\langle T \rangle . \tag{4.165}$$

The ground state of Coulomb systems is always negative (Thirring 1980). For the special case of the hydrogen atom, these relations are known already from the work by Niels Bohr. The most essential feature of the Bohr theory was the finding that the atomic energy has a lower bound. The electron cannot fall into the Coulomb singularity, that is, atoms are asymptotically stable. The important question, whether this is still true for arbitrary finite systems of N nuclei and $N_e = ZN$, forming possibly N atoms could be answered only half a century later by Dyson and Lenard (1967) and Dyson and Lenard (1968) (see also Kraeft et al. 1986; Thirring 1980). Dyson and Lenard (1967, 1968) have shown that there exists indeed a lower bound proportional to N_e:

$$E_0 \leq -CN = -C_e N_e ; \qquad 0 < C < \infty; \quad 0 < C_e < \infty . \tag{4.166}$$

The proof by Lenard and Dyson for the lower bound is a mathematical *tour de force*. Some simple arguments for the correctness of these inequalities were given already in Sect. 2.3.2. More arguments can be found below, for a proof see (Lieb and Simon 1973; Lieb and Thirring 1975; Thirring 1980).

The existence of a lower bound of the ground state energy proportional to the particle number, which is called *H-stability* of the system is of fundamental importance for all matter. It was proven that the Fermi character of the electrons is both sufficient and necessary for the *H*-stability of Coulomb systerns. In more detail, it is the fast increase of the kinetic energy with $n_e^{5/3}$ which is the essential reason for the stability of matter, that is, the Fermi pressure prevents the collapse. In order to demonstrate the importance of the Pauli principle for the stability of matter, Dyson and Lenard (1968) wrote that in the absence of the Pauli principle "we (can) show that not only individual atoms but matter in bulk would collapse into a condensed high-density phase. The assembly of any two macroscopic objects would release energy comparable to that of an atomic bomb."

Virial Theorem The essential part of the proof about stability of Coulomb systems are lower bounds based on a special form of the virial theorem providing a fixed relation between potential and kinetic energy, which is characteristic for Coulomb systems. We remind that Clausius' original virial theorem was related to the pressure of gases and look for more general quantum statistical virial theorems. Following Zubarev (1974) we find the pressure virial from scaling properties of the Hamiltonian. Let $H(q, p)$ be a general Hamiltonian (in the quantum case the corresponding operator). The pressure is then defined as the mean value of a generalized force, which is the derivative with respect to the volume, V:

$$p = -\left\langle \frac{\partial H(q, p)}{\partial V} \right\rangle . \tag{4.167}$$

In order to describe changes of the volume, we introduce a scaling parameter, λ, for linear dimensions such that

$$q = \lambda q'; \quad p = \lambda^{-1} p'; \quad V = \lambda^3 V' . \tag{4.168}$$

This gives (Zubarev 1974)

$$-\frac{\partial H}{\partial V} = -\frac{1}{3V} \left. \frac{\partial H(\lambda q, p/\lambda)}{\partial \lambda} \right|_{\lambda=1} . \tag{4.169}$$

For a Coulomb sytem, we obtain

$$-\frac{\partial H}{\partial V} = \frac{2}{3V} \sum_i \frac{p_i^2}{2m_i} + \frac{1}{3V} U . \tag{4.170}$$

As final result we obtain for Coulomb quantum systems the virial relation for the pressure (Bobrov and Trigger 2014; Subarew 1976; Zubarev 1974):

$$pV = \frac{2}{3} \langle H_{\text{kin}} \rangle + \frac{1}{3} \langle H_{\text{pot}} \rangle . \tag{4.171}$$

4.6 Second Quantization

Green's functions are a concept of mathematical physics which is very useful in our context and Bethe-Salpeter equations are some generalization of the Schrödinger equation for particles, pairs of particles etc., imbedded into a surrounding. These concepts are not simple, however, we need them in the subsequent text. We will try to explain these concepts as clear and at the same time as simple as possible.

4.6.1 Occupation Number Representations

Planck described the state of radiation in a black body by the numbers of energy quanta of oscillators with the discrete energies $n\hbar\omega$, $n = 0, 1, 2 \ldots$. Einstein used a corresponding description by oscillator quanta in his model for the excitations in solid bodies. We used similar representations in the theory of Bose and Fermi gases.

In all these cases, we operate with an integer number, n, indicating the number of elementary quanta of energy. The general idea of occupation number representations is to count the numbers, n_i, of particles, photons, phonons, etc in the energy states, ε_i. The observable is in these cases an occupations number. Following the general schema, any observable should correspond to a certain Hermitian operator. This idea will lead us to the occupation number operators. However describing this procedure in full generality, let us explain the procedure for the special case of one-dimensional oscillators.

Occupation Numbers for One-Dimensional Oscillators The idea to use occupation numbers for the representation of the state comes from oscillator theory and existed already at the very beginning of the quantum statistics, starting with the work by Planck and Einstein. The following set of relations describe a quantum mechanical harmonic oscillator:

$$\hat{P} = \frac{\hbar}{i} \frac{\partial}{\partial \hat{Q}} ; \quad \left[\hat{Q}, \hat{P}\right] = i\hbar ; \quad \left[\hat{P}, \hat{Q}\right] = \frac{\hbar}{i} ; \quad \hat{H} = \frac{1}{2m}\hat{P}^2 + \frac{1}{2}m\omega^2\hat{Q}^2 .$$

$$(4.172)$$

We introduce a formal operator, $\hat{a}$, and the adjunct operator, $\hat{a}^\dagger$,

$$\hat{a} = \frac{m\omega\hat{Q} + i\hat{P}}{\sqrt{2m\hbar\omega}} ; \qquad \hat{a}^\dagger = \frac{m\omega\hat{Q} - i\hat{P}}{\sqrt{2m\hbar\omega}} ,$$

$$(4.173)$$

and find the useful relations

$$\hat{Q} = \sqrt{\frac{\hbar}{2m\omega}} \left(\hat{a} + \hat{a}^\dagger\right) ; \qquad \hat{P} = i\sqrt{\frac{m\hbar\omega}{2}} \left(\hat{a}^\dagger - \hat{a}\right) .$$

$$(4.174)$$

The Hamiltonian can than be expressed by $\hat{a}$ and $\hat{a}^\dagger$:

$$\hat{H} = \frac{\hat{P}^2}{2m} + \frac{m}{2}\omega^2\hat{Q}^2 = \frac{\hbar\omega}{2} \left(\hat{a}^\dagger\hat{a} + \hat{a}\hat{a}^\dagger\right) .$$

$$(4.175)$$

Problem 13 Show that the commutator obeys the permutation rule

$$\left[\hat{a}^\dagger, \hat{a}\right] = -1$$

$$(4.176)$$

and show that by introducing the permutation rule the Hamiltonian assumes the form

$$\hat{H} = \hbar\omega \left(\hat{a}^\dagger\hat{a} + \frac{1}{2}\right) .$$

$$(4.177)$$

According to

$$\hat{H}\,|\Psi\rangle = \left(n + \frac{1}{2}\right)\hbar\omega\,|\Psi\rangle\,, \qquad (4.178)$$

the operator $\hat{a}^\dagger\hat{a}$ should have the eigenvalues n with $n = 0, 1, 2\ldots$ The number operator, $\hat{N} = \hat{a}^\dagger\hat{a}$, obeys the eigenvalue equation

$$\hat{N}\,|n\rangle = n\,|n\rangle \qquad (4.179)$$

and, therefore, we interprete $\hat{N}$ as the operator of the number of oscillation quanta in the *Fock space*, $\mathcal{F}$, of phonons (oscillation quanta). We introduce now occupation number representations for N-particle systems in a more systematic way (Blochinzew 1953). Consider an N-particle system with the one particle wave functions $\Psi_n(q,t)$ and the Schrödinger-equation

$$i\hbar\frac{\partial}{\partial t}\Psi(q,t) = \hat{H}\Psi(q,t). \qquad (4.180)$$

The one-particle wave function can be represented by eigenfunctions as

$$\Psi(q,t) = \sum_n c_n\psi_n(q)\,. \qquad (4.181)$$

and the corresponding expansion for N-particle wave functions is

$$\begin{aligned}
&\Psi\,(q_1\,q_2\ldots q_N, t) \\
&= \sum_{n_1}\sum_{n_2}\cdots\sum_{n_N} c\,(n_1\,n_2\ldots n_n, t)\,\psi_{n_1}(q_1)\,\psi_{n_2}(q_2)\ldots\psi_{n_N}(q_N)\,. \qquad (4.182)
\end{aligned}$$

By introducing this representation into the N-particle Schrödinger equation, we obtain (Blochinzew 1953)

$$i\hbar\frac{\partial}{\partial t}\Psi\,(q_1\ldots q_N, t) = \left(\sum_{k=1}^{N}\hat{H}\,(q_k) + \sum_{k<j}^{N}U\,(q_k, q_j)\right)\Psi(q,t)\,. \qquad (4.183)$$

Thus, we find

$$\begin{aligned}
&i\hbar\frac{\partial}{\partial t}c\,(n_1\ldots n_n, t) \\
&= \int dq_1\ldots dq_N\,\psi_{n_1}^*\ldots\psi_{n_N}^*\left(\sum_{k=1}^{N}\hat{H}\,(q_k) + \sum_{k<j}^{N}U\,(q_k, q_j)\right) \\
&\qquad\qquad \sum_{n'}c\,(n_1'\ldots n_N', t)\,\psi_{n_1'}\,(q_1)\ldots\psi_{n_N'}(q_N) \qquad (4.184)
\end{aligned}$$

and in matrix notation:

$$i\hbar \frac{\partial}{\partial t} c\, (n_1\, n_2 \ldots n_n, t) = \sum_{k=1}^{N} \sum_{n_k} H_{m_k n_k} c\, (n_1\, n_2 \ldots n_k \ldots n_n, t)$$

$$+ \sum_{k<l}^{N} \sum_{n_k, n_l} U_{m_k m_l n_k n_l} c\, (n_1 \ldots n_k \ldots n_l \ldots n_n, t) \,. \tag{4.185}$$

Here we used the definitions

$$H_{m_k, n_k} = \int dq_k \psi_{m_k}^* \, (q_k)\, \hat{H}\, (q_k)\, \psi_{n_k} \, (q_k) \tag{4.186}$$

$$U_{m_k, m_l, n_k, n_l} = \int dq_k \int dq_l \left[\psi_{m_k}^* \, (q_k)\, \psi_{m_l}^* \, (q_l) \right.$$

$$\left. U\, (q_k, q_l)\, \psi_{n_k}^* \, (q_k)\, \psi_{n_k}^* \, (q_k) \right] \,. \tag{4.187}$$

For the case of bosons, the coefficients $c\,(n_1' \ldots n_N', t)$ are symmetrical functions and for the case of fermions antisymmetric functions. In order to make the transition to occupation numbers, we perform the operations (Abrikosov et al. 1965; Blochinzew 1953)

$$c\, (n_1 \ldots n_N, t) = A\, c\, (N_{\alpha_1}\, N_{\alpha_2} \ldots N_{\alpha_N}, t) \,, \tag{4.188}$$

where A is a normalization constant. Since $c\,(n_1, n_2 \ldots, t)$ are symmetric or antisymmetric functions, they depend only on the number of arguments which are equal or different, respectively. In other words, they depend only on the occupation numbers, N_{α_1}, N_{α_2}, $\ldots$, of the one-particle states, α_1, α_2, $\ldots$. The probability of a certain occupation is

$$\left| c\, (N_{\alpha_1}\, N_{\alpha_2}, \ldots, t) \right|^2 = \sum_{m_1, m_2, \ldots} \left| c\, (m_1\, m_2, \ldots, t) \right|^2 \,, \tag{4.189}$$

where the summation is performed over all combinations of $m_1\, m_2 \ldots m_N$. The number of possibile combinations is $\frac{N!}{N_{\alpha_1}! N_{\alpha_2}! \ldots}$, therefore,

$$\left| c\, (N_{\alpha_1}\, N_{\alpha_2} \ldots N_{\alpha_N}, t) \right|^2 = \frac{N!}{N_{\alpha_1}! N_{\alpha_2}! \ldots} \left| c\, (m_1\, m_2 \ldots m_N, t) \right|^2 \tag{4.190}$$

and

$$c\, (N_{\alpha_1}\, N_{\alpha_2}, \ldots, t) = \sqrt{\frac{N!}{N_{\alpha_1}! N_{\alpha_2}! \ldots}}\, c\, (m_1\, m_2 \ldots m_N, t) \,. \tag{4.191}$$

Here, N_{α_1} is the occupation number of state α_1, N_{α_2} is the occupation number of state α_2, etc. This way we find the Schrödinger-equation

$$i\hbar \frac{\partial}{\partial t} c\left(N_{\alpha_1} N_{\alpha_2}, \ldots, t\right) = \hat{H}\left(N_{\alpha_1} N_{\alpha_2}, \ldots, t\right) . \tag{4.192}$$

and, eventually, the dynamics in occupation number representation in explicite form:

$$i\hbar \frac{\partial}{\partial t} c(N_1 N_2, \ldots, t)$$

$$= \sum_{n,m} \sqrt{N_m (N_n + 1)} \; H_{mn} \; c(N_1, \ldots, N_m - 1, \ldots, N_n + 1, \ldots, t)$$

$$+ \frac{1}{2} \sum_{\substack{nn' \\ mm'}} \sqrt{N_m N_{m'} (N_n + 1)(N_{n'} + 1)} \; U_{mm'nn'}$$

$$\times c(N_1, \ldots, N_m - 1, \ldots, N_{m'} - 1, \ldots, N_n + 1, \ldots, N_{n'} + 1, \ldots, t) . \tag{4.193}$$

4.6.2 Second Quantization

We define pairs of operators, $\hat{a}_{\alpha_k}$ (creation operator) and $\hat{a}^{\dagger}_{\alpha_k}$ (annihilation), which operate in the Fock space and have the properties

$$\left[\hat{a}_k, \hat{a}^{\dagger}_l\right] = \delta_{kl} \tag{4.194}$$

$$\hat{a}^{\dagger}_k c(N_1 \ldots N_k \ldots, t) = \sqrt{N_k + 1} \; c(N_1 \ldots N_k + 1 \ldots, t) \tag{4.195}$$

$$\hat{a}_k c(N_1 \ldots N_k \ldots, t) = \sqrt{N_k} \; c(N_1 \ldots N_k + 1 \ldots, t) \tag{4.196}$$

$$\hat{a}^{\dagger}_k c(N_1 N_2 \ldots 0 \ldots, t) = 0 . \tag{4.197}$$

These operators obey the set of relations

$$\hat{a}^{\dagger}_k \hat{a}_k c(N_1 \ldots N_k \ldots, t) = N_k \; c(N_1 \ldots N_k \ldots, t) \tag{4.198}$$

$$\hat{a}^{\dagger}_k \hat{a}_k = N_k ; \qquad \hat{a}_k \hat{a}^{\dagger}_k = N_k + 1 \tag{4.199}$$

$$\left[\hat{a}_k, \hat{a}^{\dagger}_k\right] = 1 ; \qquad \left[\hat{a}_k \hat{a}^{\dagger}_l\right] = \delta_{kl} . \tag{4.200}$$

Further we can show

$$\sqrt{N_m (N_n + 1)} \; c \, (N_1 \ldots N_m - 1 \ldots N_n + 1 \ldots, t)$$
$$= \hat{a}_n^\dagger \hat{a}_m \; c \, (N_1 \ldots N_m, \ldots N_n \ldots N_N, t) \; . \qquad (4.201)$$

By using the new creation and annihilation operators, $\hat{a}^\dagger$ and $\hat{a}$, the Hamilton operator and the equations of motion assume the form

$$\hat{H} = \sum_{mn} H_{mn} \, \hat{a}_n^\dagger \hat{a}_m + \frac{1}{2} \sum_{mm'nn'} U_{mm'nn'} \, \hat{a}_n^\dagger \hat{a}_{n'}^\dagger \hat{a}_m \hat{a}_{m'} \qquad (4.202)$$

and

$$i\hbar \frac{\partial}{\partial t} c \, (N_1 \, N_2 \ldots, t) = \hat{H} \, c \, (N_1 \, N_2 \ldots, t) \; . \qquad (4.203)$$

We notice the remarkable simplicity of the Hamilton operator in the new representation by means of creation and annihilation operators which we call *second quantization*. The method of second quantization which originally was developed in quantum field theory is nowadays also one of the basic tools of quantum statistics (Abrikosov et al. 1965).

The creation and annihilation operators satisfy for bosons the relations

$$\hat{a}_m \hat{a}_n^\dagger - \hat{a}_n^\dagger \hat{a}_m = \delta_{mn}, \qquad (4.204)$$

and

$$\hat{a}_n \hat{a}_n^\dagger = 1 + N_n \; . \qquad (4.205)$$

For fermions we have the permutation rules

$$\hat{a}_m \hat{a}_n^\dagger + \hat{a}_n^\dagger \hat{a}_m = \delta_{mn} \qquad (4.206)$$

and the particle number operator, $\hat{N} = \hat{a}_n^\dagger \hat{a}_n$, is

$$\hat{a}_n \hat{a}_n^\dagger = 1 - \hat{N}_n \; . \qquad (4.207)$$

The Pauli principle requires

$$\hat{a}_n^\dagger c \, (N_1 \, N_2 \ldots 1_n \ldots, t) = 0 \qquad (4.208)$$

$$\hat{a}_n c \, (N_1 \, N_2 \ldots 0_n \ldots, t) = 0 \qquad (4.209)$$

$$\hat{a}_n^\dagger c \, (N_1 \, N_2 \ldots 0_n \ldots, t) = \pm c \, (N_1 \, N_2 \ldots 1_n \ldots, t) \qquad (4.210)$$

$$\hat{a}_n^\dagger c \, (N_1 \, N_2 \ldots 1_n \ldots, t) = \pm c \, (N_1 \, N_2 \ldots 0_n \ldots, t) \; . \qquad (4.211)$$

Just as the Hamiltonian, all other operators of physical quantities can be represented by $\hat{a}$ and $\hat{a}^\dagger$. In analogy to quantum field theory, we define the field operator, $\hat{\Psi}$, and the conjugated field operator, $\hat{\Psi}^\dagger$, by the expansions (Abrikosov et al. 1965)

$$\hat{\Psi}(q,t) = \sum_n \hat{a}_n(t)\psi_n(q); \qquad \hat{\Psi}^\dagger(q,t) = \sum_n \hat{a}_n^\dagger(t)\psi_n^*(q). \qquad (4.212)$$

The permutation rules for bosons read

$$\hat{\Psi}(q,t)\hat{\Psi}^\dagger(q',t) - \hat{\Psi}^\dagger(q',t)\hat{\Psi}(q,t) = \sum_{m,n} \left(\hat{a}_n\hat{a}_m^\dagger - \hat{a}_m^\dagger\hat{a}_n\right)\psi_n(q)\psi_m^*(q')$$

$$= \sum_{m,n} \delta_{mn}\psi_m^*(q')\psi(q) = \sum_n \psi_n^*(q')\psi_n(q) = \delta(q'-q). \qquad (4.213)$$

For fermions we find in a similar way

$$\hat{\Psi}(q,t)\hat{\Psi}^\dagger(q',t) + \hat{\Psi}^\dagger(q',t)\hat{\Psi}(q,t) = \delta(q'-q). \qquad (4.214)$$

Using these new relations we can express the Hamiltonian by means of the operator $\hat{\Psi}(q,t)$:

$$\hat{H} = -\frac{\hbar^2}{2m} \int \hat{\Psi}^\dagger(q,t)\Delta\hat{\Psi}(q,t)\,dq$$

$$+ \frac{1}{2} \int\int \hat{\Psi}^\dagger(q,t)\hat{\Psi}^\dagger(q',t)U(q,q')\hat{\Psi}(q,t)\hat{\Psi}(q',t)\,dq\,dq'. \qquad (4.215)$$

The permutation rules for the operators $\hat{\Psi}$ and the representation of $\hat{H}$ can be interpreted as a kind of quantization of the wave function, $\Psi(q,t)$. This seems to be the origin of the denotation as second quantization.

4.6.3 Klimontovich Operator in Second Quantization

In second quantization, the operator of the particle number has the representation

$$\hat{N} = \int \hat{\Psi}^\dagger(q,t)\hat{\Psi}(q,t)dq \qquad (4.216)$$

and the corresponding operator of particle number density is defined by Kadanoff and Baym (1962)

$$\hat{n}(q,t) = \hat{\Psi}^\dagger(q,t)\hat{\Psi}(q,t). \qquad (4.217)$$

The introduction of an operator of particle density in coordinate/spin space can be extended to the phase space. In his textbook on statistical physics (Klimontovich 1982, 1986) and in further papers, Klimontovich developed a method which generalizes the Wigner method and his method of random densities in phase space on the basis of the second quantization method. Klimontovich's operator of phase density, (called by Kubo *Klimontovich* operator) is defined by Klimontovich (1982, 1986)

$$\hat{N}(\mathbf{r}, \mathbf{p}, t) = \frac{\hbar^3}{V} \int d\mathbf{g} \, \Psi^* \left(\mathbf{r} - \frac{\hbar}{2} \mathbf{g} \right) \Psi \left(\mathbf{r} + \frac{\hbar}{2} \mathbf{g} \right) e^{-i\mathbf{g} \cdot \mathbf{p}} . \tag{4.218}$$

The dynamics of the Klimontovich operator is given by

$$\frac{\partial \hat{N}(\mathbf{r}, \mathbf{p}, t)}{\partial t} + \frac{\mathbf{p}}{m} \cdot \frac{\partial \hat{N}(\mathbf{r}, \mathbf{p}, t)}{\partial \mathbf{r}} = \frac{i}{\hbar(2\pi)^3} \int \left[\hat{U} \left(\mathbf{r} - \frac{\hbar}{2} \mathbf{g} \right) - \hat{U} \left(\mathbf{r} + \frac{\hbar}{2} \mathbf{g} \right) \right]$$

$$\hat{N}(\mathbf{r}, \mathbf{p}', t) \, e^{i\mathbf{g} \cdot (\mathbf{p} - \mathbf{p}')} \, d\mathbf{g} \, d\mathbf{p}' , \tag{4.219}$$

where Klimontovich introduced the operator of potential energy (here U_0 is an external potential)

$$\hat{U}(\mathbf{r}, \mathbf{p}, t) = U_0(\mathbf{r}) + \int \Phi(\mathbf{r}, \mathbf{r}') \hat{N}(\mathbf{r}', \mathbf{p}', t) \frac{d\mathbf{r}' \mathbf{p}'}{(2\pi\hbar)^3} . \tag{4.220}$$

The Klimontovich equation for the operator of particle density in phase space is in full analogy to the Wigner equation. An essential difference is, however, that the new equation is not an approximation valid only for one-particle problems but we have now a field operator equation for systems of interacting particles. In Chap. 8 we will come back to Klimontovich's kinetic equation.

4.7 Green's Functions

4.7.1 Definition and Relations for Green's Functions

The method of second quantizations allows us to introduce another important concept of modern quantum statistics, the method of Green's functions whose origin is mainly in quantum field theory (Abrikosov et al. 1965; Bonch-Bruevich and Tyablikov 1961; Fetter and Walecka 1971; Fick 1981; Kadanoff and Baym 1962; Martin and Schwinger 1959; Matsubara 1955; Montroll and Ward 1958). The advanced and the retarded Green's functions are defined by

$$G^> (1, 1') = \frac{1}{i\hbar} \text{Tr} \left\{ \hat{\rho} \hat{\Psi}(1) \hat{\Psi}^\dagger (1') \right\} \tag{4.221}$$

$$G^< (1, 1') = \begin{cases} +\frac{1}{i\hbar} \text{Tr} \left\{ \hat{\rho} \hat{\Psi}^\dagger (1') \hat{\Psi} (1) \right\} \text{ for bosons} \\ -\frac{1}{i\hbar} \text{Tr} \left\{ \hat{\rho} \hat{\Psi}^\dagger (1') \hat{\Psi} (1) \right\} \text{ for fermions}, \end{cases} \tag{4.222}$$

where the 1, or $1'$ correspond each to a full set of coordinates, spin and time. The symbol Tr expresses here a trace which is performed over a Fock space. Causal Green's functions are defined by

$$G (1, 1') = \frac{1}{i\hbar} Tr \left\{ \hat{\rho} \hat{T} \hat{\Psi} (1) \hat{\Psi}^\dagger (1') \right\}, \tag{4.223}$$

with a time ordering operator, $\hat{T}$, which acts as follows:

$$\hat{T} \hat{\Psi} (1) \hat{\Psi}^\dagger (1') = \begin{cases} \hat{\Psi} (1) \hat{\Psi}^\dagger (1') & t_1 > t_1' \\ +\hat{\Psi}^\dagger (1') \hat{\Psi} (1) & t_1 < t_1' \quad \text{for bosons} \\ -\hat{\Psi}^\dagger (1') \hat{\Psi} (1) & t_1 < t_1' \quad \text{for fermions}. \end{cases} \tag{4.224}$$

The advanced Green's function describes a particle which is created at time t_1' and annihilated at time $t_1 > t_1'$. Correspondingly, a retarded Green's function describes a particle which is annihilated at t_1 and created again at $t_1' > t_1$. The Green's function $G (1, 1') = G_1 (1, 1')$ is called one-particle Green's function. Correspondingly, we can define many-particles Green's functions by

$$G_n (1 \ldots n, 1' \ldots n') = \frac{1}{(i\hbar)^n} Tr \left\{ \hat{\rho} \hat{T} \hat{\Psi}(1) \ldots \hat{\Psi}(n) \hat{\Psi}^\dagger(1') \ldots \hat{\Psi}^\dagger(n') \right\}. \tag{4.225}$$

The generalized operator $\hat{T}$ contains a factor $\lambda (P_n)$ which is defined by the number of permutation operations, P, which are needed

$$\text{for bosons} \quad \lambda (P) = 1$$

$$\text{for fermions} \quad \lambda (P) = \begin{cases} 1 & \text{even} \\ -1 & \text{odd}. \end{cases}$$

Relations to the density operators (matrices) are given by

$$\rho (x, x', t) = i\hbar \, G^< (1, 1') \big|_{t_1 = t_{1'}} = i\hbar G_1 (1, 1')_{t'=t+0} \tag{4.226}$$

and the relations to two-time density matrices are given by

$$\rho (x, t, x't') = i\hbar G_1 (1, 1') \quad \text{if} \quad t' > t. \tag{4.227}$$

The Green's functions for free particles are denoted by $G_1^{(0)}(1, 1')$. In a similar way as the density operators of different order are coupled by the Bogolyubov hierarchy, the Green's functions for different particle numbers are also related through a hierarchy:

$$\left(i\hbar\frac{\partial}{\partial t_1} + \frac{\hbar^2}{2m}\nabla_1^2\right) G_1\left(1, 1'\right) = \delta\left(1 - 1'\right) \pm i \int d\tilde{2}'\, V\left(1, \tilde{2}'\right) G_2\left(1, 2, 1', \tilde{2}'\right).$$

$$(4.228)$$

$$\left(i\hbar\frac{\partial}{\partial t_1} + \frac{\hbar^2}{2m}\nabla_1^2\right) G_2\left(121'2'\right)$$

$$= \delta\left(11'\right) G_1\left(22'\right) \pm G_1\left(21'\right) \delta\left(12'\right) \pm i \int d\bar{3}V\left(1\bar{3}\right) G_3\left(1231'2'\bar{3}^+\right).$$

$$(4.229)$$

$$\delta\left(1 - 1'\right) = 0 \quad \text{for} \quad x_1 \neq x_1', \quad \text{or} \quad t_1 - t_1'.$$

$$(4.230)$$

The δ-functions are needed in order to fullfil the correct initial conditions for $t_1 = t_1'$, $x_1 = x_1'$.

The dynamic equations can be solved by a perturbation expansion with respect to the order in the interactions,

$$G_n = G_n^{(0)} + G_n^{(1)} + G_n^{(2)} + G_n^{(3)} + \dots$$

$$(4.231)$$

For the case of free particles (no interactions, $V = 0$), we obtain

$$G_2^{(0)} = \begin{vmatrix} G_1^0\left(11'\right) & G_1^0\left(12'\right) \\ G_1^0\left(21'\right) & G_1^0\left(22'\right) \end{vmatrix}_{\pm},$$

$$(4.232)$$

corresponding to the correct symmetry relations. Starting from the free Green's function, by iterations we find a perturbation series

$$G_3^{(0)} = \begin{vmatrix} G_1^0\left(11'\right) & G_1^0\left(12'\right) & G_1^0\left(13'\right) \\ G_1^0\left(21'\right) & G_1^0\left(22'\right) & G_1^0\left(23'\right) \\ G_1^0\left(31'\right) & G_1^0\left(32'\right) & G_1^0\left(33'\right) \end{vmatrix}_{\pm}.$$

$$(4.233)$$

Following the perturbation expansion given above, the Green's functions of any order can be calculated by iterations and the terms can be represented by Feynman graphs as shown in Fig. 4.1.

Figure 4.2 shows the Feynman diagrams corresponding to one-particle Green's functions up to the first order in perturbation theory and Fig. 4.3 corresponding the Feynman diagrams for two-particle Green's functions. Following the schemata

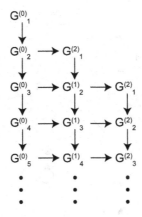

Fig. 4.1 Symbolic representation of the hierarchy of Green's functions

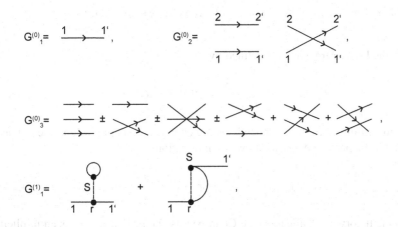

Fig. 4.2 Feynman diagram representation of the one-particle Green's function up to first order in perturbation theory

given here, the Green's functions can be calculated in escalating order after transforming them into the corresponding integrals.

4.7.2 Thermodynamics and Green's Functions

Green's functions contain many informations about the many-particle system. As an example, we consider first the momentum distributions which can be expressed by the one-particle functions

$$n(p_1) = i\hbar G_1^<(p_1, \tau = 0) , \qquad (4.234)$$

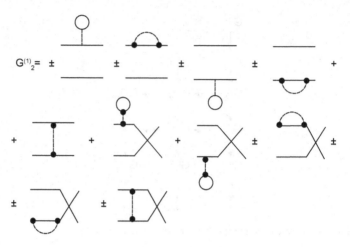

Fig. 4.3 Feynman diagram representation for the two-particle Green's function

where the Fourier transform is defined by

$$G^<(\mathbf{r}_1 - \mathbf{r}_1', \tau) = \frac{-i}{\hbar} \int \frac{d\mathbf{p}_1}{(2\pi)^3} e^{i\mathbf{p}_1 \cdot (\mathbf{r}_1 - \mathbf{r}_1')} G^<(\mathbf{p}_1, \tau) . \qquad (4.235)$$

The momentum distribution depends on the chemical potential, μ, and the temperature, T. The particle density follows by integration

$$n = \int d^3 p_1 \, n(p_1, \mu) . \qquad (4.236)$$

As in the theory of ideal gases (see Chap. 3) this can be considered as an implicit representation of the chemical potential, $n = n(\mu, T)$. The chemical potential can be obtained by inversion, $n = \mu(n, T)$ (Kraeft et al. 1986). This way, we find the Gibbs free energy, $G = N\mu$, and the other thermodynamic potentials. As an alternative way which is in some cases of advantage, we can start from the pressure expressed by two-particles Green's functions. In the grand canonical ensemble, the pressure is given by

$$pV = k_B T \ln \Xi ; \qquad \Xi = \mathrm{Trace} \left[e^{\beta \mu N - \beta \hat{H}_N} \right] \qquad (4.237)$$

$$\hat{H}_N = \hat{T} + \hat{H}_{\mathrm{int}} . \qquad (4.238)$$

The Trace is performed over the Fock space. Again, we use the idea of Debye to introduce a charging parameter, λ, which allows to change the strength of interaction gradually to its full value, $\lambda = 1$, that is, $\hat{H}_N(\lambda) = \hat{T} + \lambda \hat{H}_{\mathrm{int}}$. Then $\lambda = 1$ corresponds to the full Hamilton operator, $\hat{H}(1)$, and $\lambda = 0$ to the case of the ideal gas, $\hat{H}(0)$, that is, without interaction. All functions of the Hamilton operators are now functions of λ. By charging from 0 to 1—*per definitionem* this is the Debye charging—follow the real thermodynamics functions, in particular also pV:

$$\frac{\partial}{\partial \lambda}(pV) = -k_B T \frac{1}{\Xi} \text{Trace}\left[H_{\text{int}} e^{\beta \mu N - \beta \hat{H}_N}\right] \tag{4.239}$$

$$pV = (pV)_{\text{id}} - \int_0^1 d\lambda \frac{1}{\Xi} \text{Trace} H_{\text{int}}\, e^{\beta \mu N - \beta \hat{H}_N} = (pV)_{\text{id}} - \int_0^1 \frac{d\lambda}{\lambda}\left\langle \hat{H}_{\text{int}}\right\rangle, \tag{4.240}$$

with

$$\left\langle \hat{H}_{\text{int}}\right\rangle = \frac{\text{Trace}\left(H_{\text{int}} e^{\beta \mu N - \beta \hat{H}_N}\right)}{\text{Trace}\left(e^{\beta \mu N - \beta \hat{H}_N}\right)} = \frac{1}{2}\int d1\, d2\, V(1,2)\, G_2\big(1,2,1',2'\big), \tag{4.241}$$

where $G_2\left(1,2,1',2'\right)$ is the two-particle Green's function. This leads to the following final expression for the pressure (in the grand canonical ensemble):

$$pV = (pV)_{\text{id}} - \frac{1}{2}\int_0^1 \frac{d\lambda}{\lambda}\int d1\, d2\, V(1,2)\, G_2\left(1,2,1',2';\lambda\right). \tag{4.242}$$

This way we have two expressions for the thermodynamic functions by Green's functions. In our view the second method based on the two-particle functions is preferable for several reasons for deriving the equation of state, in particular in the chemical picture. Thus, we will give some preference to Eq. (4.242) and equivalent representations for calculating the pressure. Note that by combination of Eq. (4.242) with the Feynman diagrams shown in Fig. 4.3, there follows a diagrammatic representation of the pressure and we see clearly, which integrals have to be calculated in given order.

On the other hand, the method based on one-particle functions and Eq. (4.236) was also quite successfully applied to equation-of-state problems, in particular in combination with self energy concepts closely related to the work of a group at Rostock University (Kremp et al. 2005). This way an approach to the equation of state can be given based on the self-energy, Σ_c in combination with the related spectral function, A_c ($c = e, p$), starting, e.g., from the normalization condition for the total density

$$n_c(\beta, \mu_e, \mu_p) = \frac{1}{\Omega}\sum_1 \int_{-\infty}^{\infty} \frac{d\omega}{2\pi} f_c(\omega) A_c(1, \omega), \tag{4.243}$$

where $1 = \{\mathbf{p}_1, \sigma_1\}$ denotes momentum and spin, and $f_c(\omega) = [\exp(\beta\omega - \beta\mu_c) + 1]^{-1}$ is the Fermi distribution function. We take periodic boundary conditions with respect to the normalization volume, Ω, leading to discrete values of the momentum, $\mathbf{p}$. The transition from the sum to an integral can be performed as usual in quantum statistics. The spectral function

$$A_c(1, \omega) = \frac{2\mathrm{Im}\, \Sigma_c(1, \omega - i0)}{\left[\omega - \frac{p_1^2}{2m_c} - \Sigma_c(1, \omega)\right]^2 + [\mathrm{Im}\, \Sigma_c(1, \omega - i0)]^2} \qquad (4.244)$$

is related to the self-energy, $\Sigma_c(1, z)$, defined in the complex z plane. This way, approximations in the equation of state are traced back to approximations for the self-energy, which can be evaluated at the complex Matsubara frequencies using diagram techniques (Matsubara 1955). Note that thermodynamic potentials are obtained from the equation of state, Eq. (4.243), by relations such as an expression for the pressure, p, and the free energy density, f,

$$p(T, \mu) = \int_{-\infty}^{\mu} n(\beta, \mu')d\mu'; \qquad f(T, n) = \int_{0}^{n} \mu(\beta, n')dn'. \qquad (4.245)$$

The chemical picture can also be derived by performing a cluster expansion of the self-energy and the bound-state parts of the few-particle T-matrices occurring in this cluster expansion are taken into account (Kraeft et al. 1986; Kremp et al. 2005). On the other hand, the low-density expansion of the equation of state is obtained by considering the two-particle contributions to the self-energy, which in general will contain two-particle bound and scattering states. As long as $\mathrm{Im}\, \Sigma_c(1, \omega - i0)$ can be considered as a small quantity, an expansion of the spectral function can be performed which gives in addition to the quasiparticle δ-like structure also the contribution from the two-particle states. This way, a new version of the Beth and Uhlenbeck formula can be derived which provides new expression for the second virial coefficient in terms of the two-particle bound state energy and phase shifts (Kraeft et al. 1986; Röpke et al. 2013).

The original version of the Beth-Uhlenbeck formula and its application to gases will be explained in detail in Chap. 5 and the formulation for Coulomb systems in Chap. 6. Here we discuss a generalized Beth-Uhlenbeck formula which is based on the Green's function method and relates the densities of the constituent particles in the physical picture to the chemical potentials, including medium effects on the mean-field level (Blaschke et al. 2014; Ebeling et al. 2009, 2010; Röpke et al. 2013).

We start from a relation for the total electron density with given spin orientation (we will explicitly mark the spin orientation, assuming full degeneration such that $n_e^\uparrow = n_e^\downarrow = n_e/2$)

$$n_e^\uparrow(\beta, \mu_e, \mu_p) = \frac{1}{\Omega} \sum_p f_e\left[E^{\mathrm{qp}}(p)\right] + \frac{2}{\Omega} \sum_{P,n}^{\mathrm{bound}} g_{ep}(E_{P,n})$$

$$+ \frac{2}{\Omega} \sum_{P,n} \int \frac{dE}{\pi} g_{ep} \left(\frac{P^2}{2m_p} + E\right) \frac{d}{dE}\left\{\delta_{P,n}(E) - \frac{1}{2}\sin[2\delta_{P,n}(E)]\right\},$$

$$(4.246)$$

where $f_e(E^{\mathrm{qp}}(p))$ is the Fermi distribution of electrons with the quasiparticle dispersion relation

$$E^{\mathrm{qp}}(p) = \frac{p^2}{2m_e} + \mathrm{Re}\,\Sigma_e\left[p, E^{\mathrm{qp}}(p)\right] \qquad (4.247)$$

and

$$g_{ep}(E_{P,n}) = \frac{1}{e^{\beta(E_{P,n} - \mu_e - \mu_p)} - 1} \qquad (4.248)$$

denotes the Bose function for the electron-proton states with total momentum P and internal quantum number n. The spin degrees of freedom are treated explicitly. Disregarding hyperfine splitting into singlet and triplet states, the summation over the spin of the proton gives the factor 2. In general, also the electron-electron interaction channel should be considered which contributes to the scattering part of the second virial coefficient. In contrast to the simple Beth-Uhlenbeck formula, the free single-particle energy dispersion relation is replaced by the quasiparticle dispersion relation. The two-particle bound state energies, $E_{P,n}$, and scattering phase shifts, $\delta_{P,n}(E)$, contain medium effects to be discussed in the following subsection. The term $\sin[2\delta_{P,n}(E)]$ avoids double counting of contributions which are already taken into account in the quasiparticle shift. The microscopic origin of the concept of excluded volume is the Pauli principle which is due to the anti-symmetrization of fermionic wave functions. One of the consequences are contributions to the effective interaction between composed particles which have to be taken into account when a chemical picture is introduced. As an example, the short-range repulsion between atoms or molecules is caused by Pauli blocking of overlapping spin parallel electron orbitals. Here, however, we focus on the interaction between hydrogen atoms and free electrons. We present a microscopic treatment based on the underlying Pauli exclusion principle. The extension to effective interactions between other components of the chemical picture is straightforward but will not be studied here.

4.8   Pair Bound States and Bethe-Salpeter Equation

The bound states of pairs of particles play a major rôle for the properties of gases and plasmas. They are responsible for the formation of atoms, molecules, clusters etc. Let us first consider the treatment of a pair bound state in the case that the pair is isolated from the surrounding. This way we have a standard two-body problem, $N = 2$, with a central force potential, $U = U(r)$, with $r = |\mathbf{r}_1 - \mathbf{r}_2|$. The wave function is $\Psi(q, t) = \Psi(\mathbf{r}_1, \sigma_1, \mathbf{r}_2, \sigma_2, t)$, and the Schrödinger equation reads in the 6-dimensional space

$$i\hbar \frac{\partial \Psi}{\partial t} = \left(-\frac{\hbar^2}{2m_1} \Delta_1 - \frac{\hbar^2}{2m_2} \Delta_2 \right) \Psi + U(r)\Psi . \tag{4.249}$$

Using the simple transformation to relative x, y, z and center of mass coordinates X, Y, Z, we find (see Chap. 2):

$$i\hbar \frac{\partial \Psi}{\partial t} = -\frac{\hbar^2}{2M} \Delta_{\mathbf{r}} \Psi - \frac{\hbar^2}{2\mu} \Delta_{\mathbf{R}} \Psi + U(r)\Psi . \tag{4.250}$$

Decoupling the free translation of the center of mass and the contribution of the spins, we obtain

$$\Psi(q, t) = \chi(\sigma_1, \sigma_2) \, e^{-\frac{i}{\hbar} \mathbf{P} \mathbf{R}} e^{-\frac{i}{\hbar} E t} \, \psi(x, y, z) , \tag{4.251}$$

where $\mathbf{R}$ is the position and $\mathbf{P}$ the momentum of the center of mass. The Schrödinger equation is decomposed as

$$E = \frac{P^2}{2M} + \varepsilon ; \qquad \frac{\hbar}{2\mu} \Delta_{\mathbf{r}} \psi(\mathbf{r}) + [\varepsilon - U(r)] \, \psi(\mathbf{r}) = 0 , \tag{4.252}$$

where ε is the separate energy of relative dynamics, $\varepsilon = E - P^2/2M$. For the Coulomb problem with $N = 2$, we obtain the Schrödinger equation for the stationary bound states

$$\frac{\hbar^2}{2\mu} \Delta_{\mathbf{r}} \psi_n(\mathbf{r}) + \frac{e^2}{r} \psi_n(\mathbf{r}) = E_n^0 \psi_n(\mathbf{r}) . \tag{4.253}$$

As well known from standard quantum mechanics, the energy eigenvalues are

$$E_n^0 = -\frac{\mu e^4}{2\hbar^2 n^2} \tag{4.254}$$

and the energy eigenfunctions are given by Legendre polynomials. The ground state energy and the corresponding eigenfunction read

$$E_1^0 = -\frac{\mu e^4}{2\hbar^2}; \qquad \psi_0(r) = const.\, e^{-\frac{r}{2a_B}}. \tag{4.255}$$

We will use in the following, for convenience, Rydberg units with $m_e/m_p \ll 1$, $m_e = 1/2$, $\hbar = 1$, $e^2 = 2$. The binding energies of the isolated hydrogen atom are then $E_{P,n}^0 = P^2/2m_p + E_n^0$ with $E_n^0 = -1/n^2$. Let us study now an effective wave equation.

Effective Schrödinger Equation of Coulomb Pairs in a Plasma
The Schrödinger equation of the electron-proton system reads in momentum representation

$$p^2\phi_n(p) - \sum_q V(q)\phi_n(p+q) = E_n^0\phi_n(p). \tag{4.256}$$

Using the Fourier form of the Coulomb interaction, $V(q) = 4\pi/q^2$, the normalized wave function for the ground state $(n = 1)$ is

$$\phi_1(p) = \frac{8\sqrt{\pi}}{(1+p^2)^2}; \qquad \sum_p |\phi_1(p)|^2 = \int \frac{d^3p}{(2\pi)^3}|\phi_1(p)|^2 = 1. \tag{4.257}$$

For the normalization we assumed a periodic volume, $\Omega = 1$.

When imbedding the hydrogen atom into a plasma environment, the additional interactions with the medium can be described within a quantum statistical approach by introducing concepts such as self-energy, dynamical screening, and spectral function. This leads to the effective wave equation for plasmas given by Ebeling et al. (1977):

$$[\varepsilon_e(p_1) + \varepsilon_i(p_2) - E]\, \psi(p_1, p_2)+$$

$$+ \int \frac{dp_1'dp_2'}{(2\pi)^6} V_{ei}^s\, \delta(p_1 + p_2 - p_1' - p_2')\, \psi(p_1', p_2') = 0, \tag{4.258}$$

where ε_a are the one-particle energies and V^s is the screened potential defined by

$$\varepsilon_a = \frac{p_1^2}{2m} + \Sigma_a(p_1); \qquad V_{ei} = V_{ei} + \sum V_{ac}\, \Pi_c\, V_{cb}^s \tag{4.259}$$

(Σ_a, Π_c - self-energy and polarization function). In early publications, a simple static approximation was used (Ebeling et al. 1977) following the intuitive approach by Ecker and Weizel (1956). In later work on effective wave equations several dynamic approximations have been introduced which reflect various processes in plasmas (Kraeft et al. 1986; Kremp et al. 2005; Röpke et al. 1978; Zimmermann 1988; Zimmermann et al. 1978) which improved the effective wave equation considerably. By now, the approach is considered as a standard tool in plasma physics (Kraeft et al. 1986; Kremp et al. 2005; Zimmermann et al. 1978). A more recent form is

$$p^2 \psi_n(p) - \sum_q V(q)\psi_n(p+q) + \sum_q H^{\mathrm{pl}}(q)\psi_n(p+q) = E_n \psi_n(p), \quad (4.260)$$

assuming the adiabatic limit $m_e/m_p \ll 1$, such that the center of mass motion, $\mathbf{P}$, can be neglected. In general, the plasma Hamiltonian, $H^{\mathrm{pl}}(q)$, depends also on $\mathbf{P}$ and on the energy, if dynamical and retardation effects are taken into account. The plasma Hamiltonian shifts the energy eigenvalues, $E_n = E_n^0 + \Delta E_n$, and modifies the wave functions, $\psi_n(p)$. With increasing density, the influence of the plasma increases and the binding energies can merge into the continuum, such that bound states disappear. This breakup of bound states is called Mott effect and has important consequences for the macroscopic properties of the plasma. By means of perturbation theory, different medium effects can be considered to contribute additively to the plasma Hamiltonian,

$$H^{\mathrm{pl}}(q) = H^{\mathrm{Hartr}} + H^{\mathrm{Fock}} + H^{\mathrm{Pauli}} + H^{\mathrm{MW}} + H^{\mathrm{polp}} + H^{\mathrm{vdW}} + \ldots. \quad (4.261)$$

The first three contributions are of first order with respect to the interaction and determine the mean-field approximation, which is instantaneous in time and does not contain dynamic contributions:

$$\sum_q \left[H^{\mathrm{Hartr}}(q) + H^{\mathrm{Fock}}(q) + H^{\mathrm{Pauli}}(q) \right] \psi_n(p+q) =$$

$$\sum_{p'} V(0) \left[2f_e(p') - 2f_i(p') \right] \psi_n(p) - \sum_{p'} V(\mathbf{p'} - \mathbf{p}) f_e(p') \psi_n(p)$$

$$+ \sum_{p'} V(\mathbf{p'} - \mathbf{p}) f_e(p) \psi_n(p'). \quad (4.262)$$

These contributions are of similar structure and have to be considered simultaneously in order to find consistent approximations. Note that the electron-electron interaction is repulsive ($V(q)$), while the electron-ion interaction is attractive ($-V(q)$). The Hartree term contains the factor 2 due to spin summation (for abbreviation, only the momentum is given in the Fermi distribution function). This contribution vanishes, however, for neutral plasmas where the densities $n_c = \sum_{p'} 2f_c(p')$ of the charge carriers, $c = \{e, i\}$, compensate one another. The Fock term as well as the Pauli blocking term describe exchange terms and refer only to the interaction between particles of identical species and spin. The origin of the Pauli blocking term is the phase space occupation by free electrons according to the distribution function, $f_e(p)$. This phase space cannot be used to form a bound state. Thus, the interaction of the electron with the ion is blocked if the final state is already occupied by a free electron of the same spin orientation. The following two terms of the plasma Hamiltonian are the Montroll-Ward term giving the dynamical screening of the interaction in the self-energy, and the (dynamical) screening of the interaction between the bound particles. These contributions are related to the polarization function and are of particular interest for plasmas due to the long-range character of the Coulomb interaction. In a consistent description, both terms should be treated simultaneously. A more detailed discussion has been given by Kraeft et al. (1986), Kremp et al. (2005) and Röpke et al. (2013). The remaining contributions to the plasma Hamiltonian are of at least second order with respect to the interaction, such as the polarization potential, describing the interaction of a bound state with free charge carries, as well as the van-der-Waals interaction, describing the interaction of a bound state with another bound state (Kraeft et al. 1986; Röpke et al. 2013).

A comparison of the density dependence according to a numerical evaluation of the shifts following Ebeling et al. (2009) and Redmer et al. (2010) is shown in Fig. 4.4 for $T = 5000$ K. We see that the agreement with the data is for this temperature quite reasonable; we mention that the temperature dependence in the region of interest $(5000 \ldots 15{,}000)$ K is rather weak.

In the density range studied here, the total energy shift of the ground state is always positive. Looking at the asymptotic expressions for the Fermi limit, we see,

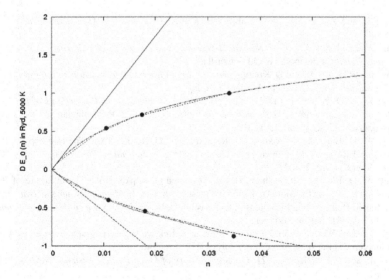

Fig. 4.4 Results of a numerical evaluation (Ebeling et al. 2009) of the Pauli shift (above) and the Fock shift (below) for the temperature $T = 5000$ K in comparison to an interpolation formula based on the asymptotic representations (the smooth lines) and the linear approximation (the cone defined by the outer straight lines). The results of the numerical evaluation are given by dots connected by the straight lines

however, that at the highest density the shifts can become negative (Arndt et al. 1996; Redmer 1997). The remaining shifts in Eq. (4.261) are smaller and may be neglected in a first approach, we refer for further discussion to the literature (Arndt et al. 1996; Kremp et al. 2005; Redmer 1997; Redmer et al. 2010; Röpke et al. 1978, 2013).

References

Abrikosov, A.A., L.P. Gorkov, and I.E. Dzyaloshinskii. 1965. *Methods of Quantum Field Theory in Statistical Physics*. Oxford: Pergamon Press.

Arndt, S., W.D. Kraeft, and J. Seidel. 1996. Two-Particle Energy Spectrum in Dense Electron–Hole Plasmas. *Physica Status Solidi (B)* 194: 601–617.

Ashcroft, N., and N.D. Mermin. 1976. *Solid State Physics*. Boston: Cengage Learning.

Blaschke, D., M. Buballa, A. Dubinin, G. Röpke, and D. Zablocki. 2014. Generalized Beth-Uhlenbeck Approach to Mesons and Diquarks in Hot Dense Quark Matter. *Annals of Physics* 348: 228–255.

Blochinzew, D.J. 1953. *Quantenmechanik*. Berlin: Deutscher Verlag der Wissenschaften.

Bobrov, V.B., and S.A. Trigger. 2014. Virial Theorem and Gibbs Thermodynamic Potential for Coulomb Systems. *Physics of Plasmas* 21: 100703.

Bogolyubov, N.N. 2005–2009. *Collected Papers*. Vol. 1–12. Moscow: Fizmatlit.

Bonch-Bruevich, V.L., and S.V. Tyablikov. 1961. *Method of Green Functions in Statistical Physics*. Moscow: Nauka.

Dirac, P.A.M. 1932. *Principles of Quantum Mechanics*. Oxford: Clarendon Press.

Dyson, F.J., and A. Lenard. 1967. Stability of Matter I. *Journal of Mathematical Physics* 8: 423–434.

Dyson, F.J., and A. Lenard. 1968. Stability of Matter. II. *Journal of Mathematical Physics* 9: 698–711.

Ebeling, W., and I. Sokolov. 2005. *Statistical Thermodynamics and Stochastic Theory of Nonequilibrium Systems*. Singapore: World Scientific.

Ebeling, W., W.D. Kraeft, and D. Kremp. 1976. *Theory of Bound States and Ionisation Equilibrium in Plasmas and Solids*. Berlin: Akademie-Verlag.

Ebeling, W., W.D. Kraeft, and D. Kremp. 1977. Nonideal Plasmas. In *Proceedings of the XIIIth International Conference on Phenomena in Ionized Gases*, ed. P. Bachmann, 73–90. Berlin: Physikalische Gesellschaft der DDR.

Ebeling, W., D. Blaschke, R. Redmer, H. Reinholz, and G. Röpke. 2009. The Influence of Pauli Blocking Effects on the Properties of Dense Hydrogen. *Journal of Physics A: Mathematical and Theoretical* 42: 214033.

Ebeling, W., D. Blaschke, R. Redmer, H. Reinholz, and G. Röpke. 2010. The Influence of Pauli Blocking Effects on the Mott Transition in Dense Hydrogen. In *Metal-to-Nonmetal Transitions*, ed. Ronald Redmer, Friedrich Hensel, and Bastian Holst. *Springer Series in Materials Science*, vol. 132, 37–61. Berlin: Springer.

Ecker, G., and W. Weizel. 1956. Zustandssumme und effektive Ionisierungsspannung eines Atoms im Inneren des Plasmas. *Annalen der Physik* 452: 126–140.

Fetter, A.I., and J.D. Walecka. 1971. *Quantum Theory of Many Particle Systems*. New York: Mc Graw Hill.

Feynman, R.P. 1972. *Statistical Mechanics*. Reading: Benjamin.

Feynman, R.P., and A.R. Hibbs. 1965. *Quantum Mechanics and Path Integrals*. New York: McGraw-Hill.

Fick, E. 1981. *Einführung in die Grundlagen der Quantentheorie*. Leipzig: Akademischer Verlag.

Huang, K. 2001. *Introduction to Statistical Physics*. London: Taylor & Francis.

Ichimaru, S. 1992. *Statistical Plasma Physics*. Redwood: Addison-Wesley.

Jaynes, E.T. 1985. Macroscopic Prediction. In *Complex Systems – Operational Approaches in Neurobiology, Physics, and Computers*, ed. Hermann Haken, vol. 31, 254–269. *Springer Series in Synergetics*. Berlin: Springer.

Kadanoff, L.P., and G. Baym. 1962. *Quantum Statistical Mechanics*. New York: Benjamin.

Karasiev, V.V., T. Sjostrom, and S.B. Trickey. 2012. Comparison of DFT and Finite Temperatature HF Approximation. *Physical Review E* 86: 056704.

Klimontovich, Yu.L. 1982. *Statistical Physics (in Russian)*. Moscow: Nauka.

Klimontovich, Yu.L. 1986. *Statistical Physics*. New York: Harwood.

Kohn, W., and L.J. Sham. 1965. Self-Consistent Equations Including Exchange and Correlation Effects. *Physical Review* 140: A1133–A1138.

Kraeft, W.D., D. Kremp, W. Ebeling, and G. Röpke. 1986. *Quantum Statistics of Charged Particle Systems*. Berlin: Akademie-Verlag.

Kremp, D., M. Schlanges, and W.D. Kraeft. 2005. *Quantum Statistics of Nonideal Plasmas*. Berlin: Springer.

Landau, L.D., and E.M. Lifshitz. 1976. *Statistical Physics (Part I)*. Moscow: Nauka.

Landau, L.D., and E.M. Lifshitz. 1980. *Statistical Physics*. Oxford: Butterworth-Heinemann.

Lieb, E.H., and B. Simon. 1973. Thomas-Fermi Theory Revisited. *Physical Review Letters* 31: 681–683.

Lieb, E.H., and W.E. Thirring. 1975. Bound for the Kinetic Energy of Fermions Which Proves the Stability of Matter. *Physical Review Letters* 35: 687–689.

Martin, P.C., and J. Schwinger. 1959. Theory of Many-Particle Systems I. *Physical Review* 115: 1342–1373.

Matsubara, T. 1955. A New Approach to Quantum-Statistical Mechanics. *Progress of Theoretical Physics* 14: 351.

Montroll, E., and J. Ward. 1958. Quantum Statistics of Interacting Particles. *The Physics of Fluids* 1: 55–72.

Parr, R.G., and W. Yang. 1989. *Density-Functional Theory of Atoms and Molecules*. New York: Oxford University Press.

Pines, D. 1963. *Elementary Excitations in Solids. Lecture Notes and Supplements in Physics*. New York: Benjamin.

Pines, D., and P. Nozieres. 1966. *The Theory of Quantum Liquids*. New York: Benjamin.

Redmer, R. 1997. Physical Properties of Dense, Low-Temperature Plasmas. *Physics Reports* 282: 35–157.

Redmer, R., B. Holst, and F. Hensel, eds. 2010. *Metal-to-Nonmetal Transitions. Springer Series in Materials Science*. Berlin: Springer.

Röpke, G., K. Kilimann, D. Kremp, and W.D. Kraeft. 1978. Two-Particle Energy Shifts in Low Density Non-ideal Plasmas. *Physics Letters A* 68, 329–332.

Röpke, G., N.-U. Bastian, D. Blaschke, T. Klähn, S. Typel, and H.H. Wolter. 2013. Cluster-Virial Expansion for Nuclear Matter within a Quasiparticle Statistical Approach. *Nuclear Physics A* 897: 70–92.

Stratonovich, R.L. 1975. *Information Theory (in Russian)*. Moscow: Sovetskoe Radio.

Subarew, D.N. 1976. *Statistische Thermodynamik des Nichtgleichgewichtes*. Berlin: Akademie-Verlag.

ter Haar, D. 1995. *Elements of Statistical Mechanics*. Oxford: Butterworth-Heinemann.

Thirring, W.E. 1980. *Lehrbuch der Mathematischen Physik*. Vol. 3: Quantenmechanik von Atomen und Molekülen, Quantenmechanik großer Systeme. Berlin: Springer.

Toda, M., R. Kubo, and N. Saito. 1983. *Statistical Physics*. Vols. I and II. Berlin: Springer.

Tolman, R. 1938. *The Principles of Statistical Physics*. Oxford: Oxford University Press.

von Neumann, J. 1932. *Mathematische Grundlagen der Quantenmechanik*. Berlin: Springer.

Zimmermann, R. 1988. *Many Particle Theory of Highly Excited Semiconductors*. Leipzig: Teubner.

Zimmermann, R., K. Kilimann, W.D. Kraeft, D. Kremp, and G. Röpke. 1978. Dynamical Screening and Self Energy of Excitons in the Electronhole Plasma. *Physica Status Solidi (B)* 90: 175–187.

Zubarev, D.N. 1974. *Nonequilibrium Statistical Thermodynamics*. New York: Consultants Bureau.

Zubarev, D.N., V. Morozov, and G. Röpke, eds. 1996. *Statistical Mechanics of Nonequilibrium Processes*. Vol. 1: Basic Concepts, Kinetic Theory. Weinheim: Wiley-VCH.

Zubarev, D.N., V. Morozov, and G. Röpke, eds. 1997. *Statistical Mechanics of Nonequilibrium Processes*. Vol. 2: Relaxation and Hydrodynamic Processes. Weinheim: Wiley-VCH.

Chapter 5
Real Gas Quantum Statistics

5.1 Cluster Expansions for Real Gases

From the classical kinetic theory of gases we know the equation of state of the ideal gas, $\beta p = n$ (see Chap. 1). For real gases, the interaction forces between the molecules lead to corrections to the ideal gas equation of state. We mention the classical theory by van der Waals and the systematic expansions with respect to density, called virial expansions (Friedman 1962; Hirschfelder et al. 1954).

The virial expansion for a classical molecular gas is a representation of the interaction part of the free energy by a series in powers of density:

$$
\begin{aligned}
F(T, V, N) = F_{\text{id}} + F_{\text{int}} &= N k_{\text{B}} T \left[\ln \left(n \Lambda^3 \right) - 1 \right] \\
&\quad - k_{\text{B}} T V \left[n^2 B_2^{\text{cl}}(T) + n^3 B_3^{\text{cl}}(T) + \dots \right].
\end{aligned}
\tag{5.1}
$$

By differentiation with respect to the volume, V, that is, $p = -\partial F / \partial V$, follows the equation of state:

$$
\beta p = -\beta \left(\frac{\partial F}{\partial V} \right)_{T,N} = n \left[1 - n B_2^{\text{cl}}(T) - 2n^2 B_3^{\text{cl}}(T) - \dots \right].
\tag{5.2}
$$

Initially, this expansion was developed for the classical case, however, the same expansion with modified coefficients is valid for the quantum case as well. The extension of the classical formula to the quantum case was derived by Uhlenbeck and Beth (1936). In the classical case, the first ideal term in the expansion is the ideal Boltzmann contribution. In the quantum case, it is the Fermi–Dirac or Bose–Einstein expression known from Chap. 3. We see that not only interactions are responsible for deviations from the ideal term, that is, from the simple law $\beta p = n$, but there are also quantum effect contributing to the virial coefficients. For ideal

© Springer Nature Switzerland AG 2019

W. Ebeling, T. Pöschel, *Lectures on Quantum Statistics*,

Lecture Notes in Physics 953, https://doi.org/10.1007/978-3-030-05734-3_5

quantum gases, the classical ideal gas law, $\beta p = n$, is not valid, but we have to distinguish between Bose gases and Fermi gases (see Chap. 3):

$$\beta p = (2s + 1)\Lambda^{-3} f_{5/2}(z) \qquad\qquad \text{for Fermi gases} \qquad (5.3)$$

$$\beta p = \frac{2s + 1}{V} \ln(1 - z) + (2s + 1)\Lambda^{-3} g_{5/2}(z) \qquad \text{for Bose gases} . \qquad (5.4)$$

Here z is the fugacity which is related to density via

$$n = z \frac{\partial(\beta p)}{\partial z} . \qquad (5.5)$$

For small density, we expand Eqs. (5.3) and (5.4) to obtain in first approximation

$$\beta p = n \left[1 \pm \frac{1}{2^{5/2}} \left(\frac{n\Lambda^3}{2s + 1} \right) + \mathcal{O}\left(n^2\right) \right] , \qquad (5.6)$$

where the upper sign $(+)$ is valid for fermions and the lower one $(-)$ for bosons. The parameter of the expansion is $n\Lambda^3$. We define $n\Lambda^3 = 1$ as the degeneration line. In log-scale we obtain the straight line $\log n = \frac{3}{2} \log T + \text{const}$. For fixed density, the degeneration starts at the characteristic temperature (see Fig. 5.1)

$$T_{\text{qs}} = \frac{h^2 n^{\frac{2}{3}}}{2\pi m k_{\text{B}}} . \qquad (5.7)$$

To find the virial functions including interaction effects, we start from the binary correlation functions, F_{ab}. In first classical approximation, the binary correlations

Fig. 5.1 Logarithmic diagram of the density-temperature plane showing the characteristic density and temperature where degeneration effects become significant

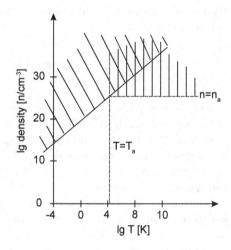

are given by a Boltzmann factor and deliver the first order mean potential energy

$$F_{ab}(1, 2) = e^{-\beta V_{ab}(1,2)} \tag{5.8}$$

$$U = \langle V \rangle = \frac{1}{2} \sum_{a,b} n_a n_b \int d\mathbf{r}_1 \, d\mathbf{r}_2 \, V_{ab}(1, 2) e^{-\beta V_{ab}(1,2)} .$$

Since the internal energy, U, and the free energy, F, are related via the thermodynamic expression $U = \partial(\beta F)/\partial\beta$, we find by integration for the second virial coefficient

$$B_2^{\mathrm{cl}} = \frac{1}{2} \int d\mathbf{r} \left(e^{-\beta V(\mathbf{r})} - 1 \right) . \tag{5.9}$$

For the higher order virial coefficients, we can derive corresponding expressions (Hirschfelder et al. 1954). In the quantum case, some modifications are needed. The mean potential energy is given by a trace

$$U = \langle V \rangle = \frac{1}{2} \sum_{a,b} n_a n_b \mathrm{Tr}[V_{ab}(1, 2) F_{ab}(1, 2)] . \tag{5.10}$$

A simple way to proceed is using coordinate representations. The pair probability given classically by a Boltzmann factor has to be replaced by its quantum statistical counterpart, the Slater sum of pairs introduced in Chap. 4. In this way, as shown by Uhlenbeck and Beth (1936), the classical expressions remain valid when replacing the Boltzmann factor by binary Slater sums,

$$e^{-\beta V(\mathbf{r})} \rightarrow S_2(\mathbf{r}) = \mathrm{const.} \sum_n e^{-\beta E_n} \phi_n(\mathbf{r}) \, \phi_n^*(\mathbf{r}) . \tag{5.11}$$

Beth and Uhlenbeck (1937) derived also other useful representations. In particular they studied the rôle of bound and scattering states and introduced phase shifts. We will discuss this in detail later.

Let us now introduce some models of interactions which are used in the field (Hirschfelder et al. 1954). Note that there are attractive forces such as van der Waals forces, chemical forces etc. and repulsive forces due to Coulomb repulsion between charges and the Pauli principle (Slater 1939). The typical form is a decaying function with a minimum. Characteristics of the potential are the energy in the minimum, D, the distance, r_0, where the potential energy changes from positive to negative values, and the location of the minimum, σ. Some typical data are given in the table below

Substance	D [eV]	r_0 [Å]
H_2	4.5	0.75
O_2	5.1	1.20
C_2	5.6	1.31
Cl_2	2.5	1.98

The simplest potential follows from quantum mechanical first order perturbation theory which yields an exponential repulsive potential describing the force due to the Pauli exclusion for overlapping atomic core wave functions:

$$V(r) = V_0 e^{-b(r-\sigma)} . \qquad (5.12)$$

For the limit $b \to \infty$ this potential yields the hard core potential,

$$\Phi(r) = \begin{cases} \infty \text{ for } r < \sigma \\ 0 \text{ for } r \geq \sigma . \end{cases} \qquad (5.13)$$

With some modification we obtain the square-well potential:

$$\Phi(r) = \begin{cases} \infty \text{ for } r < \sigma \\ -\varepsilon \text{ for } \sigma \leq r < b \\ 0 \text{ for } r \geq b . \end{cases} \qquad (5.14)$$

The Morse potential (Slater 1939) is another interaction potential based on quantum mechanical calculations:

$$U_M(r) = D \left(e^{-2b(r-\sigma)} - 2e^{-b(r-\sigma)} \right) . \qquad (5.15)$$

Its positive part describes the repulsive forces due to the Pauli exclusion for overlapping atomic core wave functions. Its negative part models qualitatively the attraction due to induced quantum mechanical dipole-dipole forces.

The Toda model is an exponential potential with an additional (nonphysical) linear attraction leading to a minimum of the potential at $r = \sigma$:

$$U_T(r) = -D + \frac{a_T}{b_T} \left[e^{-b_T(r-\sigma)} - 1 + b_T(r - \sigma) \right] . \qquad (5.16)$$

The term $-D$ provides the minimum at $r = \sigma$. Further important physical informations are contained in the frequency of oscillations around the minimum, $m\omega_0^2 = 2Db^2$, and in the stiffness of the potential which is proportional to the parameter b. The Toda potential is unphysical for larger distance but describes well the force for small distances. Its great advantage is that it allows fully analytical

Fig. 5.2 Interaction potential of atoms: the available knowledge about depth, frequency, and stiffness of the interaction of atoms can be fitted by different formulae. Here we show the Toda potential (upper curve), the Morse potential (middle curve), and the (r^{-12}, r^{-6}) Lennard–Jones potentials suitably scaled around the minimum to have identical values at the minimum and of the second derivative (frequency) and the third derivative (stiffness)

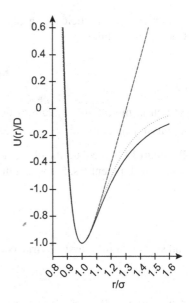

calculations. Figure 5.2 shows that the Toda potential and the Morse potential nicely agree at smaller distances up to the minimum and a bit beyond. Note, that a good fit of Toda and Morse potential near to the minimum is obtained if the Toda parameters are related to the Morse parameters by (Chetverikov et al. 2009)

$$a_T = \frac{2}{3}bD \, ; \qquad b_T = 3b \, . \tag{5.17}$$

One of the most popular models is the Lennard–Jones (6–12)-potential which is semi-empirical with respect to the repulsive part, but well founded by quantum calculations for the attractive part. We scale such that the minimum, $(-D)$, is at $r = \sigma$:

$$U_L(r) = D \left(\frac{\sigma^{12}}{r^{12}} - 2\frac{\sigma^6}{r^6} \right) . \tag{5.18}$$

A more general form is the Lennard–Jones $(n - m)$ potential,

$$U_L^{nm}(r) = \frac{D}{n - m} \left(n\frac{\sigma^m}{r^m} - m\frac{\sigma^n}{r^n} \right) . \tag{5.19}$$

Note that there is a large number of other potential models in the literature which are adapted to practical applications (Hirschfelder et al. 1954).

For charged particles in a plasma, the basic interaction is described by the Coulomb-Potential:

$$\Phi(r) = \pm \frac{e^2}{\varepsilon_r r},\tag{5.20}$$

written in the rational system of units. To obtain its representation in the international SI-system of units, we replace $e^2 \rightarrow \frac{e^2}{4\pi\varepsilon_0}$.

Problem 14 Calculate the classical second virial coefficient for hard-core and for square-well potentials. Show that the equation of state for hard sphere gases is

$$\beta p = n \left(1 + \frac{2}{3}\pi n a^3\right)\tag{5.21}$$

and show that the classical virial coefficients for Coulomb potential is divergent.

5.2   Slater Functions and Virial Coefficients

5.2.1   General Density Expansions

To provide a more general derivation of virial expansions for gases, we start from the free energy of a N-particle system (Feynman 1972):

$$F(T, V, N) = -k_B T \ln Q_N(T, V),\tag{5.22}$$

with

$$Q_N(T, V) = \mathrm{Tr}_{(1,\dots,N)}\left(e^{-\beta\hat{H}_N}\right).\tag{5.23}$$

Using Slater functions or Slater sums, respectively, the quantum statistical problem can be mapped to a classical expression. From the formula

$$Q_N(T, V) = \sum_{\sigma_1,\dots,\sigma_N} \int d\mathbf{r}_1 \dots d\mathbf{r}_N$$
$$\times \left\langle \mathbf{r}_1 \dots \mathbf{r}_N \sigma_1 \dots \sigma_N \left| e^{-\beta\hat{H}_N} \right| \mathbf{r}_1 \dots \mathbf{r}_N \sigma_1 \dots \sigma_N \right\rangle\tag{5.24}$$

and

$$Q_N(T, V) = Q_{N,\mathrm{id}}\, Q_{N,\mathrm{int}},\tag{5.25}$$

where the ideal part is defined by classical Boltzmann statistics,

$$Q_{N,\text{id}} = \frac{(2s+1)^N \ V^N}{N! \ \Lambda^{3N}} \tag{5.26}$$

follows for the non-ideal part

$$Q_{N,\text{int}} = \frac{N! \ \Lambda^{3N}}{(2s+1)^N \ V^N} \sum_{\sigma_1,\ldots,\sigma_N} \int d\mathbf{r}_1 \ldots d\mathbf{r}_N$$

$$\times \left\langle \mathbf{r}_1 \ldots \mathbf{r}_N \sigma_1 \ldots \sigma_N \left| e^{-\beta \hat{H}_N} \right| \mathbf{r}_1 \ldots \mathbf{r}_N \sigma_1 \ldots \sigma_N \right\rangle . \tag{5.27}$$

By means of Stirling's formula, $\ln N! = (N \ln N - N)$, we obtain for the free energy

$$F(T, V, N) = F_{\text{id}} - k_B T \ln Q_{N,\text{in}} = N k_B T \left[\ln \left(\frac{n\Lambda^3}{2s+1} \right) - 1 \right]$$

$$\times k_B T \ln \left[\frac{N! \Lambda^{3N}}{(2s+1)^N V^N} \sum_{\sigma_1,\ldots,\sigma_N} \int d\mathbf{r}_1 \ldots d\mathbf{r}_N \right.$$

$$\left. \times \left\langle \mathbf{r}_1 \ldots \mathbf{r}_N \sigma_1 \ldots \sigma_N \left| e^{-\beta \hat{H}_N} \right| \mathbf{r}_1 \ldots \mathbf{r}_N \sigma_1 \ldots \sigma_N \right\rangle \right] . \tag{5.28}$$

This representation of the free energy has already a classical form and can be treated, e.g., by the well known technique of Mayer expansions (Friedman 1962). Following this route, as proposed by Morita (1959) and Slater (1939) we introduce the Slater sums or Slater functions. The N-particle Slater sums/Slater functions are defined by

$$S^{(N)}(\mathbf{r}_1, \ldots, \mathbf{r}_N) = \frac{N! \Lambda^{3N}}{(2s+1)^N} \sum_{\sigma_1,\ldots,\sigma_N}$$

$$\times \left\langle \mathbf{r}_1 \ldots \mathbf{r}_N \sigma_1 \ldots \sigma_N \left| e^{-\beta \hat{H}_N} \right| \mathbf{r}_1 \ldots \mathbf{r}_N \sigma_1 \ldots \sigma_N \right\rangle . \tag{5.29}$$

At small density, we may use a pair approximation

$$S^{(N)}(\mathbf{r}_1, \ldots, \mathbf{r}_N) = \prod_{i<j}^{N} S^{(2)}(\mathbf{r}_i, \mathbf{r}_j) , \tag{5.30}$$

where *per definitionem*

$$S^{(2)}(\mathbf{r}_i, \mathbf{r}_j) = e^{-\beta U_{ij}(\mathbf{r}_i,\mathbf{r}_j)} = \frac{2!\Lambda^6}{(2s+1)^2} \sum_{\sigma_1 \sigma_2} \left\langle \mathbf{r}_1\mathbf{r}_2 \left| e^{-\beta \hat{H}_2} \right| \mathbf{r}_1\mathbf{r}_2 \right\rangle . \tag{5.31}$$

The quantum statistical partition function corresponds directly to the classical form and, consequently, also the quantum statistical second virial coefficient represented by

$$B_2^{\mathrm{qm}}(T) = \frac{1}{2V} \int d\mathbf{r}_1 d\mathbf{r}_2 \left[S^{(2)}(\mathbf{r}_i, \mathbf{r}_j) - 1 \right] . \tag{5.32}$$

5.2.2   Slater Sums for Pair Correlations

To calculate the binary Slater sum of the two-particle problem for particles of different species, a and b, and spin, s_a and s_b, we write

$$S_{ab}^{(2)}(\mathbf{r}_1, \mathbf{r}_2) = (1 + \delta_{ab})! \frac{\Lambda_a^3 \Lambda_b^3}{(2s_a + 1)(2s_b + 1)}$$
$$\times \sum_{\sigma_1,\sigma_2} \left\langle \mathbf{r}_1\mathbf{r}_2\sigma_1\sigma_2 \left| e^{-\beta \hat{H}^{(2)}} \right| \mathbf{r}_1\mathbf{r}_2\sigma_1\sigma_2 \right\rangle . \tag{5.33}$$

We define the energy eigenfunctions of two-particles belonging to species a and b, denoting the spin functions as χ_a and χ_b and the relative wave functions depending on distance $\mathbf{r}$ by $\psi_n(\mathbf{r})$ by

$$\psi_\alpha(\mathbf{r}_1, \mathbf{r}_2, \sigma_1, \sigma_2) = \frac{1}{\sqrt{V}} \chi_a(\sigma_1)\chi_b(\sigma_2) e^{-\frac{i}{\hbar}\mathbf{P}\mathbf{R}} \psi_n(\mathbf{r}) . \tag{5.34}$$

Here we used the center of mass and relative coordinates, $\mathbf{R}$ and $\mathbf{r} = \mathbf{r}_1 - \mathbf{r}_2$, the energy, $E_{\alpha=(\mathbf{P},n)} = E_n + \frac{P^2}{2M_{ab}}$, and the total mass, $M_{ab} = m_a + m_b$. We obtain

$$S_{ab}^{(2)}(\mathbf{r}) = (1 + \delta_{ab})! \frac{\Lambda_a^3 \Lambda_b^3}{(2s_a + 1)(2s_b + 1)}$$
$$\times \frac{1}{V} \sum_{\mathbf{P}} e^{\frac{-\beta P^2}{2M_{ab}}} \sum_n e^{-\beta E_n} |\psi_n(\mathbf{r})|^2 . \tag{5.35}$$

The contribution of the center of mass momenta can be explicitly calculated:

$$\frac{1}{V} \sum_{\mathbf{P}} e^{\frac{-\beta P^2}{2M_{ab}}} = \int \frac{d\mathbf{P}}{h^3} e^{-\frac{P^2}{2M_{ab}k_{\mathrm{B}}T}} = \frac{1}{\Lambda_{ab}^3} = \frac{(2\pi M_{ab}k_{\mathrm{B}}T)^{\frac{3}{2}}}{h^3} . \tag{5.36}$$

The wave functions for radial-symmetric potentials are products of radial and angular parts (Blochinzew 1953; Landau and Lifshitz 1980),

$$\psi_\alpha(\mathbf{r}) = R_{s\ell}(r)Y_{\ell m}(\vartheta, \varphi), \tag{5.37}$$

with the standard set of quantum numbers and the standard radial and momentum eigenfunctions, $R_{s\ell}(\mathbf{r})$ and $Y_{\ell m}(\vartheta, \varphi)$. For pairs of different species there are no exchange terms and we find

$$S_{ab}^{(2)}(\mathbf{r}) = \frac{h^3}{(2\pi m_{ab}k_{\mathrm{B}}T)^{3/2}} \left(\sum_{s,\ell,m} e^{-\beta E_{s\ell}} \left| R_{s\ell}^2 \right| |Y_{\ell m}(\vartheta, \varphi)|^2 + \ldots \right), \tag{5.38}$$

where the dots stand for the scattering state contributions. With the relations

$$4\pi \sum_{m=-l}^{\ell} |Y_{\ell m}(\vartheta, \varphi)|^2 = 2\ell + 1; \; \lambda_{ab} = \frac{\hbar}{\sqrt{2m_{ab}k_{\mathrm{B}}T}}; \; m_{ab} = \frac{m_a m_b}{m_a + m_b} \tag{5.39}$$

we arrive at the final result:

$$S_{ab}^{(2)}(r) = 2\pi^{1/2}\lambda_{ab}^3$$

$$\times \left(\sum_{\ell=0}^{\infty} (2\ell + 1) \left[\sum_{s=1}^{\ell-1} e^{-\beta E_{s\ell}} |R_{s\ell}(r)|^2 + \int dk\, e^{-\frac{\beta\hbar^2 k^2}{2m_{ab}}} |R_{k\ell}|^2 \frac{d\, n_\ell(k)}{d\, k} \right] \right). \tag{5.40}$$

Here we introduced the density of states, $\frac{d\, n_\ell(k)}{d\, k}$, defined as the number of states with quantum number ℓ in the interval dk (Beth and Uhlenbeck 1937; Huang 1963, 2001; Kremp and Kraeft 1972; Uhlenbeck and Beth 1936).

5.3   The Second Virial Coefficient

5.3.1   Virial Coefficient Including Exchange Effects

For a gas of species with pair interaction, the expansion for the free energy reads

$$F(T, V, N) = F_{\mathrm{id}} - k_{\mathrm{B}}TV \left(\sum_{ab} n_a n_b\, B_{ab}(T) + \sum_{abc} n_a n_b n_c\, B_{abc}(T) + \ldots \right) \tag{5.41}$$

and the corresponding equation of state is

$$\beta p = -\beta \left(\frac{\partial F}{\partial V}\right)_{T,N} = \sum_a n_a \left(1 - \sum_b n_b \, B_{ab} - 2 \sum_{bc} n_b n_c B_{abc} - \ldots \right),$$

(5.42)

with the second virial coefficient expressed by the binary Slater function or Slater sum

$$B_{ab}(T) = \frac{1}{2} \int d\mathbf{r} \, [S_{ab}(\mathbf{r}) - 1] \,.$$

(5.43)

Representing the Slater sum as above, including the contributions of bound and scattering states we obtain for the second virial coefficient

$$B_{ab}(T) = (1 + \delta_{ab})! \, 4\pi^{\frac{3}{2}} \lambda_{ab}^3 \sum_{\ell=0}^{\infty} (2\ell + 1) \left[\sum_{s=\ell+1}^{\infty} \int_0^{\infty} dr r^2 e^{-\beta E_{s\ell}} |R_{s\ell}(r)|^2 \right.$$
$$\left. + \int_0^{\infty} dk \int_0^{\infty} dr \, r^2 e^{-\lambda_{ab}^2 k^2} \left(\frac{dn_\ell(k)}{dk} - \frac{dn_\ell^0(k)}{dk}\right) |R_{s\ell}(r)|^2 \right].$$

(5.44)

Here, $n_\ell(k)$ is the density of the states of pairs and $n_\ell^0(k)$ is the corresponding density for the non-interacting system. Using the normalization for the radial functions,

$$\int_0^{\infty} dr \, r^2 |R_{s\ell}(r)|^2 = 1 \,,$$

(5.45)

we obtain the *Beth–Uhlenbeck formula* (Beth and Uhlenbeck 1937; Uhlenbeck and Beth 1936)

$$B_2(T; a, b) = 4\pi^{\frac{3}{2}} \lambda_{ab}^3 \sum_{\ell=0}^{\infty} (2\ell + 1) \left[\sum_{s=\ell+1}^{\infty} e^{-\beta E_{s\ell}} \right.$$
$$\left. + \int_0^{\infty} dk \, e^{-\lambda_{ab}^2 k^2} \left(\frac{dn_\ell(k)}{dk} - \frac{dn_\ell^0(k)}{dk}\right) \right].$$

(5.46)

In this calculation of the second virial coefficient of quantum gases, so far we neglected the symmetry character of the two-particle wave functions. Including the symmetry we find the Slater sum with exchange effects

$$S_{ab}^{(2)}(\mathbf{r}_1, \mathbf{r}_2) = (1 + \delta_{ab})! \, \frac{\Lambda_a^3 \Lambda_b^3}{(2s_a + 1)(2s_b + 1)}$$

$$\times \sum_{\sigma_1, \sigma_2} \langle \mathbf{r}_1 \mathbf{r}_2 \sigma_1 \sigma_2 \Big|^{S,A} \Big| e^{-\beta \hat{H}^{(2)}} \Big| \mathbf{r}_1 \mathbf{r}_2 \sigma_1 \sigma_2 \rangle^{S,A}. \tag{5.47}$$

The upper indexes "S" and "A" mean that symmetrical or anti-symmetrical wave functions have to be constructed. For symmetric/anti-symmetric wave functions, we write

$$\psi^{S,A}(1, 2) = \frac{1}{\sqrt{2}} [\psi_\alpha(1, 2) \pm \psi_\alpha(1, 2)] \,. \tag{5.48}$$

Introducing the wave functions, we obtain the Slater sums

$$S_{ab}^{(2)}(\mathbf{r}) = (1 + \delta_{ab})! \, 2\sqrt{\pi} \lambda_{ab}^3 \sum_{\ell=0}^{\infty} (2\ell + 1) \left(1 \pm \delta_{ab} \frac{(-1)^\ell}{(2s_a + 1)} \right)$$

$$\times \left(\sum_{s=\ell+1}^{\infty} e^{-\beta E_{s\ell}} |R_{s\ell}(r)|^2 + \int_0^\infty dk \, e^{-\lambda_{ab}^2 k^2} \frac{dn_\ell(k)}{dk} \right) |R_{k\ell}(r)|^2 \tag{5.49}$$

and finally the virial coefficients

$$B_{ab}(T) = (1 + \delta_{ab})! \, 4\pi^{\frac{3}{2}} \lambda_{ab}^3 \sum_{\ell=0}^{\infty} (2\ell + 1) \left(1 \pm \frac{(-1)^\ell}{(2s_a + 1)} \right)$$

$$\times \left(\sum_{s=\ell+1}^{\infty} e^{-\beta E_{s\ell}} + \int_0^\infty dk \, e^{-\lambda_{ab}^2 k^2} \left[\frac{dn_\ell(k)}{dk} - \frac{dn_\ell^0(k)}{dk} \right] \right). \tag{5.50}$$

A different and sometimes more convenient way of writing the second virial coefficient is obtained from a relation between scattering phase shifts and Jost functions (Ebeling et al. 1976; Kraeft et al. 1969; Kremp et al. 1971): We express the second virial coefficient by means of the two-particle trace

$$B_{ab}(T) = \text{const. Tr} \left(e^{-\beta H_{ab}} - e^{-\beta H_{ab}^0} \right). \tag{5.51}$$

The Hamiltonian and the relative thermal wave length of pairs of species, a and b, are defined by

$$H_{ab} = -\frac{\hbar^2}{2\mu_{ab}}\Delta + V_{ab}; \qquad \lambda_{ab} = \frac{\hbar}{\sqrt{2\mu_{ab}k_BT}}. \qquad (5.52)$$

Using the resolvent representation for the exponential operator, we obtain

$$B_{ab} = \frac{4\pi^{3/2}(1+\delta_{ab})\lambda_{ab}^3}{(2s_a+1)(2s_b+1)}\frac{1}{2\pi i}\int_c e^{-\beta z}F(z)\,dz, \qquad (5.53)$$

$$F(z) = \text{Tr}\left(\frac{1}{H_{ab}-z} - \frac{1}{H_{ab}^0-z}\right). \qquad (5.54)$$

After some algebra we arrive at a complex representation by Jost functions, $D_\ell(z)$, first obtained by Kraeft et al. (1969)

$$F(z) = -\frac{(2s_a+1)(2s_b+1)}{(1+\delta_{ab})}\sum_{\ell=0}^{\infty}(2\ell+1)\left(1\pm\frac{(-1)^\ell}{2s_a+1}\right)\frac{d}{dz}\ln D_\ell(z). \qquad (5.55)$$

The Jost functions, $D_\ell(z)$, are analytical functions with poles at the bound states and a branch cut at the positive real axis defined by the scattering phase shifts. These functions as well as other scattering quantities are exactly known for Coulomb systems.

If the attracting well of the interatomic force is deep enough such that bound states are formed, the contribution of scattering states may be neglected, at least at low temperature. In the limit of low T, we obtain

$$B_{ab}(T) \approx (1+\delta_{ab})!\,4\pi^{\frac{3}{2}}\lambda_{ab}^3\,e^{-\beta E_{10}}, \qquad (5.56)$$

with the assumption $|\beta E_{10}| \gg 1$ where E_{10} is the lowest eigenvalue in the well. At low T, the contribution of the bound states increases exponentially and, therefore, the ground state dominates. Some specific problems related to this exponential increase will be discussed later. For Coulomb systems, the energy eigenvalues do not depend on the quantum number, ℓ,

$$E_{s\ell} = E_s = -\frac{m_{ab}e_a^2e_b^2}{2\hbar^2}\left(\frac{Z}{s}\right)^2. \qquad (5.57)$$

A specific problem of Coulomb systems is the divergence of the bound state contribution due to the infinite number of bound states for $s \to \infty$. For this reason, the cluster expansion is divergent and cannot be applied to Coulomb systems. This difficult problem will be studied in more detail in Chap. 6 where we include the effect of *Debye screening*.

5.3.2 Beth–Uhlenbeck Method for Non-associating Gases

As known from quantum mechanics, only deep potential wells can lead to bound states. In many cases, however, the attractive van der Waals force between atoms or molecules in a gas is too weak to allow for bound states. A common example for weak attractive forces are noble gases. In this case the second virial coefficient contains only scattering contributions. At temperature below 100 K classical calculations are insufficient and we need quantum statistical calculations based on the density of states (Costa et al. 2013; Kilpatrick et al. 1954; Kraeft et al. 1969; Rogers et al. 1970). In order to find the density of states we analyze the asymptotics of the radial wave functions. For large r we obtain

$$\lim_{r \to \infty} R_{k\ell}(r) \approx \frac{1}{r}\sqrt{\frac{2}{\pi}} \sin\left[kr - \frac{\pi}{2}\ell + \delta_\ell(k)\right], \qquad (5.58)$$

where $\delta_\ell(k)$ are the scattering phase shifts. Enclosing the system into a sphere of radius R, assuming $R_{s\ell}(r \geq R) = 0$, the relation

$$kR - \frac{\pi}{2}\ell + \delta_\ell(k) = n_\ell(k)\pi; \qquad n_\ell(k) = 0, \pm 1, \pm 2, \ldots \qquad (5.59)$$

holds. For free particles, it assumes the form

$$kR - \frac{\pi}{2}\ell = n_\ell^0(k)\pi; \qquad n_\ell^0(k) = 0, \pm 1, \pm 2, \ldots \qquad (5.60)$$

and we find

$$\frac{1}{\pi}\delta_\ell(k) = n_\ell(k) - n_\ell^0(k). \qquad (5.61)$$

In the continuous limit we write

$$\frac{dn_\ell(k)}{dk} - \frac{dn_\ell^0(k)}{dk} = \frac{1}{\pi}\frac{d\delta_\ell(k)}{dk}. \qquad (5.62)$$

Obviously, the phase shifts are required to compute the second virial coefficient, $B_2(T; a, b)$, which, in general, are found by means of numerical methods (Costa et al. 2013; Kilpatrick et al. 1954; Rogers et al. 1970). There are only a few cases where exact analytical expressions are available, e.g., for hard-sphere, square-well, and for Coulomb potentials (Beth and Uhlenbeck 1937; Kraeft et al. 1969). A quantum calculation of the phase shifts, $\delta_\ell(k)$, requires as a rule numerical integration of the radial Schrödinger equation,

$$\left[\frac{d^2}{dr^2} + k^2 - \frac{2\mu}{\hbar^2}V(r) - \frac{\ell(\ell+1)}{r^2}\right]\psi_\ell(r) = 0, \qquad (5.63)$$

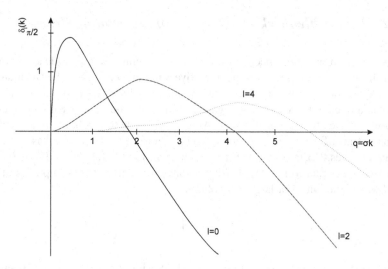

Fig. 5.3 Schematic picture of the phase shifts with $\ell = 0, 2, 4$ for He–He interaction (Kilpatrick et al. 1954)

for large distances. We find the phase shifts by comparison of the numerical solution with the asymptotic solution,

$$\psi_\ell(r) = A \sin\left(kr - \frac{\ell\pi}{2} + \delta_\ell\right).\qquad(5.64)$$

From this comparison, the phase shift can be derived as a function of k and ℓ (see Fig. 5.3). After computing the phase shifts for the given interaction potential in analytical or numerical form, the virial coefficient can be calculated as a function of temperature.

5.4   Equation of State for Gases with Deep Bound States

5.4.1   Fugacity Expansions

Let us summarize our findings about density expansions for real plasmas. The pressure of a one-component plasma can be represented by

$$\beta p = n\left[1 - nB_2(T) - 2n^2 B_3(T) + \dots\right].\qquad(5.65)$$

Fig. 5.4 Pressure of a classical gas of hard spheres. We show the result of a virial expansion with respect to density (cut after linear terms) in comparison to the Carnahan–Starling formula which is considered as nearly exact

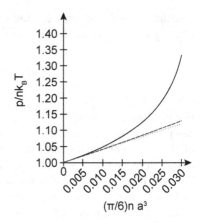

For most rare gases, e.g. hard-sphere gases, this expansion converges, since the virial terms decays rapidly. To assure convergence, thus, we need conditions like

$$n B_2(T) \ll 1, \quad n^2 B_3(T) \ll 1 ; \dots \tag{5.66}$$

For gases of spheres of diameter d, this corresponds to the condition $nd^3 \ll 1$ or $d \ll n^{-1/3}$. Thus, the effective diameter of particles, d, should be small as compared to the average distance between the particles in the gas. For gases, in most cases this condition is satisfied. In Fig. 5.4 we show the density expansion for the compressibility coefficient as a function of the dimensionless parameter, $y = \pi n d^3/6$, in comparison with the Carnahan–Starling expression, which fits well numerical MC and MD data for hard spheres. The agreement of the density expansion

$$\frac{p}{n k_B T} = 1 + \frac{2}{3} \pi n d^3 + \dots \tag{5.67}$$

and the Carnahan–Starling approximation is rather good. There are, however, physical conditions when the contribution of the second virial coefficient

$$n B_2(T) = \frac{n}{2} \int d\mathbf{r} \left(S^{(2)}(\mathbf{r}) - 1 \right) \tag{5.68}$$

increases exponentially with the inverse temperature, $\beta = 1/k_B T$. In this case, we have to search for other representations of the pressure.

From the theory of ideal quantum gases we know that the pressure can be represented also by an expansion with respect to the fugacity, z. A corresponding expansion for a classical hard-sphere gas is given also for comparison in Fig. 5.4 (Friedman and Ebeling 1979). We see that a density expansion cutted after the linear terms describes the real hard-sphere gas rather well. Following the pioneering work by Hill (1956), we have a different situation in gases with deep bound states, revealing association and reaction effects (see also Ebeling 1974; Friedman and Ebeling 1979). Remind that in systems with deep bound states we have

$$B_2(T) \approx 8\pi^{\frac{3}{2}}\lambda^3 \sum_{s\ell} e^{\frac{-E_{s\ell}}{k_B T}} . \tag{5.69}$$

For the case that the ground state energy, $|E_{10}| > k_B T$, the exponential function can give a large contribution. Let us assume, for example, a typical binding energy of $0.1\,\text{eV}$ which corresponds to 10^3 Kelvin. Then for temperature below room temperature we obtain so large values of $B_2(T)$ that the pressure in the virial series becomes negative and the expansion loses physical meaning.

In order to check the power of the fugacity expansion introduced by Hill (1956), which is by now considered as the appropriate tool, we remember the basic formulae for the grand canonical ensemble given in Chap. 2. In the grand ensemble, the basic independent quantity is the chemical potential or the fugacity, $z = e^{\beta\mu}$, instead of the density. The density operator is given by

$$\hat{\rho} = \Xi^{-1} \exp\left(\beta\hat{N}\mu - \beta\hat{H}_N\right) , \tag{5.70}$$

and the grand canonical partition function is

$$\Xi = \sum_{N=0}^{\infty} e^{\beta\mu N} \text{Tr}_{\mathcal{H}_N} e^{-\beta\hat{H}_N} = \text{Trace}_{\mathcal{F}}\left(e^{-\beta\mathcal{H}}\right) . \tag{5.71}$$

Here we introduced the trace in the N-particle space of the Hamiltonian, H_N, and the generalized trace denoted by $\text{Trace}_{\mathcal{F}}$ which operates in the Fock space. The Fock space is a generalization of the Hilbert space, including the particle numbers, N, as an additional (discrete) variable. Symbolically, it can be written as a sum of operators:

$$\mathcal{F} = \mathcal{H}^{(0)} + \mathcal{H}^{(1)} + \mathcal{H}^{(2)} + \cdots + \mathcal{H}^{(N)} . \tag{5.72}$$

The trace of the density operator is normalized to unity,

$$\text{Trace}\left(\hat{\rho}\right) = 1 , \tag{5.73}$$

and the partition function is given by

$$\Xi = \text{Trace}_{\mathcal{F}}\left(e^{\beta\hat{N}\mu - \beta\hat{H}_N}\right) = \sum_{N=0}^{\infty} e^{\beta\mu N} \text{Tr}_{\mathcal{H}_N}\left(e^{-\beta\hat{H}_N}\right)$$

$$= \sum_{N=0}^{\infty} z^N \text{Tr}_{\mathcal{H}_N}\left(e^{-\beta\hat{H}_N}\right) . \tag{5.74}$$

The thermodynamic functions follow from

$$pV = -k_B T \ln \Xi \ ; \qquad \langle N \rangle = \frac{\partial}{\partial \mu}(pV) = \beta z \frac{\partial}{\partial z}(pV) \,. \tag{5.75}$$

These formulae provide an expansion of the pressure with respect to the fugacity. Just for convenience we introduce the quasi-density

$$\tilde{z} = \Lambda^{-3} z \,, \tag{5.76}$$

where Λ is the *thermal wave length*. The quasi-density, $\tilde{z}$, has the dimension of density and approaches the density, n, in the classical limit ($\mu = k_B T \ln n \Lambda^3$). Further we define a new partition function,

$$\tilde{Q}_N(T, V) = N! \Lambda^{3N} \mathrm{Tr}_{\mathcal{H}_N} \left(e^{-\beta \hat{H}_N} \right) \,. \tag{5.77}$$

In this convenient notation, we obtain a relation that resembles the corresponding relation from classical statistics:

$$\Xi = \sum_{N=0}^{\infty} \frac{\tilde{z}^N}{N!} \tilde{Q}_N(T, V) \,. \tag{5.78}$$

For the mean particle number we find, correspondingly,

$$\langle N \rangle = \frac{\partial}{\partial \mu}(pV) = \beta z \frac{\partial}{\partial z}(pV) = \beta \tilde{z} \frac{\partial}{\partial \tilde{z}}(pV) = \frac{\displaystyle\sum_{N=0}^{\infty} N \frac{\tilde{z}^N}{N!} \tilde{Q}_N}{\displaystyle\sum_{N=0}^{\infty} \frac{\tilde{z}^N}{N!} \tilde{Q}_N} \,. \tag{5.79}$$

These formulae provide expansions of the pressure and the density with respect to the fugacity, $\tilde{z}$:

$$\beta pV = \ln \sum_{N=0}^{\infty} \frac{\tilde{z}^N}{N!} \tilde{Q}_N(T, V) \,. \tag{5.80}$$

We obtain a virial expansion with respect to the quasi-density, $\tilde{z}$, which is related to the previous density expansion. The virial coefficients follow from the comparison of the coefficients on both sides of the series expansions: By using the expansion of the log-function, $\ln(1 + x) = x - \frac{x^2}{2} + \frac{x^3}{3} - \ldots$, we find

$$\beta p = \frac{1}{V} \sum_{j=1}^{\infty} b_j \tilde{z}^j = \frac{1}{V} \ln \left(1 + \tilde{Q}_1 \tilde{z} + \frac{1}{2!} \tilde{Q}_2 \tilde{z}^2 + \frac{1}{3!} \tilde{Q}_3 \tilde{z}^3 + \ldots \right)$$

$$= \frac{1}{V} \left(\tilde{Q}_1 \tilde{z} + \frac{1}{2!} \tilde{Q}_2 \tilde{z}^2 + \ldots \right) - \frac{1}{2V} \left(\tilde{Q}_1 \tilde{z} + \frac{1}{2!} \tilde{Q}_2 \tilde{z}^2 \ldots \right)^2$$

$$+ \frac{1}{3V} \left(\tilde{Q}_1 \tilde{z} + \frac{1}{2!} \tilde{Q}_2 \tilde{z}^2 \ldots \right)^3 . \tag{5.81}$$

Comparing equal powers, $\tilde{z}^k$, we obtain the virial coefficients

$$b_1 = \frac{1}{V} \tilde{Q}_1 \tag{5.82}$$

$$b_2 = \frac{1}{2! \, V} \left(\tilde{Q}_2 - \tilde{Q}_1^2 \right) \tag{5.83}$$

$$b_3 = \frac{1}{3! \, V} \left(\tilde{Q}_3 - 3 \tilde{Q}_1 \tilde{Q}_2 + 2 \tilde{Q}_1^3 \right) . \tag{5.84}$$

General relations can be obtained by using the rules of Thiele's semi invariants or the equivalent methods of cumulant expansions. For the first terms in the series we find

$$\beta p = b_1 \tilde{z} + b_2 \tilde{z}^2 + b_3 \tilde{z}^3 + \ldots \tag{5.85}$$

$$n = \tilde{z} \frac{\partial}{\partial \tilde{z}} (\beta p) = b_1 \tilde{z} + 2 b_2 \tilde{z}^2 + 3 b_1 \tilde{z}^3 + \ldots \tag{5.86}$$

In principle, this virial expansion is equivalent to the density expansion, however as we conclude from Fig. 5.5, the convergence can be rather different. We will address this point in more detail later. Explicitly, we find

$$b_1(T) = \frac{1}{V} \tilde{Q}_1 = \frac{\Lambda^3}{V} \mathrm{Tr}_{\mathcal{H}_1} \left(e^{-\beta \hat{H}_1} \right) = 1 \tag{5.87}$$

$$b_2(T) = \frac{1}{2V} \left(\tilde{Q}_2 - \tilde{Q}_1^2 \right) = \frac{1}{V} \Lambda^6 \mathrm{Tr}_{(12)} \left(e^{-\beta \hat{H}_2} - e^{-\beta \hat{H}_2^{(0)}} \right) . \tag{5.88}$$

By comparison with Eq. (5.43) follows

$$b_2(T) = B_2(T) . \tag{5.89}$$

For higher orders, the relations appear more complicated.

Fig. 5.5 Pressure of hydrogen gas as a function of temperature calculated from the fugacity expansion (curve a) and the density expansion (curve b)

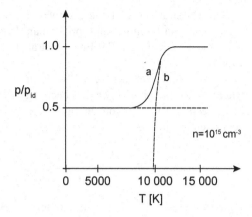

5.4.2   Chemical Picture

Let us analyze the simple model case of a gas where only the second virial coefficient is relevant:

$$\beta p = \tilde{z} + B_2(T)\,\tilde{z}^2 \tag{5.90}$$

$$n = \tilde{z} + 2B_2(T)\,\tilde{z}^2 . \tag{5.91}$$

We consider Eq. (5.90) as an implicite relation and eliminate the fugacity

$$2B_2(T)\tilde{z}^2 + \tilde{z} - n = 0 \tag{5.92}$$

$$\tilde{z}^2 + \frac{1}{2B_2(T)}\tilde{z} - \frac{n}{2B_2(T)} = 0 . \tag{5.93}$$

With the physically relevant solution of this quadratic equation,

$$\tilde{z} = \frac{1}{4B_2(T)}\left(\sqrt{1 + 8nB_2(T)} - 1\right), \tag{5.94}$$

we find the pressure

$$\beta p = \frac{1}{2}n + \frac{1}{2}\tilde{z} = \frac{n}{2} + \frac{1}{8B_2(T)}\left(\sqrt{1 + 8nB_2} - 1\right). \tag{5.95}$$

This equation which contains only the second virial coefficient has a rather different analytical form than the corresponding density expansion containing only second virial contributions. An application of Eq. (5.65) to the pressure of hard sphere gases is shown in Fig. 5.4. When comparing to the nearly exact Carnahan–Starling expression, we can show that Eq. (5.95) is worse than the density expansion. The advantage, that the pressure can never assume negative values, becomes relevant if $B_2(T)$ assumes very large positive values. This result suggests, that fugacity expansions can be more appropriate than density expansions for gases with deep bound states (Ebeling 1974; Ebeling et al. 1976; Friedman and Ebeling 1979; Hill 1956).

As a second example we study atomic hydrogen gas in the parameter region where the formation of molecules becomes relevant. Assuming a bound state energy of molecules (dissociation energy) of $D = 4.7\,\mathrm{eV}$, we find the second virial coefficient of the gas:

$$B_H^{(2)}(T) \approx \frac{h^3}{(2\pi m_H k_B T)^{3/2}} \exp\left(\frac{4.7\,\mathrm{eV}}{T[\mathrm{eV}]}\right). \tag{5.96}$$

Figure 5.5 shows the pressure as a function of temperature for a hydrogen gas. The full lines show the fugacity expansion and the dashed lines the density expansion including only the second virial coefficient. The curves show in a convincing way that the fugacity representation describes the physics of an hydrogen gas. There are two limiting cases:

1. At low temperature, with $p \rightarrow \frac{1}{2} n k_B T$ for $T \rightarrow 0$, we see that the effective particle density is about 50% of the density at high temperature. This is the region where molecules are formed. The number of hydrogen molecules, H_2, per cm^3 is half of the total density of atoms.
2. At high temperature, there are no molecules and the effective particle density equals the total density of atoms.

We see that the fugacity expansion describes in a correct way the physics of the equilibrium between atoms and molecules in the gas with deep bound states.

So far, we did not use any explicit chemical notation for the description of the system, but only physical quantities such as virial coefficients. How would a chemist describe the system we study? In first approximation, a chemist would consider the system as a mixture of atoms at density n_A and molecules at density n_M. Due to the assumption of ideality, the pressure is

$$\beta p = n_A + n_M. \tag{5.97}$$

In a chemical description, the equilibrium between atoms and molecules would be described by a mass action law,

$$\frac{n_M}{n_A^2} = K(T) \tag{5.98}$$

with the mass action constant

$$K(T) \sim e^{-\frac{E_0}{k_B T}}, \tag{5.99}$$

where E_0 is the binding energy of a molecule forming from two atoms. Since the total density of the reacting mixture is given by $n = n_A + 2n_M$, we arrive at the equation of state

$$\beta p = n_A + n_M; \qquad\qquad \beta p = n_A + n_A^2 K(T) \qquad\qquad (5.100)$$

$$n = n_A + 2n_M; \qquad\qquad n = n_A + 2n_A^2 K(T). \qquad\qquad (5.101)$$

We obtain full equivalence of the physical description given by Eqs. (5.90) and (5.91) and the chemical description given by equations above. The descriptions are identical for

$$\tilde{z} = n_A \qquad\qquad (5.102)$$

$$K(T) = B_2(T). \qquad\qquad (5.103)$$

Consequently, we interprete $\tilde{z}$ as a kind of density of free atoms and $B_2(T)\,\tilde{z}^2$ as the density of free molecules.

Summarizing our findings we arrive at the conclusions:

1. For systems with attracting forces leading to deep bound states between the particles and $(e^{-E_{10}/k_{\mathrm{B}}T} \gg 1)$, the method of density expansions does not converge and can lead even to negative values of pressure.
2. Fugacity expansions describe the correct physics for the entire range of temperature.

In real systems, at the same time we can have strong repulsive forces with hard sphere core and attracting forces leading to deep bound states. In this case we need possibly a description which is somehow between fugacity expansions and density expansions and is based on a more complicated quasi-particle picture (Ebeling 1974; Hill 1956). In particular the class of systems with Coulomb interaction which is in part repulsive and in part attractive needs more sophisticated descriptions between density and fugacity expansions (Ebeling et al. 1976, 2012; Kremp et al. 2005).

5.5   Helium and Other Quantum Gases at Low Temperature

5.5.1   Virial Expansion for Helium

At high temperature, the properties of gases are only weakly influenced by the spin. This is different at low temperature, say $T < 100\,\mathrm{K}$, where the spin of the atoms (or molecules) plays an important rôle. Note that H-atoms with the spin $s = 0$ and ^{4}He atoms with the spin $s = 1$ are bosons in difference to ^{3}He atoms which have $s = \frac{1}{2}$ and are, therefore, fermions. This leads to essential differences between the properties of the fermionic ^{3}He gas and bosonic ^{4}He gas. The second virial coefficients of ^{3}He and ^{4}He have been calculated at closely spaced temperature over

the range 0.3 K to 60 K using the Lennard–Jones 12-6 potential with parameters
given by de Boer and Michels and based on a calculation of the phase shifts sketched
in Fig. 5.3 (Kilpatrick et al. 1954). A more recent calculation of the virial expansion
for He$_4$ in the region 3 K $< T <$ 100 K was given by Costa et al. (2013). Following
these authors we calculate

$$B_2(T) = 8\pi^{\frac{3}{2}}\lambda^3 \left[\frac{1}{16} + \left(e^{-\frac{E_{10}}{k_B T}} - 1 \right) + \frac{2}{\pi q_0^2} \int e^{-\frac{q^2}{q_0^2}} G(q) q\, dq \right], \qquad (5.104)$$

where

$$G(q) = \sum (2\ell + 1)\delta_\ell(q); \quad q^2 = \frac{2\mu E}{\hbar^2}; \quad q_0^2 = \frac{2\mu\sigma^2 k_B T}{\hbar^2}, \qquad (5.105)$$

with $\sigma \simeq 2.556\,\text{Å}$. Figure 5.6 shows the virial coefficients of He$_4$ gas, given by
Costa et al. (2013). The ideal quantum term contributes approximately 11% at 3 K
and less than 1% at 100 K. The He$_4$ diatomic molecule has a rather weak bound state
which contributes less than 0.1% at 3 K and even much less at higher temperature.
In most cases, this contribution can be neglected. The dominant contribution comes
from the scattering states.

5.5.2   Phase Transitions in Low-Temperature Gases

In the previous section we have shown that the properties of rare He$_4$ gas can be
well described by a density expansion and the Beth–Uhlenbeck representation of

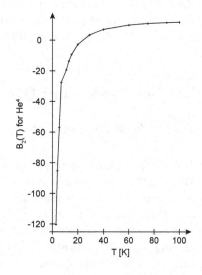

Fig. 5.6 Second virial
coefficient of Helium-4 in the
temperature range
$1, \ldots, 100$ K (Costa et al.
2013)

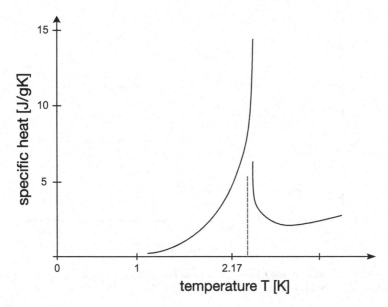

Fig. 5.7 Specific heat of ^{4}He over temperature with a λ point at $T = 2.17$ K (after Huang 2001)

the second virial coefficient. At very low temperature and high density, the virial expansion fails (Huang 2001). This becomes evident when looking at measurements of the specific heat in the region $1\,\text{K} < T < 5\,\text{K}$, as performed, e.g., by Hill (1956) shown in Fig. 5.7, (see also Costa et al. 2013).

The data show a λ curve for the specific heat. The curve resembles the specific heat of a Bose gas shown in Fig. 3.6 but in detail it is too different to come to the conclusion that He_4 behaves like an ideal Bose gas. The measurements show that He_4 in the region of the phase transition at

$$T_c = 2.172\,\text{K}; \qquad n_c = 2.16 \cdot 10^{22}\,\text{cm}^{-3}; \qquad v_c = 46.2\,\overset{\circ}{\text{A}}{}^3/\text{atom} \qquad (5.106)$$

is a Bose gas under strong influence of interactions. We see also that the simple Beth–Uhlenbeck theory is unable to describe the behavior near the phase transition. In the given region, He_4 is a non-ideal Bose gas which is outside the scope of this book. The tools developed here, such as, the Beth–Uhlenbeck method cover only the weakly non-ideal region. They do, however, not cover such interesting phenomena as the phonon and roton excitations shown in Fig. 5.8. For the theoretical interpretation of these spectra and the related phenomenon of superfluidity the reader is referred to the special literature explaining e.g. Landau's model and other more advanced theories of strongly non-ideal Bose gases (Bobrov and Trigger 2018; Landau and Lifshitz 1976; Pines and Nozieres 1966).

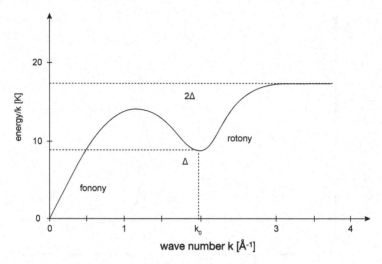

Fig. 5.8 Schema of phonon and roton excitations in Landau's model for the dispersion curve of He$_4$. The wave number, k, is given in Å^{-1} (after Huang 2001)

5.6 Weakly Interacting Quantum Gases

5.6.1 Bloch Equation

We consider now interacting quantum gases in the region of weak correlations, but without any restriction with respect to degeneracy. This means we consider the border region near ideality where weak interactions are taken into account in linear approximation, that is, in Hartree–Fock approximation. Our basic assumption about the potential is that it has a Fourier transform, however, the Coulomb case is excluded so far. As a standard example for a gas with weak interactions we consider a quantum gas of Yukawa particles with the interaction

$$V_{ab}^Y(r) = g_{ab}\frac{e^{-\eta r}}{r} . \tag{5.107}$$

This potential was developed by Yukawa (1935) in order to describe forces in elementary particle physics, mediated by the exchange of massive particles. The potential plays now a paradigmatic rôle in statistical physics, as it found applications in many fields (Langin et al. 2016; Ott et al. 2014). The corresponding Fourier transform is well defined and reads

$$\tilde{V}_{ab}^Y(\mathbf{t}) = \frac{4\pi g_{ab}}{t^2 + \eta^2} . \tag{5.108}$$

In the approximation of the second virial coefficient we have (Beth and Uhlenbeck 1937; Uhlenbeck and Beth 1936)

$$F = F^{\text{id}} - V k_B T \left(\frac{1}{2} \sum_{ab} n_a n_b \int dr_1 dr_2 \left[S_{ab} (r_1, r_2) - 1 \right] + \ldots \right). \quad (5.109)$$

Following Kelbg and Hoffmann (1964) we proceed by systematic expansions of the Slater sums (Slater functions) with respect to the interaction parameter. We denote this expansion by

$$S_{ab} = S_{ab}^{(0)} + S_{ab}^{(1)} + S_{ab}^{(2)} + \ldots \quad (5.110)$$

Instead of using the representation of the Slater sums by energy eigenfunctions we use here an alternative representation by plane waves which was introduced already in the original papers by Beth and Uhlenbeck (1937) and Uhlenbeck and Beth (1936). Following again early work by the Kelbg school we use the representation by plane wave eigenfunctions (see Ebeling et al. 1967),

$$S_{ab}(r) = A \int d\mathbf{k} e^{-\mathbf{k} \cdot \mathbf{r}} e^{-\beta \hat{H}_{ab}} e^{+i \mathbf{k} \cdot \mathbf{r}} . \quad (5.111)$$

Here, the operator of relative motion is

$$\hat{H}_{ab} = -\frac{\hbar}{2 m_{ab}} \Delta + V_{ab} , \quad (5.112)$$

with the relative mass m_{ab}. We introduce a Fourier representation in the distance coordinate space

$$\tilde{S}_{ab}^{(n)} (\mathbf{t}) = \int d\mathbf{r} \, S_{ab}^{(n)} (\mathbf{r}) e^{i \mathbf{t} \cdot \mathbf{r}} . \quad (5.113)$$

Since we need only an integral over the Slater sum, it is sufficient to find this Fourier transform in the zero point, $t = 0$. We find the expansion with respect to the interaction parameter:

$$F = F^{\text{id}} - \frac{1}{2} V \sum_a n_a \sum_b n_b \sum_{p \geq 1} \tilde{S}_{ab}^{(p)} (1, 2)(t = 0) + \ldots \quad (5.114)$$

Following Kelbg and Hoffmann (1964), we see that the averaging carried out with the canonical density operator leads to terms in the free energy related to the contribution of $t = 0$ of the Slater function for two particles. In order to calculate the Slater function we solve the Bloch equation iteratively (Beth and Uhlenbeck 1937;

Ebeling et al. 1967; Uhlenbeck and Beth 1936). To this end, we define the functions

$$W_{ab} = e^{-\beta \hat{H}_{ab}} e^{i\mathbf{k}\cdot\mathbf{r}} .$$ (5.115)

By differentiation with respect to the reciprocal temperature, β, we find for W_{ab} the more convenient differential equation

$$\frac{\partial}{\partial \beta} W_{ab} = -\hat{H}_{ab} e^{-\beta \hat{H}_{ab}} e^{+i\mathbf{k}\cdot\mathbf{r}} = -\hat{H}_{ab} W_{ab} .$$ (5.116)

This is the *Bloch equation* named after Felix Bloch (1905–1983) which reads in standard form

$$\frac{\partial W_{ab}}{\partial \beta} + \hat{H}_{ab} W_{ab} = \frac{\hbar^2}{2m_{ab}} \Delta W_{ab} - V_{ab} W_{ab} = 0 .$$ (5.117)

The Bloch equation is related to the inhomogeneous heat equation and diffusion equation (Fourier equation):

$$\frac{\partial}{\partial t} T - \kappa \Delta T = q .$$ (5.118)

There are standard methods for the solution of the Fourier equation. Here we use an iterative method which yields a series with respect to the interaction parameter. In zeroth approximation, we have

$$\frac{\partial W_{ab}^0}{\partial \beta} = \frac{\hbar}{2m_{ab}} \Delta W_{ab}^0 ; \qquad W_{ab}^0 = \exp\left(ikr - \frac{\beta \hbar^2}{2m_{ab}} k^2\right) .$$ (5.119)

Higher approximations follow from iterations of the equation (Ebeling et al. 1967)

$$\frac{\partial W_{ab}^{n+1}}{\partial \beta} - \frac{\hbar^2}{2m_{ab}} \Delta W_{ab}^{n+1} = V_{ab} W_{ab}^n .$$ (5.120)

It is convenient to introduce a new function and the corresponding Fourier transform by

$$v_{ab}(r) = e^{-i\mathbf{k}\mathbf{r}} W_{ab} ; \qquad v_{ab}(\mathbf{k}, \mathbf{t}) = \int d\mathbf{r}\, e^{i\mathbf{t}\cdot\mathbf{r}} v_{ab}(k, r) .$$ (5.121)

Then the first two iterations give

$$v_{ab}^{(1)}(\mathbf{k}, \mathbf{t}) = -\beta \int_0^\beta d\beta'\, \tilde{V}_{ab}(\mathbf{t}) \exp\left[-\lambda_{ab}^2(\mathbf{t}^2 - 2\mathbf{k}\cdot\mathbf{t})\left(1 - \frac{\beta'}{\beta}\right)\right]$$ (5.122)

$$v_{ab}^{(2)}(\mathbf{k}, t) = \frac{1}{\pi^3 8} \int_0^\beta d\beta' \int_0^{\beta'} d\beta'' \int dt' \, \tilde{V}_{ab}(t - t') \tilde{V}_{ab}(t') \tag{5.123}$$

$$\times \exp\left\{-\lambda_{ab}^2\left[(t^2 - 2\mathbf{k}\cdot\mathbf{t})\left(1 - \frac{\beta'}{\beta}\right) + (t'^2 - 2\mathbf{k}\cdot\mathbf{t}')\left(\frac{\beta'}{\beta} - \frac{\beta''}{\beta}\right)\right]\right\}.$$

On the grounds of these first two approximations for the Bloch functions we are able to calculate the free energy up to the second order in the coupling, g_{ab}, without further approximations (Kraeft et al. 1986).

5.6.2 Slater Function and Free Energy

The corresponding Fourier transforms of the Slater functions are

$$\tilde{S}_{ab}(t) = \frac{\lambda_{ab3}}{\pi^{3/2}} \int d\mathbf{k} e^{-\lambda^2 k^2}\left[v(\mathbf{k}, t, \beta) + \frac{(-1)^{2s_a}}{2s_a + 1}\delta_{ab} \, v(\mathbf{k}, t + 2\mathbf{k}, \beta)\right]. \tag{5.124}$$

Explicit calculations for the Coulomb case will be given in Chap. 6 following Ebeling et al. (1967) and Hoffmann and Ebeling (1968a,b). For Yukawa particles, the first (linear) approximation of the Slater function reads (after solving the integrals in Eq. (5.122))

$$\tilde{S}_{ab}^{(1)}(t) = -\beta\frac{4\pi g_{ab}}{t^2 + \eta^2}e^{-\frac{1}{2}\lambda_{ab}^2 t^2} \cdot {}_1F_1\left(\frac{1}{2}, \frac{3}{2}, \frac{1}{2}\lambda_{ab}^2 t^2\right)$$

$$+ \frac{(-1)^{2s_a}\delta_{ab}}{(2s_a + 1)}4\pi\beta g_{ab}\int_0^1 d\alpha \int \frac{d\mathbf{k}}{\eta^2 + (2\mathbf{k} + \mathbf{t})^2}e^{-\lambda_{ab}^2[k^2 + \alpha\mathbf{t}\cdot(\mathbf{t} + 2\mathbf{k})]}, \tag{5.125}$$

where $_1F_1$ is the standard degenerated hypergeometric function (Kummer function). After these calculations we find the free energy by performing the limit $t \to 0$. No divergency problems arrive up to the second order, neither for $r \to 0$ nor for $r \to \infty$. In the classical limit, $\hbar = 0$, the free energy is

$$F = F^B - V k_B T \sum_a n_a \sum_b n_b\left(-4\pi\frac{\beta g_{ab}}{\eta^2} + 2\pi\frac{\beta^2 g_{ab}^2}{\eta}\right). \tag{5.126}$$

Including quantum effects leads to more complicated integrals, e.g., in second order

$$\int_0^\infty dt \frac{t^2 e^{-bt^2}}{(\eta^2 + t^2)^2} = -\frac{\sqrt{\pi b}}{2} + \frac{\pi}{\eta}\left(1 + 2b\eta^2\right)\text{erfe}\left(\sqrt{b}\eta\right), \tag{5.127}$$

where erfe(z) is a special version of the error function, erfc(z),

$$\text{erfe}(z) = e^{z^2}\text{erfc}(z); \qquad \text{erfc}(z) = \frac{2}{\sqrt{\pi}}\int_z^\infty dt\, e^{-t^2}. \qquad (5.128)$$

Using Eq. (5.127) there remains only one integral over an algebraic expression in the variable α:

$$\tilde{S}_{ab}^{(2)}\,(\mathbf{t}=0) = \frac{\pi}{\eta}\beta^2 g_{ab}^2 - \frac{\pi\sqrt{\pi}}{2}\lambda_{ab}\beta^2 g_{ab}^2 + \frac{\pi}{\eta}\beta^2 g_{ab}^2\, f_2(\zeta). \qquad (5.129)$$

Here the first two terms originate from integrals where the pure classical and the pure quantum part were taken explicitly and the mixed contributions given by the integral f_2 can be estimated by using expansions or a median value approximation,

$$f_2(y) = \int_0^1 d\alpha\left[\left(1 + \sqrt{\alpha(1-\alpha)}y^2\right)\text{erfe}\left(y\sqrt{\alpha\,(1-\alpha)}\right) - 1\right]$$

$$\simeq \left(1 + \frac{1}{2}y^2\right)\left[\text{erfe}\left(\frac{y}{2}\right) - 1\right]. \qquad (5.130)$$

Correspondingly, we obtain for the free energy containing the direct and the exchange contributions respectively, the first two orders in the coupling, g_{ab},

$$F = F^{\text{id}} - Vk_BT\sum_{ab}n_a n_b 2\pi\lambda_{ab}^3\left[Q(\gamma_{ab}, \zeta) + \frac{(-1)^{2s_a}}{2s_a + 1}E(\gamma_{ab}, \zeta)\right]. \qquad (5.131)$$

Here the dimensionless coupling parameter, γ, and the decay parameter, ζ, are defined by

$$\gamma_{ab} = \beta\frac{g_{ab}}{\lambda_{ab}}; \qquad \zeta = \eta\lambda_{ab}. \qquad (5.132)$$

The *virial functions*, Q, for the contribution of Heisenberg quantum effects (direct terms, not including exchange) and E for the exchange term are in principle defined by infinite series in the coupling g. Based on Eq. (5.129) the first two terms of each of the series read

$$Q(\gamma, \zeta) = \frac{2\gamma}{\zeta^2} + \frac{1}{2}\pi^{3/2}\gamma^2 + \frac{\gamma^2}{\zeta}[1 + f_2(\zeta)] \qquad (5.133)$$

$$E(\gamma, \zeta) = \frac{\sqrt{\pi}}{4} + \frac{\gamma}{2}\left[1 - \frac{\zeta}{2}\text{erfc}\left(\frac{\zeta}{2}\right)\right]. \qquad (5.134)$$

In the limit $\lambda \to 0$, this result converges to the result for the classical Yukawa gas, Eq. (5.126). The other limit, $\eta \to 0$, does not exist but for small η we see that the result for Coulomb plasmas (this is the second term in Eq. (5.129)) appears which was obtained for Coulomb systems long ago (Ebeling et al. 1967, 1976). This special case will be discussed in detail in Chap. 6. However, as discussed above, the Yukawa potential model is of value by its own. The model is nowadays considered as paradigmatic and is used to approximately describe a wide range of physical systems, especially strongly coupled plasmas (Kilimann and Ebeling 1990; Rogers et al. 1970) and plasmas of charged particles on a background as, e.g., dusty plasmas (Langin et al. 2016; Ott et al. 2014). Note that all physical properties in Yukawa one-component plasmas are expected to be universal in η when expressed in appropriately scaled units (Langin et al. 2016).

References

Beth, E., and G.E. Uhlenbeck. 1937. The Quantum Theory of the Non-Ideal Gas. II. Behaviour at Low Temperatures. *Physica* 4: 915–924.

Blochinzew, D.J. 1953. *Quantenmechanik.* Berlin: Deutscher Verlag der Wissenschaften.

Bobrov, V.B., and S. Trigger. 2018. Bose–Einstein Condensate and Singularities of the Frequency Dispersion of the Permittivity in a Disordered Coulomb System. *Theoretical and Mathematical Physics* 194: 404–414.

Chetverikov, A.P., W. Ebeling, and M.G. Velarde. 2009. Local Electron Distributions and Diffusion in Anharmonic Lattices Mediated by Thermally Excited Solitons. *European Physical Journal B* 70: 217–227.

Costa, É., N.H.T. Lemes, M.O. Alves, R.C.O. Sebastião, and J.P. Braga. 2013. Quantum Second Virial Coefficient Calculation for the ^{4}He Dimer on a Recent Potential. *Journal of the Brazilian Chemical Society* 24: 363–368.

Ebeling, W. 1974. Statistical Derivation of the Mass Action Law or Interacting Gases and Plasmas. *Physica* 73: 573–584.

Ebeling, W., H.J. Hoffmann, and G. Kelbg. 1967. Quantenstatistik des Hochtemperatur-Plasmas im Thermodynamischen Gleichgewicht. *Contributions to Plasma Physics* 7: 233–248.

Ebeling, W., W.D. Kraeft, and D. Kremp. 1976. *Theory of Bound States and Ionisation Equilibrium in Plasmas and Solids.* Berlin: Akademie-Verlag.

Ebeling, W., W.D. Kraeft, and G. Röpke. 2012. On the Quantum Statistics of Bound States within the Rutherford Model of Matter. *Annals of Physics* 524: 311–326.

Feynman, R.P. 1972. *Statistical Mechanics.* Reading: Benjamin.

Friedman, H.L. 1962. *Ionic Solution Theory.* New York: Interscience.

Friedman, H.L., and W. Ebeling. 1979. Theory of Interacting and Reacting Particles. *Rostocker Physikalische Manuskripte* 4: 330–348.

Hill, T.L. 1956. *Statistical Mechanics.* New York: McGraw Hill.

Hirschfelder, J.O., C.F. Curtis, and R.B. Bird. 1954. *Molecular Theory of Gases and Liquids.* New York. Wiley.

Hoffmann, H.J., and W. Ebeling. 1968a. On the Equation of State of Fully Ionized Quantum Plasmas. *Physica* 39: 593–598.

Hoffmann, H.J., and W. Ebeling. 1968b. Quantenstatistik des Hochtemperatur-Plasmas im Thermodynamischen Gleichgewicht. II. Die Freie Energie im Temperaturbereich 10^6 bis 10^8 °K. *Contributions to Plasma Physics* 8 (1): 43–56.

Huang, K. 1963. *Statistical Mechanics.* New York: Wiley.

Huang, K. 2001. *Introduction to Statistical Physics.* London: Taylor & Francis.

Kelbg, G., and H.J. Hoffmann. 1964. Quantenstatistik Realer Gase und Plasmen. *Annalen der Physik* 469: 310–318.

Kilimann, K., and W. Ebeling. 1990. Energy Gap and Line Shifts for H-Like Ions in Dense Plasmas. *Zeitschrift für Naturforschung* 45a: 613–617.

Kilpatrick, J.E., W.E. Keller, E.F. Hammel, and N. Metropolis. 1954. Second Virial Coefficients of He^3 and He^4. *Physical Review* 94: 1103–1110.

Kraeft, W.D., W. Ebeling, D. Kremp. 1969. Complex Representation of the Quantumstatistical Second Virial Coefficient. *Physics Letters A* 29: 466–467.

Kraeft, W.D., D. Kremp, W. Ebeling, and G. Röpke. 1986. *Quantum Statistics of Charged Particle Systems*. Berlin: Akademie-Verlag.

Kremp, D., and W.D. Kraeft. 1972. Analyticity of the Second Virial Coefficient as a Function of the Interaction Parameter and Compensation Between Bound and Scattering States. *Physics Letters A* 38: 167–168.

Kremp, D., W.D. Kraeft, and W. Ebeling. 1971. Quantum-Statistical Second Virial Coefficient and Scattering Theory. *Physica* 51: 146–164.

Kremp, D., M. Schlanges, and W.D. Kraeft. 2005. *Quantum Statistics of Nonideal Plasmas*. Berlin: Springer.

Landau, L.D., and E.M. Lifshitz. 1976. *Statistical Physics (Part I)*. Moscow: Nauka.

Landau, L.D., and E.M. Lifshitz. 1980. *Statistical Physics*. Oxford: Butterworth-Heinemann.

Langin, T.K., T. Strickler, N. Maksimovic, P. McQuillen, T. Pohl, D. Vrinceanu, et al. 2016. Demonstrating Universal Scaling for Dynamics of Yukawa One-Component Plasmas After an Interaction Quench. *Physical Review E* 93: 023201.

Morita, T. 1959. Equation of State of High Temperature Plasma. *Progress in Theoretical Physics* 22: 757–774.

Ott, T., M. Bonitz, L.G. Stanton, and M.S. Murillo. 2014. Coupling Strength in Coulomb and Yukawa One-Component Plasmas. *Physics of Plasmas* 21: 113704.

Pines, D., and P. Nozieres. 1966. *The Theory of Quantum Liquids*. New York: Benjamin.

Rogers, F.J., H.C. Graboske, and D.J. Harwood. 1970. Bound Eigenstates of the Static Screened Coulomb Potential. *Physical Review A* 1: 1577–1586.

Slater, J.C. 1939. *Introduction to Chemical Physics*. New York: Mc Graw Hill.

Uhlenbeck, G.E., and E. Beth. 1936. The Quantum Theory of the Non-Ideal Gas I. Deviations From the Classical Theory. *Physica* 3: 729–745.

Yukawa, H. 1935. On the Interaction of Elementary Particles. I. *Proceedings of the Physico-Mathematical Society of Japan* 17: 48–57.

Chapter 6
Quantum Statistics of Dilute Plasmas

6.1 Basic Physics of Plasmas

6.1.1 Screening and Lattice Formation in Coulomb Systems

The development of a systematic statistical theory for systems with Coulomb interactions is related to characteristic problems:

1. Debye's screening problem,
2. Wigner's problem of lattice formation,
3. Herzfeld's bound state problem.

These problems are more or less due to the unique properties of the Coulomb potential and its long-range character. The law of forces between two charges e_a and e_b is long-range and the integral over its potential diverges, that is, the field energy does not exist and the correlations between two charges decay infinitely slow. Since plasmas and electrolytes as well as metals and semiconductors consist of charges, the statistical theory has to overcome these problems in a proper way. In the course of research on systems with Coulomb interactions which started at the beginning of the twentieth century, it became more clear what are the most essential divergencies. We will start with a discussion of the screening problem which is connected to the work by Milner and the Debye school and continue then with the discussion of the divergence of the Bohr atom studied by Herzfeld and Planck. What is the essence of the screening problem? Around the turn from the nineteenth to the twentieth century, experimental studies provided evidence that Coulomb systems such as plasmas and electrolytic solutions show essential deviations from gases with respect to the form of the thermodynamic functions. In particular, the pressure or its analogue in solutions, the osmotic pressure, cannot be expanded into Taylor series with respect to density. This was first understood by Milner (1912) and presented in a statistical approach by Debye and Hückel (1923). Debye and Hückel developed in fact the modern concept of screening. An important contribution was added later

© Springer Nature Switzerland AG 2019
W. Ebeling, T. Pöschel, *Lectures on Quantum Statistics*,
Lecture Notes in Physics 953, https://doi.org/10.1007/978-3-030-05734-3_6

by Bogolyubov (see Bogolyubov 1946, 2005–2009) and Mayer (1950). Following these ideas, the long-range divergencies in Coulomb systems are regularized by replacing long-range Coulomb potentials by screened potentials. In other words, the Coulomb potential for the interactions of two charges belonging to the species a and b, given above takes into account screening by the surrounding plasma to be replaced by the Debye potential (Falkenhagen 1971; Falkenhagen and Ebeling 1971)

$$V_{ab}^{D} = \frac{e_a e_b e^{-r/r_D}}{\varepsilon_r r} \, ; \qquad r_D^{-2} = \kappa^2 = \frac{4\pi\beta}{\varepsilon_r} \sum_a n_a e_a^2 \, , \tag{6.1}$$

with $\beta = 1/(k_B T)$. To derive this effective potential, Debye used a generalized Poisson equation. As an alternative approach, Bogolyubov (1946) proposed to solve the integral equation

$$V_{ab}^{D}(1, 2) = V_{ab}(1, 2) + \beta \sum_c n_c \int dr_3 V_{ac}(1, 3) V_{bc}^{D}(2, 3) \, . \tag{6.2}$$

The radial distribution and the correlation function g_{ab} are according to Bogolyubov related to the screened potential via

$$F_{ab} = 1 + g_{ab} \, ; \qquad g_{ab} = -\beta V_{ab}^{D} \, . \tag{6.3}$$

The mean Coulomb energy is then according to Debye given by

$$U_c = \frac{V}{2} \sum_{ab} \int d\mathbf{r} V_{ab}(r) F_{ab}(r) = -V \frac{\kappa^3}{8\pi} \, . \tag{6.4}$$

This means that the energy per particle increases with the root of density,

$$-\frac{U_c}{N} \sim \sqrt{n} \, . \tag{6.5}$$

The free energy can be obtained from Debye's charging procedure, by integrating with respect to e^2, which leads to the pre-factor 3/2. Note that, strictly speaking, there is also a contribution linear in e^2. Carrying out first the summation, the linear term results in a zero contribution due to electro neutrality. However, carrying out first the integration, we find a sum of infinities which makes no sense. However this term cancels out due to electro-neutrality. We will come back to this problem in Sect. 6.6.2. A different procedure to derive the free energy which is due to Mayer (1950) and Macke (1950) is connected with the summation of chain or ring diagrams

Fig. 6.1 Peter Debye who developed 1924 with Erich Hückel the concept of screening. Source: http://doi.org/10.3932/ethz-a-000045888

of Coulomb lines. For the classical case, by using the charging procedure for a pure classical Coulomb system we obtain the same result as Debye (Fig. 6.1)

$$F_D = \frac{V}{2} \sum_{ab} \int de^2 \int d\mathbf{r} V_{ab}(r) \left[1 + g_{ab}\left(r; e^2\right) \right] \tag{6.6}$$

$$F = F_{id} + F_D ; \qquad F_D = -k_B T V \frac{\kappa^3}{12\pi} . \tag{6.7}$$

Further important contributions to the classical statistical theory of screening effects were given later by Bogolyubov, Mayer, Meeron, Zubarev, Yukhnovsky, and Friedman (Friedman 1962). The quantum theory of screening was developed by Macke, Gell-Man, Brueckner, Montroll, Ward, Vedenov, Larkin, DeWitt, Kelbg and others. Most important results in the present context were obtained in different scientific centers. Vedenov and Larkin (1959) in Moscow obtained first quantum corrections to Debye's results by using field theoretical methods. Another center was founded by Kelbg in Rostock (for early results see, e.g., Ebeling et al. 1976; Kelbg 1963a; Kraeft et al. 1986). The Rostock school used first the method of effective potentials to include quantum corrections and later other more refined methods as Green functions (Ebeling et al. 1976; Kelbg 1963a).

A different approach based on the Montroll-Ward method was developed by DeWitt et al. (for main results see DeWitt 1962; DeWitt et al. 1995; Montroll and Ward 1958). A third important approach was based on the Feynman-Kac methods and developed by Alastuey and Perez (1992) and an approach based on the effective field methods of quantum field theory was developed by Brown and Yaffe (2001) All these quite different methods lead at lower density essentially to the same results (Alastuey et al. 2015) (Fig. 6.2).

So far we discussed for plasmas at low density the classical case and the first quantum corrections. At high density, the behavior of Coulomb systems is determined by lattice effects as shown first in the 30th by the young Hungarian physicist Eugene Wigner (1902–1995: Fig. 6.2). Wigner enrolled in 1921 at the

Fig. 6.2 Eugene Wigner
visiting as young scientist
Berlin and other scientific
centers. Source: https://
thespectrumofriemannium.
files.wordpress.com/2013/04/
wigner.jpg

Technische Hochschule in Berlin (today Technische Universität Berlin), where he studied chemical engineering, similar to his fellow from Budapest grammar school Johann von Neumann (1903–1957). They attended also the famous Wednesday afternoon colloquia of the German Physical Society. These colloquia featured such luminaries as Max Planck, Max von Laue, Rudolf Ladenburg, Werner Heisenberg, Walther Nernst, Wolfgang Pauli, and Albert Einstein. No doubt, these colloquia influenced Wigner and von Neumann very much as well as other young physicists. For example, Wigner met here another physicist from Hungary, Leó Szilard, who at once became Wigner's closest friend. Later Wigner worked at the Kaiser Wilhelm Institute for Physical Chemistry and Electrochemistry (now the Fritz Haber Institute), and there he met Michael Polanyi, who became Wigner's teacher. Polanyi supervised Wigner's dissertation "Formation and Decay of Molecules". After a stay in Budapest, in 1926 Wigner returned to Berlin and worked at the Kaiser Wilhelm Institute for Physics in Berlin on problems of X-ray crystallography. Six months later Wigner started working with Richard Becker at the Technical University. Note that his old friend Johann von Neumann was from 1928 to 1933 the youngest Privatdozent at the Berlin University, Unter den Linden, und wrote his ground-breaking book on the mathematical foundation of quantum mechanics and quantum statistics (von Neumann 1932). Under these circumstances, Wigner explored quantum mechanics, studying the work by Erwin Schrödinger and others. This way Wigner was prepared to become one of the most brilliant scientists in the field of the analytical behavior of plasma thermodynamic functions. He was the first who understood, that at high density the analytical behavior of Coulomb systems changes and approaches a regime where the Coulomb energy is determined by lattice effects and increases more slowly as in the Debye regime by Wigner (1934)

$$- \frac{U_c}{N} \sim n^{1/3} \,. \tag{6.8}$$

A system of electrons with a positive smeared out background form a bcc-lattice, a so-called Wigner lattice. We switch now to a problem posed by the new quantum theory of the Bohr atom, which was overlooked by Eggert and Saha but was solved a few years later in a first approach by Planck and worked out by Leon Brillouin (1889–1969), one of the most influential French pioneers of quantum theory.

6.1.2 The Divergence of the Partition Function

The model by Bohr was a true revolution for the understanding of the structure of atoms. Note that the Austrian physicist Karl Herzfeld who was influenced by Otto Stern and Friedrich Hasenöhrl developed in parallel to Bohr a similar model, which he investigated in detail between 1912 and 1916. He detected that the Bohr model had a serious deficiency, namely the statistical partition function is divergent. This fundamental problem will be discussed now and in more detail at various places of this book. Simply ignoring the open problems with the partition function, Eggert and Saha succeeded to find the solution of the most urgent problems of astrophysics, the ionization phenomena (Eggert 1919; Saha 1920). In their derivation Eggert and Saha used theoretical tools as Sackur-Tetrode's formula of the chemical constant for calculating the chemical potential of electrons. On the other hand, they missed evidently the significance of the partition function in the chemical potential of the atoms. In fact the atomic partition function is divergent for Coulomb atoms. The problem of the divergence of the partition function of the Bohr atom was seen already by Bohr but studied then in detail by Herzfeld and later by many others including Planck. For an atom with energy levels E_{sl} depending on the main quantum numbers s, l, the atomic partition function is defined by

$$\sigma(T) = \sum_{s,l} e^{-\beta E_{sl}} . \tag{6.9}$$

For hydrogen withe $E_{sl} = E_s = -I/s^2$ we obtain

$$\sigma(T) = \sum_{s=1}^{\infty} s^2 e^{-\beta E_s} = \sum_{s=1}^{\infty} s^2 e^{\frac{I}{s^2 k_B T}} . \tag{6.10}$$

The terms in the atomic partition function for hydrogen according to the definition, Eq. (6.10), diverge as s^2 (see Fig. 6.3). The problem of the divergency of the atomic partition function was seen already by Bohr and in particular by Herzfeld but it took a long time before the mathematical background for a serious treatment of this problem was available. To discuss this in detail, is one of the tasks of this book, we will come back to this soon. However the people urgently needed to solve the ionization problems in astrophysics and plasma physics had no time to wait for strict solutions. As we have pointed out above, for the most urgent calculations the

Fig. 6.3 The terms in the atomic partition function (6.10) in comparison to the renormalized formula Eq. (6.16). Up to terms of order $E_s^{max} \simeq k_B T$ the terms decrease monotonically and then they start to increase as s^2. For comparison we show the Brillouin-Planck-Larkin renormalized partition function

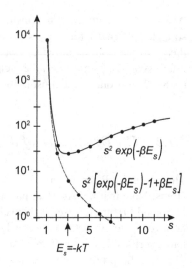

problem was simply ignored and the partition function was estimated by the first term of the sum or a few terms.

The solution is connected with the so-called bound state problem for Coulomb systems. From the point of view of statistical physics, the major problem is that the partition function of Bohr atoms is infinite. The reason is that the terms in the atomic partition function for hydrogen and correspondingly for other plasmas according to the definition, Eq. (6.10), diverges as s^2 (see Fig. 6.3). For the calculations, the problem was simply ignored and the partition function was estimated by the first term of the sum or a few terms. One should admit that there were some good mathematical reasons for such a procedure. Looking at the terms in Fig. 6.3 we see a strong similarity to the so-called asymptotic series, the terms first decrease and then increase again. In such cases mathematical theories suggest to cut the series at the smallest terms. In our case this is the term with $|E_s| \simeq k_B T$. A more physical argument was, that terms with energies smaller than the thermal one $|E_s| \simeq k_B T$ are not sufficiently stable and cannot contribute (Riewe and Rompe 1938). A first reasonable solution of the divergence problem of the Coulomb partition function was first given by Planck (1924). Planck divided the atomic partition function Σ as demonstrated in Fig. 1.4 into three parts

$$\Sigma = \Sigma_1 + \Sigma_2 + \Sigma_3 , \tag{6.11}$$

where Σ_1 is the contribution of the scattering states, Σ_2 is the contribution of the high bound states near to the series limit, and Σ_3 is the contribution of lowest bound states. The border between the lower and the higher bound states is according to Planck's estimate at $-\varepsilon'$ where the border energy is given by

$$\frac{e^2}{d} < \varepsilon' < k_B T . \tag{6.12}$$

Thus, the border is an energy between the thermal energy and the Wigner energy. In a next step, Planck estimated the sum of the first two contributions by an semiclassical estimate and found it equal to the sum over states for free particles

$$\Sigma_1 + \Sigma_2 \simeq \frac{V}{\Lambda^3} . \tag{6.13}$$

In other words, the interaction parts (which both are divergent but of opposite sign) compensate each other. Due to the compensation we have

$$\Sigma = \frac{V}{\Lambda^3} + \Sigma_3 . \tag{6.14}$$

Since the free contribution is not counted as an internal atomic contribution, we get

$$\sigma(T) = \Sigma_3 = \sum_{sl}^{E_{sl} < -\varepsilon'} e^{-\beta E_{sl}} . \tag{6.15}$$

In the more general case that the energy levels depend on density, which will be discussed later, we need an appropriate density and temperature dependent choice of the upper border $\varepsilon'(n, T)$. At low density, $\varepsilon'(n, T)$ should be near to the terms in the partition function where is the minimum of the terms in the sum.

A disadvantage of the procedure by Planck and generalizations of it working with upper limits of the s-summations is the possibility of discontinuities, when levels cross with changing density the upper limit. A procedure which at least for hydrogen-like plasmas avoids this problem was worked out by Brillouin (1931a,b). Much later, a more strict foundation was worked out in the 60th in a letter by Larkin (1960) and in extended work by one of the present authors in collaboration with Kremp and Kraeft (Ebeling 1967, 1968a; Ebeling et al. 1976). We do not find this procedure in Planck's paper but it was worked out by Brillouin (1931a,b) with reference to Planck's work) and much later by Larkin (1960). The regularization is based on a Taylor expansion of the partition function for hydrogen levels and subtracting the first divergent terms,

$$\sigma(T) = \sum_{s=1}^{\infty} s^2 \left(e^{-\beta E_s} - 1 + \beta E_s \right) . \tag{6.16}$$

The convergent (regularized) form of the atomic partition function given by Eq. (6.16) is nowadays called the Planck-Brillouin-Larkin (PBL) partition function. The convergence is due to omitting in the series the divergent contributions up to $\mathcal{O}\left(e^4\right)$. The practical equivalence of the original Planck procedure and the PBL-procedure at least in the region of lower temperature (and small density) is seen from Fig. 6.3. A deeper physical justification for the procedure in Eq. (6.16) is based on the fact that the contributions up to $\mathcal{O}\left(e^4\right)$ are influenced by screening effects and have to be treated separately. An alternative way to express this is the observation that there are compensation effects between contributions of the terms just below and just above the series limit (see Fig. 1.4). Due to these and other physical effects, the sum of contributions to any thermodynamic quantity is always finite (see Ebeling et al. 1976). The important work by Planck (1924) on the partition function and the subsequent work by Brillouin (1931a,b) was largely ignored for about 30 years. Then the formula Eq. (6.16) appeared in a letter by Larkin (1960) as a side product of a strict field-theoretical statistical thermodynamics by Larkin (1960) and Vedenov and Larkin (1959). Larkin obtained Eq. (6.16) but evidently he was not aware of Planck's and Brillouin's contributions and did not give any exhaustive discussion. The whole complexity of the problem and the connection with the Eggert-Saha equation became clear in later work of the Rostock school (Ebeling 1967, 1968a, 1974) and was summarized later in several books (Ebeling et al. 1976; Kraeft et al. 1986; Kremp et al. 2005).

We will come back to the PBL-regularization procedure in Chap. 6, but let us explain here the physical arguments behind omitting terms, as proposed by Planck and in final form first by Brillouin (1931a,b). Let us assume for a moment the formula Eq. (6.16) as the result of Planck's and Brillouin's great intuition and study the consequences for a simple plasma consisting of N_e free electrons, N_i free ions, and N_b bound electrons. According to Planck, the contributions of the bound states to the free energy of a simple plasma are

$$F_b = -k_B T N_b \ln \sigma(T), \tag{6.17}$$

where N_b is here the number of particles in bound states, that is, the number of atoms. What are the other contributions beyond the ideal ones? At first we have for the contribution of free charges the Debye term, F_D, according to Eq. (6.7). The next correction we express in a second order virial approximation by integrals corresponding to ion-ion-, electron-electron- and electron-ion pairs in classical approximation (Ebeling 1967, 1968a). For equally charged pairs we obtain

$$\delta F_{ii} = \delta F_{ee}$$

$$= \frac{N_e^2}{2V} \int_0^\infty dr \, 4\pi r^2 \left[\exp\left(-\frac{e^2}{k_B T r}\right) - 1 + \frac{e^2}{k_B T r} - \frac{1}{2} \left(\frac{e^2}{k_B T r}\right)^2 \right]. \tag{6.18}$$

The difference to standard second order virial coefficients (see Chap. 5) is only that the linear and quadratic orders in e^2 were subtracted. This is necessary to avoid double-counting, since these terms were already taken into account for the screening procedure. The contribution of the electron-ion interaction gives in same approximation

$$\delta F_{\text{ie}} = \frac{N_e N_i}{2V} \int_0^\infty dr\, 4\pi r^2$$

$$\times \left[\exp\left(+\frac{e^2}{k_B T r} \right) - 1 - \frac{e^2}{k_B T r} - \frac{1}{2}\left(\frac{e^2}{k_B T r} \right)^2 - S_\pm^b(r) \right]. \tag{6.19}$$

Here we have an additional term subtracted which accounts for Planck's partition function which corresponds in classical approximation to the subintegral terms

$$S_\pm^b(r) = \frac{4\pi}{(2\pi m_e k_B T)^{3/2}}$$

$$\int_0^{p_0} dp\, p^2 \left[\exp\left(-\frac{p^2}{2m k_B T} + \frac{e^2}{k_B T r} \right) + 1 - \frac{p^2}{2m k_B T} + \frac{p^2}{2m k_B T} \right]. \tag{6.20}$$

The momentum p_0 separates the negative from the positive pair energies:

$$E(p_0, r) = \frac{p_0^2}{2m} - \frac{e^2}{r} = 0\,; \qquad p_0 = \sqrt{\frac{2me^2}{r}}\,. \tag{6.21}$$

Calculating all the integral contributions to the free energy given above, we find the sum of all corrections in the classical second virial approximation,

$$\delta F_2 = \delta F_{\text{ii}} + \delta F_{\text{ee}} + 2\delta F_{\text{ie}}\,. \tag{6.22}$$

After a long and tedious but fully analytical calculation we find the surprising result (Ebeling 1967)

$$\delta F_2 = 0\,. \tag{6.23}$$

From this, we may formulate a theorem: With the Planck-Brillouin-Larkin choice of regularization, the corrections to Debye's law due to bound states are in classical approximation zero, only higher order in density corrections to the limiting law are possible. This statement can be summarized in the formula

$$F = F_{\text{id}} - k_B T \left[\frac{\kappa^3}{12\pi} + n_e n_i \Lambda^3 \sigma(T) + \mathcal{O}\left(n^{5/2} \right) \right] \tag{6.24}$$

There are recent papers claiming that the PBL-partition function is wrong and should be replaced by other choices. For example Starostin et al. (2003) obtained a different renormalized partition function and arrived at the statement, that the PBL-partition function is not correct (Starostin et al. 2003). Following Lars Onsager's views (see Hemmer et al. 1996), we are convinced that statements about correctness of regularized partition functions do not have a direct physical meaning since the choice of Coulomb partition functions is to some extent arbitrary (Ebeling and Norman 2003; Ebeling et al. 1976). There is a well known statement by Onsager about a related problem in electrolyte theory, which is applicable to our problem. Onsager said in a discussion remark at the electrolyte conference held 1968 in Montpellier with respect to the choice of the mass action constant for ion association (see Falkenhagen 1971; Falkenhagen and Ebeling 1971) (page 249 of Hemmer et al. 1996): *Bjerrum's choice is good but we could vary it within reason. In a complete theory this would not matter; what we remove from one side of the ledger would be entered elsewhere with the same effect.* The problem of defining the mass action constant for electrolytic association is completely equivalent to our problem of the choice of a plasma partition function and the corresponding mass action constant. Applying Onsager's theorem to our problem we come to the conclusion that there is indeed some freedom in defining the partition function, but in a complete theory this would not matter, "what we remove from one side of the ledger would be entered elsewhere with the same effect". In correct theories, which consider both "sides of the ledger" (see Hemmer et al. 1996), the different choices of partition functions can be used (in some limits) on equal rights. Following Onsager, we would prefer, to apply the terms *correct* or *wrong* not to the bound state partition function but only to expression for complete thermodynamic functions, containing a sum of bound state and free state contributions. Following Onsager, the statement, a partition function of bound states (or the corresponding mass action constant) is correct or is wrong is meaningless in the context of a correct imbedding theory, which includes bound and free quantum states.

6.2   Pair Correlations on Non-degenerate Plasmas

6.2.1   Density Matrix of Pairs

According to Langmuir, a plasma is a system with free electrons. This definition includes a wide range of physical systems, examples are discharge plasmas, plasmas in solids, and astrophysical plasmas. Here we consider mainly plasmas consisting of electrons, e, and ions, i, or protons, p. In logarithmic scale the straight line, $n_e \Lambda_e^3 = 1$, in the temperature-density diagram (see Fig. 6.4) separates degenerate from non-degenerate electron-components in plasmas. The thermal De Broglie wavelength,

$$\Lambda_e = \frac{h}{\sqrt{2\pi m_e k_B T}} \, , \tag{6.25}$$

corresponds to the thermal momentum and n_e is the density of electrons in the plasma. The straight line $n_p \Lambda_p^3 = 1$ separates in hydrogen plasmas the regions of non-degenerate protons from the region of degenerate protons. In non-degenerate

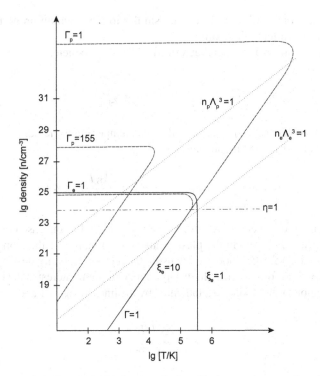

Fig. 6.4 Characteristic lines in the log-log $n - T$ plot of hydrogen plasmas separating regions of different physical conditions

plasmas, the bound states between electrons and the positive charges play a particular rôle. If I is the ionization energy, then as a rule of thumb, the bound states are essential if $\frac{I}{k_B T} > 10$. The ionization energy, I, is the ground state energy of the atoms in the plasma, $I = |E_{10}|$, or possibly the ground state of outer electron shells. Measuring T in electron Volt, we have $1\,\text{eV} \approx 10^4$ K, and obtain according to our rule, that for hydrogen with $I = 13.6\,\text{eV}$ bound states are to be expected for $T < 15,000$ K.

In Fig. 6.4 we show in the $n - T$ plane straight lines (in log-log scales) separating different regions for H-plasmas. First we consider the separation of degenerate and non-degenerate subsystems. Due to the different masses of electrons and protons, $m_p \simeq 1840\, m_e$, we have

$$\Lambda_e \gg \Lambda_p \quad \text{with} \quad \Lambda_e = \frac{h}{\sqrt{2\pi m_e k_B T}}; \quad \Lambda_p = \frac{h}{\sqrt{2\pi m_p k_B T}}. \quad (6.26)$$

The light electrons degenerate at much smaller density than the protons while in most cases the protons behave rather classical. The quantum properties of gas atoms which are also contained in a plasma are similar to the quantum properties of the free protons according to their similar masses.

The state of non-ideality is characterized by the Γ-parameter,

$$\Gamma_e = \frac{e^2}{\Theta_e d} \ ; \qquad \Gamma_p = \frac{e^2}{\Theta_p d} \ . \tag{6.27}$$

Here we defined generalized electron and proton temperatures including degeneracy by using the Fermi-Dirac kinetic energies

$$\Theta_e = \frac{2k_B T}{n_e \Lambda_e^3} I_{3/2}(\alpha_e) \ ; \qquad \Theta_p = \frac{2k_B T}{n_p \Lambda_p^3} I_{3/2}(\alpha_p). \tag{6.28}$$

Below the line of degeneracy we have $\Theta_e = \Theta_p = k_B T$. For the case that there are also atoms with density n_0 in the plasma, above the line $\eta = 1$ the atoms become densely packed. At further increased density, the atoms are finally destroyed, since there is no space to form atomic shells anymore. Our main interest is devoted to the triangular region in the lower part, indicated by the lines $n_e \Lambda_e^3 = 1$ and $\xi_e = 1$, where

$$\xi_e = 2\sqrt{\frac{I}{\Theta_e}} \tag{6.29}$$

is the interaction parameter of the electron system.

In this triangular region, the line $\Gamma = \Gamma_e = \Gamma_p = 1$ separates the region of weakly coupled plasmas below this line from strongly coupled plasmas (above the line). An explicit treatment of systems with Coulomb interactions is in much respect similar to the theory of gases but requires in contrast to real gases the discussion of several singularities. In this respect, the Coulomb potential is a very special "singular potential" which gives rise to several divergencies in the statistical thermodynamics. The first one is related to its long range character, that is, the Coulomb potential decays slowly with r^{-1}. This property causes problems with several integrations in statistical thermodynamics, in particular the second virial coefficient defined in Chap. 5 does not converge. To overcome this problem we will introduce screening effects, following Debye and Hückel. The other singularity at $r = 0$ can be removed only by taking into account quantum effects.

In fact, two particles, i and j with masses m_i, m_j and thermal momentum corresponding to the temperature, T, do not experience one another as point particles but as charge clouds with an effective diameter given by the de Broglie wavelength of relative thermal motion,

$$\lambda_{ij} = \frac{\hbar}{\sqrt{2m_{ij}k_B T}} \ ; \qquad \frac{1}{m_{ij}} = \frac{1}{m_i} + \frac{1}{m_j} \ . \tag{6.30}$$

For interaction effects, that is, effects of non-ideality, the relative thermal wavelength is more relevant than the individual wave length which determines the ideal properties.

We start the systematic study of the Coulomb effects in plasmas with the investigation of plasmas at fairly low density and sufficiently high temperature, that is, far from the region where bound states are formed. Here the Coulomb interactions of the free charges play an important rôle and cause characteristic Coulomb correlations. Let us study the diagonal elements of the density operator of pairs,

$$\rho_{ab}^{(2)}(\mathbf{r}_1\mathbf{r}_2) = \langle \mathbf{r}_1\mathbf{r}_2 | \hat{\rho}_{ab} | \mathbf{r}_1\mathbf{r}_2 \rangle, \tag{6.31}$$

which are responsible for the thermodynamic functions and other physical properties, as we know from Chap. 4.

We represented the diagonal element of the density matrix for a pair of particles of species a and b, with $ab \in \{++, --, -+, +-\}$. In most cases, this function depends only on the distance of the two particles. In the limit of low density the pair density operator in diagonal representation corresponds to the binary Slater sum

$$\lim_{n \to 0} \rho_{ab}^{(2)}(\mathbf{r}_1\mathbf{r}_2) = S_{ab}^{(2)}(\mathbf{r}); \qquad \mathbf{r} = \mathbf{r}_1 - \mathbf{r}_2. \tag{6.32}$$

Let us further study this expression: According to Chap. 4, we have

$$\hat{\rho}_{ab} = e^{-\beta \hat{H}_{ab}}. \tag{6.33}$$

The Hamilton operator of relative motion is

$$\hat{H}_{ab} = -\frac{\hbar}{2m_{ab}}\Delta + V_{ab}; \qquad V_{ab} = \frac{e_a e_b}{\varepsilon_0 r}, \tag{6.34}$$

where m_{ab} is the reduced (relative) mass. The binary Slater function (6.31) can be written in form of the integral

$$S_{ab}(r) = A \int_{\{\mathbf{k}\}} d\mathbf{k} e^{-i\mathbf{k}\mathbf{r}} e^{-\beta \hat{H}_{ab}} e^{+i\mathbf{k}\mathbf{r}}, \tag{6.35}$$

with $\mathbf{k} = \mathbf{p}/m$ and a constant, A. For further calculations, we modify the representation and introduce Bloch equations, known from Chaps. 4 and 5. We proceed in the same way as in Chap. 5: First we define

$$W_{ab} = e^{-\beta \hat{H}_{ab}} e^{+i\mathbf{k}r} \tag{6.36}$$

and find the Bloch equation in β space,

$$\frac{\partial W_{ab}}{\partial \beta} + \hat{H}_{ab} W_{ab} = 0 \tag{6.37}$$

or more explicitly

$$\frac{\partial W_{ab}}{\partial \beta} = \frac{\hbar}{2m_{ab}} \Delta W_{ab} - V_{ab} W_{ab} \,. \tag{6.38}$$

For free particles, its solution reads

$$W_{ab}^0 = e^{ikr - \frac{\beta \hbar^2}{2m_{ab}} k^2} \,. \tag{6.39}$$

This can be confirmed by differentiation:

$$\frac{\partial W_{ab}^0}{\partial \beta} = -\frac{\hbar}{2m_{ab}} k^2 e^{ikr - \frac{\beta \hbar^2}{2m_{ab}} k^2} \tag{6.40}$$

and

$$\Delta e^{i\mathbf{k}\mathbf{r}} = -k^2 e^{i\mathbf{k}\mathbf{r}} \,. \tag{6.41}$$

The first order solution which was found for Coulomb systems by Kelbg (1963a, 1964) and Kelbg and Hoffmann (1964) follows by introducing the zeroth order into the right-hand side of the Bloch equation,

$$\frac{\partial W_{ab}^1}{\partial \beta} - \frac{\hbar^2}{2m_{ab}} \Delta W_{ab}^1 = V_{ab} W_{ab}^0 \,. \tag{6.42}$$

This partial differential equation has the structure of the heat equation

$$\frac{\partial}{\partial t} T - \kappa \Delta T = q \,, \tag{6.43}$$

therefore, we find the solution by standard methods. Note that here we use the error function, $\Phi(x)$, with the definition

$$\Phi(x) = \frac{1}{\sqrt{2\pi}} \int_0^x e^{-\frac{t^2}{2}} dt = \frac{1}{2} \mathrm{erf}\left(\frac{x}{\sqrt{2}}\right) \tag{6.44}$$

$$\mathrm{erf}(z) = \frac{2}{\sqrt{\pi}} \int_0^z e^{-t^2} dt \,. \tag{6.45}$$

The error function as well as the degenerated hypergeometric function play a central rôle in the quantum statistics of Coulomb systems (Kelbg 1963b, 1964, 1972). For later applications we note the following properties:

$$\Phi(0) = 0; \qquad \frac{d\Phi}{dx}\bigg|_{x=0} = \frac{1}{\sqrt{2\pi}}; \qquad \Phi(x) = \frac{1}{\sqrt{2\pi}}x + \mathcal{O}\left(x^2\right). \qquad (6.46)$$

Introducing our solution into the Slater sum, Eq. (6.35), we obtain finally (in Gaussian units)

$$S_{ab}(r) = 1 - \beta \frac{e_a e_b}{\varepsilon_0 r}\left(1 - e^{\left(\frac{r}{\lambda_{ab}}\right)^2} + \sqrt{\pi}\frac{r}{\lambda_{ab}}\left[1 - \Phi\left(\frac{r}{\lambda_{ab}}\right)\right]\right). \qquad (6.47)$$

This important result for the correlation function of two charges, a and b, was found by Kelbg (1963b) in first order of the (direct) Coulomb interaction.

6.2.2 Method of Effective Potentials

The pair correlation function derived above can be written in the form

$$S_{ab}(r) = 1 - \beta V_{ab}^{K}, \qquad (6.48)$$

with the Kelbg potential

$$V_{ab}^{K}(r) = V_{ab}\left(1 - e^{-\left(\frac{r}{\lambda_{ab}}\right)^2} + \sqrt{\pi}\frac{r}{\lambda_{ab}}\left[1 - \Phi\left(\frac{r}{\lambda_{ab}}\right)\right]\right). \qquad (6.49)$$

In analogy to the Boltzmann (exponential) approximation, we guess that higher correlations can be approximated by

$$S_{ab}(r) = \rho_{ab}(r, n_a = 0) \approx e^{-\beta V_{ab}^{K}(r)}. \qquad (6.50)$$

This correlation function is shown in Fig. 6.5. In contradiction to a classical Boltzmann pair correlation function for a Coulomb potential, the quantum statistical correlation is finite at $r = 0$. These results suggest that the finite potential, $V_{ab}^{K}(r)$, obtained by Kelbg (1963b) replaces the Coulomb potential for quantum statistical thermal plasmas, that is, the Kelbg potential plays the rôle of an effective potential.

The idea of effective potentials was introduced in quantum chemistry by Hellmann (1935) and Gombás (1965) and in statistical physics by Morita (1959). Since 1963 Kelbg and co-workers at Rostock University developed this method to a powerful tool in plasma physics. The idea by Günter Kelbg was to replace the

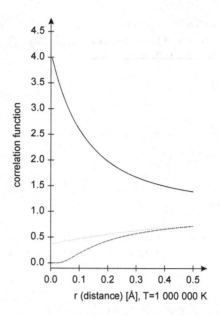

Fig. 6.5 The distribution function for (+−),(−−) and (++) pairs as a function of distance in Å for hydrogen plasmas with the parameters $T = 10^6$ K, $n = 0$

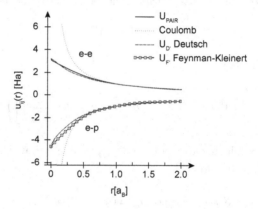

Fig. 6.6 Effective pair potential for electron-electron (e-e) and electron-proton (e-p) pairs at temperature $T = 10^6$ K in several approximations. Dotted lines: pure Coulomb interaction; solid lines: exact pair potential, U_{pair}; dashed lines: Kramers-Hellman-Deutsch potential, Eq. (6.57); open circles: U_F variational perturbative potential due to Feynman and Kleinert (Kleinert 1995). Distances are in units of $a_B = \hbar^2/m_e e^2$ and potentials are in units Ha $= e^2/a_B$

Coulomb potential by an effective potential which is finite at zero distance due to quantum effects (Kelbg 1963b) (see Fig. 6.6).

Originally, Kelbg's theory was based on quantum perturbation theory for the diagonal matrix elements (Slater sums) of the density matrix, where he found exact expressions for the first orders in e^2. His theory was then worked out by a group at Rostock University (including Ahlbehrendt, Ebeling, Hetzheim, Hoffmann, Kraeft, Kremp, Schmitz, Töwe, later joined by Röpke, Schlanges, Redmer, Bonitz, Blaschke, Reinholz and others). The Rostock School in Quantum Statistics formed by Günter Kelbg concentrated on analytical calculations of thermodynamic functions based on effective potentials (Ebeling 1967, 1968a,b; Ebeling et al. 1967, 1976; Hoffmann and Ebeling 1968a,b; Kelbg and Hoffmann 1964; Kraeft et al. 1986). Several authors calculated the two-particle density matrix from the known wave functions including numerical approaches (Barker 1969; Kalman et al. 1998; Rohde et al. 1968; Storer 1968a,b; Trubnikov and Elesin 1965). Zamalin, Norman and Filinov used effective potentials in the Monte Carlo simulations (Zamalin et al. 1977; Zelener et al. 1981). Deutsch (1977) introduced useful approximations and Hansen and McDonald (1981) and later Ortner (1999); Ortner et al. (1999, 2000) applied effective potentials in molecular dynamics simulations. Further Kleinert (1995) developed a variational approach.

We demonstrate now the calculation of analytical pair potentials from wave functions. Using our knowledge about the Slater sum representation for real gases, we write for a Coulomb system

$$S_{ab}(r) = (1 + \delta_{ab})! \pi^{3/2} \lambda_{ab}^3 \sum_{\alpha} e^{-\beta E_\alpha} |\psi(r, \vartheta, \phi)|^2 \,, \tag{6.51}$$

where the sum is performed over the entire spectrum. By using theorems for the spherical functions, we see that the result depends only on the radial eigenfunctions, $R_\alpha(r)$, and we obtain

$$S_{ab}(r) = (1 + \delta_{ab})! \, 2\sqrt{\pi} \lambda_{ab}^3 \sum_{\ell=0}^{\infty} (2\ell + 1) \left(1 \pm \delta_{ab} \frac{(-1)^\ell}{2s_a + 1} \right)$$

$$\times \left(\sum_{s=l+1}^{\infty} e^{-\beta E_{s\ell}} |R_{s\ell}(r)|^2 + \int_0^\infty dk \, e^{-\lambda_{ab}^2 k^2} \frac{d\delta_\ell(k)}{\pi dk} \right) |R_{k\ell}(r)|^2 \,. \tag{6.52}$$

Here, $\delta_\ell(k)$ denotes the phase shift of the wave function which is known exactly for Coulomb systems, as well as the discrete spectrum. This representation by integrals and sums is exact and can be used also for numerical calculations (Barker 1968, 1969; Ebeling et al. 1968; Rohde et al. 1968; Storer 1968a,b; Zamalin et al. 1977; Zelener et al. 1981). A simpler approach is based on Taylor expansions of the Slater sum (Rohde et al. 1968). Using the properties of the Coulomb wave function, we obtain for small r:

$$S_{ab}(r) = [4\sqrt{\pi} \xi_{ab} J_1 (\xi_{ab}) \, \Theta \, (\xi_{ab}) + \sqrt{\pi} \xi_{ab}^3 Z_3(\xi_{ab})$$

$$+ \delta_{ab} \frac{(-1)^{2s_a}}{2s_a + 1} 4\sqrt{\pi} \, \xi_{ab} J_1(\xi_{ab})] e^{-\xi_{ab} \left(\frac{r}{\lambda_{ab}} \right)} \,. \tag{6.53}$$

Defining the exact two particle effective potential by

$$u_{ij}^{\text{exact}}(r) = -k_B T \ln S_{ab}(r) \,, \tag{6.54}$$

we find by means of Eq. (6.53) the exact value of the effective potential in zeroth order,

$$u_{ij}^{\text{exact}}(0) = -k_B T \ln S_{ab}(r = 0) \,. \tag{6.55}$$

Remind that the effective interaction potential by Kelbg for two particles of species i, j has at zero distance the value

$$u_{ij}^{K}(0) = \frac{e_i e_j \sqrt{\pi}}{\lambda_{ij}} \,. \tag{6.56}$$

Since Kelbg's expression as well as the representations by wave functions are rather complicated, several useful approximations have been proposed. The simplest one has been elaborated by Deutsch (1977) and discussed in much detail by Baimbetov and his school (Arkhipov et al. 2000, 2011; Baimbetov et al. 1995; Ebeling 2016):

$$u_{ij} = \frac{e_i e_j}{r} \left(1 - e^{-\alpha_{ij} r}\right) \,; \qquad \alpha_{ij} \simeq \frac{1}{\lambda_{ij}} \,. \tag{6.57}$$

This potential was first used in quantum mechanics by Kramers, in quantum chemistry by Hellmann, in electrolyte theory by Glauberman and Juchnovskii and in plasma physics by Deutsch. The molecular-dynamical simulations by Hansen and McDonald (1981) were also based on this approximation. The Kramers-Hellmann-Deutsch potential has only one free parameter per pair and does not have sufficiently plasticity to fit the short and as well the long-range properties of the effective potential. The easiest way to fit the potential by Kelbg to the correct value at $r = 0$ is

$$\alpha_{ij} = \frac{\sqrt{\pi}}{\lambda_{ij}} \,. \tag{6.58}$$

Another simple approximation by Zelener et al. (1981) uses the Coulomb potential for opposite charges in the form

$$u_{+-} = \begin{cases} -\frac{e^2}{r} & \text{if } r > r_0 \\ \varepsilon & \text{if } r < r_0 \end{cases} \qquad \text{with} \quad r_0 = e^2/\varepsilon \,. \tag{6.59}$$

Extensive Monte Carlo simulations for dense plasmas were based on this simple effective potential (Trigger et al. 2003; Zamalin et al. 1977; Zelener et al. 1981).

For small distance, there are heuristic corrections (Ortner et al. 2000; Sadykova and Ebeling 2007). Kelbg's approximations (first order in e^2) is not exact for $r = 0$ but it clearly differs from the expression obtained above from the hydrogen wave functions. In contrast, the first derivative of Kelbg's potential at $r = 0$ and the asymptotic $r \to \infty$ are correct, that is, in agreement with quantum mechanics. This leads us to the idea to include the higher orders in e^2 by adapting a free parameter, γ_{ij}, formally introduced into Eq. (6.49)

$$u_{ij}(r) = \frac{e_i e_j}{r} \left\{ 1 - e^{-\frac{r^2}{\lambda_{ij}^2}} + \sqrt{\pi} \frac{r}{\gamma_{ij}\lambda_{ij}} \left[1 - \Phi\left(\gamma_{ij}\frac{r}{\lambda_{ab}}\right)\right]\right\}. \qquad (6.60)$$

With

$$\gamma_{ij} = \frac{u_{ij}^{K}(0)}{u_{ij}^{exact}(0)} \qquad (6.61)$$

and taking u_{ij}^{exact} from quantum mechanics as shown above, we can improve Kelbg's expression substantially (see Filinov et al. 2003a, 2004). This approximation will be called improved diagonal Kelbg potential (IDKP). Figure 6.7 shows the fit-parameter, γ, as a function of temperature. For $T \lesssim 10,000\,\text{K}$ the deviations from the original Kelbg potential become large. Klimontovich and Ebeling (1972) have shown, how effective potentials can be extended to non-equilibrium situations.

The method of effective potentials reduces the problem of quantum plasmas to a classical problem with temperature-dependent effective potentials. The relatively simple method of effective potentials found many applications in analytical and numerical calculations. The majority of applications are found in equilibrium statistics (Kraeft et al. 1986, 2015b).

The idea by Morita (1959) to express the diagonal density matrix in the form of Boltzmann factors with 2-particle, 3-particle etc. effective potentials was discussed in detail in Chap. 4. The advantage of this procedure is, that the calculation of the partition function can be reduced to a purely classical problem including, however,

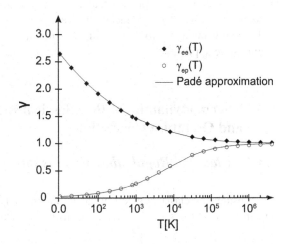

Fig. 6.7 Heuristic parameter, $\gamma(T)$, which improves the Kelbg expression by providing exact values of the potential and its derivative at $r = 0$, obtained from the exact solution of the two-particles problem

many-particle interactions. On the other hand a well-known disadvantage of this method is, that several many-particle effects, like exchange and bound state effects are not sufficiently described. Most successful applications concern the field of analytical thermodynamics. By means of cluster expansions, exact virial expansions for the thermodynamic functions of quantum plasmas could be found up to the order $n^{5/2}$ in density. This procedure will be in detail explained in the following sections. As the convergence of series is poor, the range of validity was extended by Padé methods (Kraeft et al. 1986). The available Padé formulae are widely used in applications to plasma and astrophysics (Fortov 2009, 2011).

There are several interesting applications of effective potentials within the path integral Monte Carlo (PIMC) method, which is a new promising field of computational physics (Filinov et al. 2001, 2003a,b, 2004; Trigger et al. 2003; Zamalin et al. 1977; Zelener et al. 1981). PIMC calculations are based on the high-temperature approximation for the density matrix and the Trotter-Feynman expansions of the exponential operator. Therefore, for applications we need reliable expressions for the high-temperature density matrices and, in particular, the off-diagonal elements. At this point, the formulae expressing the density matrix by effective potentials come into play. This procedure improves the convergence significantly and delivers reliable results. We mention also interesting applications of effective potentials to molecular dynamics simulations, in particular, the pioneering work by Hansen and McDonald (1981). More recent MD simulations have shown that the use of Kelbg-type effective potentials provides good thermodynamic functions, correlation functions, micro-field distributions, and, to some extent also non-equilibrium properties (Klimontovich and Ebeling 1972; Ortner 1999; Ortner et al. 2000).

By now, the method of effective potentials became a well established tool in the theory of dense quantum plasmas, and we extended the method to non-diagonal contributions. In combination with quantum Monte Carlo simulations this approach describes thermodynamic properties and spatial correlations in strongly coupled Coulomb systems rather well. For the description of dynamical processes like transport phenomena we need potentials with time-dependent parameters. Note that these potentials can be introduced into wave packet methods (Ebeling et al. 2017).

In the following, we focus on dilute (weakly non-ideal) plasmas where analytical methods are applicable.

6.3   Thermodynamics of the Classical Electron Gas and Quantum Corrections

6.3.1   Classical Bogolyubov Expansions

We consider the classical electron gas to understand screening in low Debye order and in higher orders. To this end, we investigate this simplest case and follow a procedure proposed by Bogolyubov which is an extension of the Debye-Hückel schema to higher density. In 1946, Bogolyubov (Bogolyubov 2005–2009) proposed to expand the correlation functions with respect to the so-called Bogolyubov plasma parameter, $\mu_B = e^2/k_B T r_D$. The thermodynamic functions follow from the pair correlation functions by integration. In first order, we obtain the Debye laws for the thermodynamics functions. To obtain better approximations, one has to go to higher orders in the plasma parameter and the density (Ichimaru 1992; Kalman et al. 1998; Klimontovich 1982; Pines and Nozieres 1966; Schmitz and Kremp 1967).

We consider here the one-component Coulomb gas on an neutralising background, the so-called OCP problem. In a first step we calculate the higher order distribution function for a classical OCP within the Bogolyubov method. Let us give, for example, some approximations for the pair distribution function of the OCP. In first order with respect to the plasma parameter we obtain

$$F_2(r) = 1 + g(r); \qquad g(r) = -\frac{e^2}{k_B T r} e^{-\kappa r}. \tag{6.62}$$

The first order correlation function, $g_1(r)$, is given by the Debye potential (for details of the calculation (see, e.g., Falkenhagen 1971; Falkenhagen and Ebeling 1971)). This approximation delivers only the limiting laws in thermodynamics. In order to find the free energy up to the orders n^2 and e^6, we need the second order solution which is found from the next approximation

$$F_2(1, 2) = 1 + g(1, 2) + \frac{g^2(1, 2)}{2} + n \int d\mathbf{r}_3 \left[g(1, 3)^2 g(2, 3) + g(1, 3)\, g(2, 3)^2 \right]. \tag{6.63}$$

After carrying out the Debye charging procedure, we find the result for the free energy which goes beyond the Debye expression:

$$F_{cl} = F_B - V k_B T \left\{ \frac{\kappa^3}{12\pi} + \frac{\pi}{3} n^2 \left(\beta e^2 \right)^3 \left[\ln \left(3\beta e^2 \kappa \right) + 2C - \frac{11}{6} \right] + \mathcal{O}\left(n^{5/2} \right) \right\}. \tag{6.64}$$

Introducing the Bogolyubov parameter, $\mu_B = \beta e^2 \kappa$, we obtain

$$F_{cl} = F_B - N k_B T \left(\frac{1}{3}\mu_B + \frac{1}{12}\mu_B^2 \left[\ln \mu_B + 2C + \ln 3 - \frac{11}{6} \right] + \mathcal{O}\left(\mu_B^3 \right) \right). \tag{6.65}$$

This result was obtained by several authors using different methods, e.g., by Abe, Meeron, Friedman, Yukhnovsky, Guernsey and others. The characteristic parameter of the expansion is the dimensionless parameter

$$\mu_B = \frac{e^2 \kappa}{k_B T}; \qquad \kappa^2 = 4\pi \beta n e^2. \tag{6.66}$$

To obtain the higher orders in μ_B requires considerable efforts. As far as we see, one of the most complete results was derived by Ortner (1999) using the method of collective variables. Note that in a first approach Wigner discussed already in 1934 a one-component plasma on a neutralising background. Wigner (1934) showed that the discrete charges will form a *bcc*-lattice and estimated the correlation energy as

$$\varepsilon_c = -a_{bcc}\,\Gamma\,. \tag{6.67}$$

Later studies suggested a more complicated expression (DeWitt 1976; Ortner 1999; Ortner et al. 1999; Stolzmann and Ebeling 1998). The fit formula by DeWitt for large Γ reads

$$u = u_c = -0.89461\,\Gamma + 0.9165\,\Gamma^{1/4} - 0.5012\,. \tag{6.68}$$

These analytical and numerical studies suggest that asymptotically for large Γ, the Coulomb energy per particle at finite temperature should be linear in Γ. A fractional law of this type is typical for a lattice-like structure (Ebeling 2016; Wigner 1938). Summarizing these results we found a rather complete description of the thermodynamic functions for the region were classical approximations work.

6.3.2   Quantum Corrections

To proceed to the quantum region, we use the following trick: For the electron gas we write the identity

$$F = F_{qu} = F_{cl} + \left(F_{qu} - F_{cl}\right)\,. \tag{6.69}$$

The second contribution does not cause any problems with screening and can be expressed up to second order in the density by using the real gas expression (5.28):

$$F_{qu} - F_{cl} = -\frac{1}{2}k_B V n^2 \int d\mathbf{r}_1 d\mathbf{r}_2 \left(S^{(2)}(\mathbf{r}) - S_{cl}^{(2)}\right)\,; \quad \mathbf{r} \equiv \mathbf{r}_1 - \mathbf{r}_2\,. \tag{6.70}$$

We shall evaluate this rather complicated integral. First, we write

$$F_{qu} - F_{cl} = -k_B V n^2 \,\delta B_2(T) \tag{6.71}$$

with

$$\delta B_2(T) = 2\pi \int_0^\infty dr\,r^2 \left(S_2(r) - e^{-\frac{\beta e^2}{r}}\right)\,. \tag{6.72}$$

The most convenient way to compute the integral is the expansions of the Slater sum with respect to Planck's constant. The density matrix for interactions, $V(r)$, can be represented (Landau and Lifshitz 1980) as a series in $\hbar^2$ (Kelbg and Hoffmann 1964):

$$S_2(r) = e^{-\beta V} e^{-\beta V} \left[1 + \frac{\beta \lambda^2}{6} \left(-\Delta V + \frac{\beta}{2} (\nabla V)^2 \right) + \mathcal{O}\left(\hbar^4 \right) \right], \qquad (6.73)$$

where we denote as usual $\lambda^2 = \hbar^2/(mk_B T)$. The integration can be performed easily to obtain a series in $\hbar^2$ (Hoffmann and Ebeling 1968a,b; Kelbg 1964):

$$\delta B_2(T) = -\frac{\pi}{6} \left(\beta e^2 \right)^3 \left(\frac{\lambda^2}{(\beta e^2)^2} + \frac{1}{5} \frac{\lambda^4}{(\beta e^2)^4} + \frac{4}{21} \frac{\lambda^6}{(\beta e^2)^6} + \cdots \right). \qquad (6.74)$$

This expansion works well for low temperature where the parameter $\lambda^2/(\beta e^2)^2$ is small. Evidently, this is an asymptotic series. In order to find proper Taylor expansions, we have to apply completely different methods (Ebeling 1968a; Montroll and Ward 1958) which are based on expansions with respect to the weakness of the charges e^2. For physical reasons which will become clear later, we use the parameter

$$\xi = \xi_{ee} = -\frac{\beta e^2}{\lambda}. \qquad (6.75)$$

We will show now that using expansions in ξ, the integrals in the second virial coefficient, Eq. (6.72), can be exactly evaluated without making use of any approximations (Ebeling 1968a). We will come back to this problem in later sections where the second virial coefficient for the electron gas will be obtained as a special case.

6.4 Screening and Thermodynamic Functions of Non-degenerate Plasmas

6.4.1 Debye-Hückel Screening

In linear Kelbg approximation with respect to the charge parameter, e^2, the binary density matrix is given by

$$\rho_{ab}(r) = 1 - \beta V_{ab}^K(r). \qquad (6.76)$$

We introduce a correlation function by

$$g_{ab}(r) = -\beta V_{ab}^K(r) \qquad (6.77)$$

and rewrite the binary density matrix in the form

$$\rho_{ab}(r) = 1 + g_{ab}(r). \tag{6.78}$$

A characteristic feature of Coulomb systems is the screening effect treated first successfully by Debye. The idea of screening is that in addition to the direct interaction we should include the interactions with the field of particles surrounding the pair, at least in a mean field approximation. According to Bogolyubov, for the Debye potential we find in mean field approximation

$$V_{ab}^{D}(\mathbf{r}) = V_{ab}^{K}(\mathbf{r}) + \sum_{c} \int d\mathbf{r}' \, n_{c} \, V_{bc}^{K} \left(\mathbf{r} - \mathbf{r}'\right) g_{ac}\left(\mathbf{r}'\right). \tag{6.79}$$

The Debye potential consists of a direct contribution and a mean field correction. Let us first search for a solution in the Coulomb approximation, $V^{K} \rightarrow V$. We switch to a Fourier representation of the Debye correlation function,

$$g_{ab}(q) = -\frac{4\pi \, \beta e_{a} \, e_{b}}{q^{2} + \kappa^{2}}, \tag{6.80}$$

where

$$\kappa^{2} = 4\pi \beta \sum_{a} n_{a} e_{a}^{2}. \tag{6.81}$$

After transformation back to coordinate space we obtain the Debye distribution

$$g_{ab}(r) = -\frac{\beta e_{a} e_{b} e^{-\kappa r}}{r} = -\beta \, V_{ab}^{D}. \tag{6.82}$$

This procedure corresponds to replacing the Coulomb potential by the screened Debye potential which is short-ranged with the range $r_{D} = 1/\kappa$. The solution of Eq. (6.79) in Fourier space follows by some matrix operations. The procedure is known, however, in detail technically complicated (Kelbg 1963b, 1964, 1972; Sadykova and Ebeling 2007). In the most simple approximation, we replace the Coulomb potential by a Debye potential without changing the remaining terms. In this approximation, the correlation function reads

$$g_{ab}(r) = -\beta \, V_{ab}^{D} \left\{ 1 - \exp\left(-\frac{r^{2}}{\lambda_{ab}^{2}}\right) + \sqrt{\pi} \frac{r}{\lambda_{ab}} \left[1 - \Phi\left(\frac{r}{\lambda_{ab}}\right)\right] \right\}. \tag{6.83}$$

For the calculation of the free energy we apply the Debye charging procedure which leads to the formula (see Chap. 4)

$$F = F_{id} - \frac{1}{2}k_B T \sum_{ab} n_a n_b \int_0^1 d\lambda V_{ab} g_{ab}(r; \lambda) . \tag{6.84}$$

Introducing here the Debye approximation of the correlation function we obtain the following expression for the free energy:

$$F = F_{id} + \frac{V}{2} \sum_{ab} n_a n_b \int_0^1 d\lambda \, V_{ab} V_{ab}^D(r; \lambda)$$

$$\times \left\{ 1 - \left[\exp\left(-\frac{r^2}{\lambda_{ab}^2} \right) + \sqrt{\pi} \frac{r}{\lambda_{ab}} \left(1 - \Phi\left(\frac{r}{\lambda_{ab}} \right) \right) \right] \right\} . \tag{6.85}$$

6.4.2   Ring Functions

After carrying out the technically difficult integration in Eq. (6.85), we find

$$F = F_{id} - V \frac{4\pi}{3k_B T \kappa} \sum_{ab} n_a n_b e_a e_b R (\kappa \lambda_{ab}) , \tag{6.86}$$

where

$$\lambda_{ab}^2 = \frac{\hbar^2}{2m_{ab}k_B T} \tag{6.87}$$

and the first terms of the Taylor series read

$$R(x) = 1 - \frac{3\sqrt{\pi}}{16}x + \frac{1}{4}x^2 + \dots \tag{6.88}$$

The full function can be expressed by $_1F_1$ functions which are the standard degenerated hypergeometric functions (Kummer functions)

$$R(x) =_1F_1 \left(1, \frac{5}{2}, \frac{x^2}{4} \right) + \frac{3\sqrt{\pi}}{x} \left(1 - e^{\frac{x^2}{4}} \right) \tag{6.89}$$

with the definition (Kelbg 1963b)

$$_1F_1 (\alpha, \gamma; z) = 1 + \frac{\alpha}{\gamma} \frac{z}{1!} + \frac{\alpha(\alpha+1)}{\gamma(\gamma+1)} \frac{z^2}{2!} + \frac{\alpha(\alpha+1)(\alpha+2)}{\gamma(\gamma+1)(\gamma+2)} \frac{z^3}{3!} + \dots \tag{6.90}$$

For many purposes, we need the chemical potential of a species a given by

$$\mu_a = \frac{\partial F}{\partial N_a} \, . \tag{6.91}$$

From Eq. (6.86) we find the interaction part of the chemical potential,

$$\mu_a^{\text{int}} = \frac{2\pi\beta^2 e_a^2}{\kappa} \sum_b e_b^2 \, G(\kappa\lambda_{ab}) \, , \tag{6.92}$$

where

$$G(x) = \frac{\sqrt{\pi}}{x} \left\{ 1 - e^{\frac{x^2}{4}} \left[1 - \Phi\left(\frac{x}{2}\right) \right] \right\} = 1 - \frac{\sqrt{\pi}}{4}x + \cdots \simeq \frac{1}{1 + \frac{\sqrt{\pi}}{4}x} \, . \tag{6.93}$$

This way we found in a first approximation a complete description of the thermodynamic functions of non-degenerate plasmas in the region where bound states are not relevant. Sometimes the approximation

$$G(x) \simeq \frac{1}{1 + \frac{\sqrt{\pi}}{4}x} \tag{6.94}$$

is used, which is correct up to the linear term and is analytically quite similar to the original Debye-Hückel functions (Falkenhagen 1971; Falkenhagen and Ebeling 1971). This most simple approximation is called quantum Debye-Hückel approximation (QDHA) (see Ebeling et al. 1976; Kraeft et al. 1986). We denote the functions $R(x)$ and $G(x)$ as ring functions, since they are generated from ring type Mayer or Feynman diagrams. The first approach to calculate ring functions for quantum plasmas was performed by DeWitt (1962) and DeWitt et al. (1995). There are many further approaches to improve the ring functions (DeWitt et al. 1995; Kraeft et al. 1986). All known ring functions agree in the linear term in x and disagree in higher approximations or even in the asymptotic behavior. For simple estimates the use of the QDHA is recommended.

6.5　The Pair Bound State Contributions

6.5.1　Mean-Mass Approximation—Symmetrical Plasmas

The problem of bound states in dense plasmas belongs to the difficult problems of theoretical physics. Beginning with the classical works of Planck and Brillouin and later by Bethe and Salpeter, this is still a crucial topic which gave rise to many theoretical and experimental investigations. The state of understanding the problem of bound states around the end of last century was reflected in several books (Ebeling et al. 1976; Kraeft et al. 1986; Kremp et al. 2005) and also in several review papers.

The problem of bound states in plasmas is so complicated that it is justified to investigate first a special case in which the problem appears simplified. Here, we have in mind the completely symmetrical plasma with $m_+ = m_- = m$. Strictly speaking, such plasmas do not exist in nature except in rather exotic cases such as electron-positron plasmas. In general, real plasmas are not mass-symmetrical, we will show however, that the difference of the mass plays a minor rôle and gives only small corrections. In order to avoid misunderstandings, here we do not mean the rôle of masses in the ideal terms, but only about the rôle of different masses of particles for interaction effects, that is, for collisions. In many situations we may assume that the small effects due to the difference of masses in interaction effects may be neglected and we call this the mean mass approximation (MMA).

We start this study with the simple observation that bound states are a physical effect related to high orders in e^2 and remind that already Rutherford's theory provides an order $\mathcal{O}\left(e^4\right)$ for the ground state energy. Further we have in mind the Planck-Brillouin theory. On the other hand we have seen in the previous section that screening effects are concerned with the lower orders in e^2, e.g. the Debye terms are of order $\mathcal{O}\left((e^2)^{3/2}\right)$. This leads us to the hypothesis that for the higher orders the cluster expansion for gases given in Chap. 5 is an appropriate instrument while for lower orders in e^2 we need a Debye-like screening theory as explained above. Following this argument, we combine the cluster expansion from Chap. 5 with Eq. (6.86) and obtain

$$
F = F_{\mathrm{id}} - V k_{\mathrm{B}} T \left(\frac{4\pi}{3\kappa} \sum_{ab} n_a n_b e_a^2 e_b^2 R\left(\kappa \lambda_{ab}\right) \right.
$$

$$
\left. + \frac{1}{2} \sum_{ab} n_a n_b \int d\mathbf{r} \left[S_{ab} - 1 - S_{ab}^{(1)} - S_{ab}^{(2)} - S_{ab}^{(3)} \right] + \ldots \right). \tag{6.95}
$$

In difference to real gases, the formula for the second virial coefficient contains additional terms, $S_{ab}^1, S_{ab}^2, S_{ab}^3$, in the argument of the integral. These are the lowest orders in S_{ab} which were subtracted. What is the physical meaning of these operations? The first order and the third order, $S_{ab}^{(1)}$, $S_{ab}^{(3)}$ yield odd contributions in $e_a e_b$ which sum up to zero for symmetrical plasmas. The second order, $S_{ab}^{(2)}$, was omitted since these terms are already contained in the ring contribution and have to be omitted to avoid double counting. We remind that the Debye-Kelbg terms contain already the contributions up to the orders e^4, and we have to avoid double counting of these terms. Consider now the special case of a two-component plasma with equal relative masses:

$$
n_+ = n; \quad n_- = n; \quad m_\pm = \mu; \quad \lambda_{++} = \lambda_{--} = \lambda_{+-} = \lambda. \tag{6.96}
$$

Then the sum in the ring term in Eq. (6.95) contributes in lowest order just κ^3 and the expression for the free energy density reads

$$f = f_{\text{id}} - k_{\text{B}}T \frac{\kappa^3}{12\pi} R(\kappa\lambda)$$

$$- \frac{1}{2}n^2 \int d\mathbf{r} \left(2S_{+-} + S_{++} + S_{--} - 4 - 2S_{+-}^{(2)} - S_{++}^{(2)} - S_{--}^{(2)} \right) + \dots \tag{6.97}$$

where

$$R(\kappa\lambda) = 1 - \frac{3\sqrt{\pi}}{16}\kappa\lambda + \frac{1}{4}\kappa^2\lambda^2 - \dots \tag{6.98}$$

Remember that the terms linear in $e_a e_b$ and the terms $e_a^3 e_b^3$ cancel due to the antisymmetry in the charges and the symmetry in the masses. Note that a similar expression was obtained by Planck and Brillouin. Consequently, our next step is also inspired by the ideas by Planck (1924) and Brillouin (1931a,b). We separate the contributions of the lower bound states which are contained entirely in S_{+-} from the rest of the integral. We define the bound state contribution as

$$S_{ab}^b(r) = 2\sqrt{\pi}\lambda_{ab}^3 \left(\sum_{\ell=0}^{\infty}(2\ell+1) \sum_{s=1}^{\ell-1} \left(e^{-\beta E_s} - 1 + \beta E_s \right) |R_{s\ell}(r)|^2 \right), \tag{6.99}$$

where the energy levels are as usual

$$E_s = -\frac{me^4}{2\hbar s^2}. \tag{6.100}$$

The meaning of the pre-factor in front of the wave functions, $e^{-\beta E_s} - 1 + \beta E_s$, is that all levels with $|\beta E_s| < 1$ are suppressed in the sum. In the early Planck paper we find still a discrete limit for the counting at the level with $|E_s| \simeq k_{\text{B}}T$, that is,

$$s^2 \simeq \frac{me^4}{2\hbar k_{\text{B}}T}. \tag{6.101}$$

Later, Brillouin introduced, however, the equivalent but mathematically more elegant smooth weight factor which we used here. In a next step we define the free part of the second virial coefficient, B^f, as the remaining part and obtain, finally, from Eq. (6.97)

$$f = f_{\text{id}} - k_{\text{B}}T \frac{\kappa^3}{12\pi} R(\kappa\lambda) - 4n^2\pi\sqrt{\pi}\lambda^3 \sum_{s=1}^{\infty} s^2 \left(e^{-\beta E_s} - 1 + \beta E_s \right)$$

$$- n^2 B^f(T, e^2, \hbar) + \dots \tag{6.102}$$

Here, $B^f(T, e^2, \hbar)$ denotes the contribution of the free states. Max Planck has shown that this contribution is close to zero in the classical limit, $\hbar \to 0$, and small otherwise. In Sect. 6.1, we have shown analytically that the relation $B_f(T, e^2, \hbar = 0) = 0$ is exact in the classical limit, following Ebeling (1967, 1968a). Further we can check numerically that $B_f(T, e^2, \hbar \neq 0) \simeq 0$ is true with high precision. In spite of the fact that by now we do not have a mathematical proof, we will assume $Bf(T, e^2, \hbar) = 0$ for all mass-symmetrical plasmas. The analytical proof will follow in one of the next sections. The proof is based on the cancellation of all odd power contributions in e^2 which is exact for mass-symmetrical plasmas.

Summarizing our result, we arrive at the following formulae for the free energy density and the pressure of low density mass-symmetrical plasmas (otherwise in MMA):

$$f = f_{id} - k_B T \frac{\kappa^3}{12\pi} R(\kappa\lambda) - 4n^2\pi^{3/2}\lambda^3 \sum_{s=1}^{\infty} s^2 \left(e^{-\beta E_s} - 1 + \beta E_s\right) + \mathcal{O}\left(n^{5/2}\right)$$

(6.103)

$$\beta p = \beta p_{id} - \frac{\kappa^3}{24\pi}\left[R(\kappa\lambda) + x R'(\kappa\lambda)\right]$$
$$- 4n^2\pi\sqrt{\pi}\lambda^3 \sum_{s=1}^{\infty} s^2 \left(e^{-\beta E_s} - 1 + \beta E_s\right) + \mathcal{O}\left(n^{5/2}\right).$$

(6.104)

This result is exact up to the order $\mathcal{O}\left(n^2\right)$. Without providing a full proof here, we state further that these results are excellent approximations also for unsymmetrical plasmas with masses of the positive charges larger than the electron mass, $m_+ > m_e$. The argument for this claim is quite simple—from the quantum mechanics of Coulomb systems, in particular from Bohr's theory, we know that most properties depend essentially only on the electron masses and the heavy masses give only small corrections through the replacement of electron masses by reduced masses. We will provide all necessary proofs of these statements in the next section. The price to pay are quite heavy calculations—those readers who believe in our arguments can jump immediately to the overnext section. Consequently, Eqs. (6.103) and (6.104) can be used as very good approximations for any binary plasma using $\lambda = \lambda_{+-}$ for temperature below the ionization temperature. The reason of the restriction with respect to temperature is that at very high temperature the differences of the masses plays a more important rôle in spite of the fact that then the corrections due to non-ideality remain always small.

We started from the observation that the simple partition function of Coulombic atoms is divergent. Now we see, that in a quantum statistical derivation of thermodynamic quantities, there is no divergence anymore. Instead of the simple Bohr-Herzfeld partition function we find the renormalized Planck-Brillouin-Larkin partition function

$$\sigma_{BH} \to \sigma_{BPL}$$

(6.105)

with

$$\sigma_{\text{BH}} = \sum_{s=1}^{s_{\max}} s^2 e^{-\beta E_s} \; ; \qquad \sigma_{\text{BPL}} = \sum_{s=1}^{\infty} s^2 \left(e^{-\beta E_s} - 1 + \beta E_s \right) . \qquad (6.106)$$

The maximal main quantum number, $s_{\max}$, is proportional to temperature.

The results of our quantum statistical derivation, Eq. (6.104), contain only the PBL-partition function. We arrived at the PBL procedure in a natural way, so it seems to be at least in the limit of dilute plasmas the most natural way to truncate the partition function.

Summarizing our results, we come to the important conclusion that the free energy of symmetrical plasmas, Eq. (6.103), contains the BPL-partition function and is exact up to the density order $\mathcal{O}\left(n^2\right)$ and that plasmas with larger masses of the positive charges do not lead to significantly different results. As we see, the PBL-partition function results in a quite natural way. Thus, we cannot confirm recent work by Starostin et al. (2003) and Starostin and Roerich (2006) which questions the equation of state obtained with this method. We note that at higher density, instead of the bare levels, E_s, the density and temperature dependent levels, $E_{s\ell}(n, T)$ have to be used.

6.5.2   The Second Virial Coefficient for General Plasmas

In order to evaluate the contributions of pair correlations in all orders of the interaction parameters of pairs, $e_i e_j$, we define the dimensionless interaction parameter for the general case,

$$\xi_{ij} = -\frac{e_i e_j}{k_B T \lambda_{ij}} . \qquad (6.107)$$

Note that this parameter is proportional to the root of the reduced masses and inversely proportional to the root of temperature. Bound states exist only if at least some of the interaction parameters are positive. According to Chaps. 2 and 5, the free energy is given by

$$F\left(T, V, N, \{\xi_{ij}\}\right) = F_{\text{id}}(T, V, N) - k_B T \ln Q_{N,int}\left(T, V, N, \{\xi_{ij}\}\right) \qquad (6.108)$$

$$F_{\text{id}}(T, V, N) = N k_B T \left(\ln\left[\frac{n \Lambda^3}{2s + 1}\right] - 1 \right) . \qquad (6.109)$$

Evidently this is an analytical function of the parameters $e_i\, e_j$ or—more correctly—of the dimensionless parameter defined by Eq. (6.107). We see that these interaction parameters are positive for $+-$ interactions and negative for $++$ and $--$ interactions. Our aim is to represent the thermodynamic potentials by analytical functions of the interaction parameters.

To calculate the interaction contributions, we start from the cluster expansion for real quantum gases, for the moment including only the second virial coefficient. We split the free energy into the contribution of a Boltzmann gas and an excess part, containing all correlations including the correlations due to the symmetry properties of the particles:

$$F(T, V, N) = F_{id} + F_{ex} = Nk_B T \left[\ln \left(n \Lambda^3 \right) - 1 \right]$$

$$- k_B T V \left[\frac{1}{2} \sum_{a,b} n_a n_b \int d\mathbf{r}_1 d\mathbf{r}_2 \left(S_{ab}^{(2)} - 1 \right) \right] + \dots , \qquad (6.110)$$

where the first term is the ideal free energy in Boltzmann approximation. This expansion contains all terms, however, all individual integrals are divergent due to the Coulomb singularity at large distances. If we succeed to replace all Coulomb potentials by Debye potentials in a correct way, no divergencies will appear anymore. Usually, this is achieved by a summation of an infinite number of terms. We will use here an equivalent but much simpler procedure and start from the Bogolyubov screening equation in the form

$$\beta V_{ab}(r) = -g_{ab}(r) - \beta \sum_c \int d\mathbf{r}' \, n_c \, V_{bc} \left(\mathbf{r} - \mathbf{r}' \right) g_{ac} \left(r' \right) , \qquad (6.111)$$

where now Coulomb and Debye potentials appear:

$$V_{ab} = \frac{e_a e_b}{r} ; \qquad g_{ab} = -\beta \frac{e_a e_b}{r} e^{-\kappa r} . \qquad (6.112)$$

By iteration we formally obtain an infinite series of convergent integrals

$$\beta V_{ab} = -g_{ab}(r) + \sum_c n_c \int d\mathbf{r}_3 g_{ac} \, g_{bc} - \sum_{cd} n_c n_d \int d\mathbf{r}_3 d\mathbf{r}_4 \, g_{ac} \, g_{bc} \, g_{cd} + \dots \qquad (6.113)$$

with $r \equiv |\mathbf{r}_1 - \mathbf{r}_2|$. Introducing here the Debye approximation of the correlation function, we obtain the free energy

$$F_{ex} = -\frac{1}{2} k_B T \sum_{a,b} n_a n_b \int d\mathbf{r}_1 d\mathbf{r}_2 \int_0^1 d\lambda V_{ab} g_{ab}(r; \lambda)$$

$$- k_B T V \frac{1}{2} \sum_{a,b} n_a n_b \int d\mathbf{r}_1 d\mathbf{r}_2 \left(S_{ab}^{(2)} - 1 - V_{ab} \int_0^1 d\lambda g_{ab}(r; \lambda) \right)$$

$$+ \mathcal{O} \left(n_a n_b n_c \right) . \qquad (6.114)$$

On the basis of this expression we represent the contribution of the second virial coefficients to the free energy by

$$F_{ex}^2 = F_{ring} + F_{direct}^2 + F_{exch}^2 \, . \tag{6.115}$$

Here the ring term is defined as the first term above, the direct contribution comes from the direct term in the binary Slater sum and the exchange contribution comes from the exchange part in the binary Slater sum, $S_{ab}^{(2)}$. More details will be provided in the next section.

6.6 Evaluation of the Second Virial Coefficients

6.6.1 The Exchange Contributions to the Virial Functions

One of the most important interaction contributions to the free energy of plasmas which we considered so far is the Hartree-Fock contribution (see Chaps. 3 and 5). The exchange effects play a special rôle, since they are essentially interactions of short-range nature. The exchange contribution to the thermodynamic functions is defined above. At least for small density, divergence problems due to screening effects are not relevant, thus, the free energy can be represented in the same way as for a quantum gas with short-range forces (see Chap. 5). Therefore, we write

$$F_{exch}^2 = -k_B T V \left[\frac{1}{2V} \sum_{ab} n_a n_b \int S_{ab}^{exch} (\mathbf{r}_1, \mathbf{r}_2) \, d\mathbf{r}_1 \, d\mathbf{r}_2 + \mathcal{O}\left(n^3\right) \right] . \tag{6.116}$$

Here S_{ab} is the Slater function, which is the diagonal part of the two-particle density operator

$$S_{ab} = \text{const.} \left\langle \mathbf{r}_1, \mathbf{r}_2 \left| e^{-\beta H_{ab}} \right| \mathbf{r}_2, \mathbf{r}_1 \right\rangle . \tag{6.117}$$

By using the known Coulomb wave functions having the radial parts $R(r)$, for fermions we obtain

$$S_{ab}^{exch} = -\sum_{\ell=0}^{\infty} \delta_{ab} \frac{(-1)^\ell (2\ell + 1)}{2s_a + 1} \int_0^\infty \frac{dn_k}{dk} e^{-\lambda_{ab}k^2} |R_{k\ell}(r)|^2 \, dk \, . \tag{6.118}$$

We use the method by Beth and Uhlenbeck (1937) and Uhlenbeck and Beth (1936) and exploit the fact that the density of states, dn_k/dk, is exactly known for Coulomb systems (Ebeling 1968a, 1969). By using the normalization of the radial wave functions, $R_{k\ell}$, and the known expressions for the Coulomb phase shifts by the functions $\psi(x) = (d/dx)\ln\Gamma(x)$, we obtain for the exchange free energy of electrons

$$F_{\text{exch}}^2 = -k_B T V \left\{ n_e^2 \, 2\sqrt{\pi} \, \text{Re}\left[\int_0^\infty dx \, \frac{e^{-x^2}}{x} \, f_{\text{ex}}\left(\frac{\xi_{\text{ee}}}{x}\right) \right] + \mathcal{O}\left(n_e^3\right) \right\} \tag{6.119}$$

with

$$f_{\text{ex}}(y) = \frac{1}{2}y \sum_{\ell=0}^{\infty} (-1)^\ell \, (2\ell+1) \, \psi\left(\ell+1-\frac{i}{2}y\right). \tag{6.120}$$

We introduce the complex variable $z = x^2$ and transform to an integral in the complex plane with a contour C circumventing the positive axis including the zero point in positive direction. The integral reads then (Ebeling 1968a; Kremp and Kraeft 1972)

$$F_{\text{exch}}^2 = -k_B T V n_e^2 \sqrt{\pi} \, \text{Re}\left[\int_C dz \frac{e^{-z}}{\sqrt{z}} \, f_{\text{ex}}\left(\frac{\xi_{\text{ee}}}{\sqrt{z}}\right) \right]. \tag{6.121}$$

We expand the function $\psi(x) = \frac{d}{dx}\ln\Gamma(x)$ into a Taylor series with respect to the interaction parameter, ξ, and integrate the convergent series term by term making use of

$$\frac{1}{\Gamma(s)} = \frac{1}{2\pi i}\int_C dz(-z)^{-s}\, e^{-z}. \tag{6.122}$$

This procedure delivers the exchange contributions of electrons to the free energy,

$$F_{\text{exch}}^2 = +k_B \, T \, V\pi \, n_e^2 \, \lambda_{\text{ee}}^3 \, E\,(\xi_{\text{ee}}) \,, \tag{6.123}$$

where $E(x)$ can be represented either by a complex integral as introduced above or by a converging infinite series:

$$E(x) = \sum_0^\infty \frac{\sqrt{\pi}\left(1 - 2^{2-n}\right)\zeta(n-1)\,x^n}{2^n\,\Gamma\left(1+\frac{n}{2}\right)}. \tag{6.124}$$

The coefficients in this series are determined by the Coulomb scattering phase shifts (Ebeling 1968a; Kremp and Kraeft 1972). Inserting the explicit values of the zeta and gamma functions, the first two terms in the exchange function series yield

$$E(x) = \frac{\sqrt{\pi}}{4} + \frac{x}{2} + \dots \tag{6.125}$$

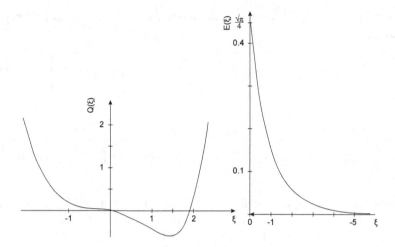

Fig. 6.8 The characteristic functions, $Q(\xi)$ and $E(\xi)$, determining the second virial coefficient of quantum plasma. The exchange function, $E(\xi)$, is shown for negative arguments only (right-hand side), since this contribution is relevant only for interacting equal charges. The Q function includes here a linear term

These terms are already known to us, since they are contained in the Fermi-Dirac and Hartree-Fock contributions. The higher order contributions are described by means of the transcendent function, $E(x)$, first derived by Ebeling (1968a) and given by infinite series representing ladder diagrams. This result was confirmed in several papers (Alastuey and Perez 1992, 1996; Brown and Yaffe 2001; Kraeft and Kremp 1968; Kraeft et al. 1986). The fact that the already known linear term obtained by Hartree-Fock calculations fits the general law of the series, can be considered as a confirmation of our derivation. The shape of the function $E(x)$ for negative arguments is shown on the right-hand side of Fig. 6.8. Note that the exchange contribution disappears rapidly (approximately exponentially) with increasing argument.

6.6.2   Direct Contributions to the Virial Functions

The calculation of the direct contribution to the second virial coefficient is a more difficult problem because of a divergency in the order e^6. Therefore, we split F^2_{direct} into the order e^6 and the remaining contributions of the orders e^{2k} with $k > 3$. These higher orders are convergent since $(e^2/r)^k$ has convergent integrals for $k > 3$. For symmetrical electro-neutral systems, the term $k = 3$ does not contribute, therefore, symmetrical systems are much easier to evaluate. Let us start with the terms of order

e^{2k} with $k > 3$. By using the Coulomb wave functions and the Beth-Uhlenbeck technique, similar as before, we have

$$S_{ab}^{\text{direct}}\left(r, e^{2k}, k > 3\right) = \sum_{\ell=0}^{\infty} \int_0^{\infty} \frac{dn_k}{dk} e^{-\lambda_{ab} k^2} |R_{k\ell}(r)|^2 \, dk. \tag{6.126}$$

The density of states, dn_k/dk, which we used already previously, is exactly known for Coulomb systems. The main difference to the previous case is that the delta-functions and the spin factors are now absent. By using the normalization of the redial wave functions, $R_{k\ell}$, and the known expressions for the Coulomb phase shifts by $\psi-$ functions, we obtain for the direct contribution of a-b pairs (Ebeling 1969)

$$F_{\text{direct}}^2\left(e^{2k}, k > 3\right) = -k_{\text{B}} \, T \, V \sum_{ab} \left\{ n_a n_b \, 2\sqrt{\pi} \, \text{Re}\left[\int_0^{\infty} dx \, \frac{e^{-x^2}}{x} \, f_{\text{dir}}\left(\frac{\xi_{ab}}{x}\right) \right] \right\} \tag{6.127}$$

with

$$f_{\text{dir}}(y) = \frac{1}{2} y \sum_{\ell=0}^{\infty} (-1)^{\ell} (2\ell + 1) \, \psi\left(\ell + 1 - \frac{i}{2} y\right) \left(e^{2k}, k > 3\right). \tag{6.128}$$

We introduce the complex variable $z = x^2$. Transforming to an integral in the complex plane with a contour, C, circumventing the positive axis including the zero point in positive direction we obtain the integral

$$F_{\text{direct}}^2\left(e^{2k}, k > 3\right) = -k_{\text{B}}T \, V \sum_{ab} \left\{ n_{\text{e}}^2 \sqrt{\pi} \, \text{Re}\left[\int_C dz \, \frac{e^{-z}}{\sqrt{z}} \, f_{\text{dir}}\left(\frac{\xi_{ab}}{\sqrt{z}}\right) \right] \right\}. \tag{6.129}$$

We expand the function $\psi(x) = \frac{d}{dx} \ln \Gamma(x)$ into a Taylor series with respect to the interaction parameter, ξ, and integrate the convergent series term by term making use of the definition of the function $1/\Gamma(s)$. This procedure provides us the following representation of the direct contributions of Coulomb interactions to the free energy in the orders e^{2k}, $k > 3$:

$$F_{\text{direct}}^2\left(e^{2k}, k > 3\right) = k_{\text{B}}T \, V\pi \sum_{ab} n_a n_b \, \lambda_{ab}^3 \, Q_4\left(\xi_{ab}\right), \tag{6.130}$$

where $Q_4(x)$ can be represented either by a complex integral as introduced above or by a converging infinite series:

$$Q_4(x) = \sum_{n=4}^{\infty} \frac{\sqrt{\pi} \, \zeta(n-2) \, x^n}{2^n \, \Gamma\left(1 + \frac{n}{2}\right)}. \tag{6.131}$$

What is still missing in our calculation are the orders $\mathcal{O}\left(e^2\right)$ and $\mathcal{O}\left(e^6\right)$. The order $\mathcal{O}\left(e^2\right)$ is not a serious problem since this term contributes zero to the free energy in electro-neutral systems. Nevertheless it has been added in several of the earliest works which contained the full series (Alastuey and Perez 1992, 1996; Brown and Yaffe 2001; Ebeling 1968a, 1969) in order to have a nice analytical representation. Taking the term from the Hartree calculations in Chap. 5, we find

$$F_{\text{direct}}^2(e^2) = -k_B T V \pi \sum_{ab} n_a n_b \lambda_{ab}^3 \frac{\xi_{ab}}{6}. \tag{6.132}$$

Introducing here the definitions of ξ and the reduced masses, we obtain

$$F_{\text{direct}}^2\left(e^2\right) = -V\pi \sum_{ab} n_a n_b \lambda_{ab}^2 e_a e_b = 0. \tag{6.133}$$

Thus, the direct linear term does not contribute to the free energy. Nevertheless, we decided to keep the term as a contribution to the full function since it provides nice analytical properties of the complete function, $Q(\xi)$, and keeps the expressions in full analogy with the structure of virial functions known from electrolyte theory. Of course, we are always free to add zero-valued terms to a function, if this appears useful. However recently, Kraeft et al. (2015a) developed a controversy: "The introduction of the linear term (they mean here $\xi/6$ in Eq. (6.132)) leads to unjustified results for Coulomb systems ..." and referred to papers by Alastuey and Perez and earlier papers of one of the present authors (W.E.). In a comment to this paper (Alastuey et al. 2015), it was shown that a careful comparison of thermodynamic functions leads to perfect agreement of all known results for the thermodynamics functions obtained by three groups of authors, up to the known orders $n^{5/2}$ in the density. Consequently, if there is any disagreement of thermodynamic function, it must be due to the higher orders terms which are, however, unknown by now. Hence, we do not see the problem with the zero-valued contribution in (6.132). Therefore, we take the liberty to keep or omit the zero-valued, in dependence of just mathematical convenience. Note that the $Q(\xi)$-function is not a measurable physical quantity but was introduced as an auxiliary function in order to abbreviate long mathematical expressions. The fact that different derivations lead to different auxiliary functions is irrelevant as far as the final results for the physical measurable quantities are not affected (Alastuey et al. 2015).

Much more difficult than the $\mathcal{O}\left(e^2\right)$ contribution, is the contribution of order $\mathcal{O}\left(e^6\right)$. This special and really difficult mathematical problem was analytically solved (Ebeling et al. 1976; Kraeft et al. 1986; Kremp et al. 2005). We use a mathematical trick based on exact calculations for the electron gas. As we have shown for the electron gas, there should be a contribution of order $n^2 e^6 \ln(\kappa)$. For compatibility with the result for the electron gas we make the ansatz

$$F_{\text{direct}}^2\left(e^{2k}, k = 3\right) = -V k_B T \sum_{ab} \frac{\pi}{3} \left(\beta\, e_a e_b\right)^3 \left[\ln\left(\kappa \lambda_{ab}\right) + A_0\right], \tag{6.134}$$

with the constant A_0 yet to be determined. In order to be compatible with Eq. (6.114) we write

$$A_0 = C + \ln 3 + \lim_{R \to \infty} \left[\ln \frac{R}{\lambda_{ab}} + \frac{6}{(\beta e_a e_b)^3} \int_0^R dr\, r^2 \, S_{ab}^{(2)} \left(r; e^6 \right) \right], \qquad (6.135)$$

where $C = 0.5772\ldots$ is the Euler constant and the contribution of the Slater sum contains only the third order in the interaction, $\mathcal{O}\left(e^6\right)$. After a long and complicated calculation we obtain the result (Ebeling et al. 1976; Hoffmann and Ebeling 1968a,b):

$$A_0 = \frac{C}{2} + \ln 3 - \frac{1}{2} \simeq 0.8872\ldots \qquad (6.136)$$

Since this result is in agreement with the previous one for the electron gas we have completed the calculations for the second virial function.

References

Alastuey, A., and A. Perez. 1992. Virial Expansion of the Equation of State of a Quantum Plasma. *Europhysics Letters* 20: 19–24.

Alastuey, A., and A. Perez. 1996. Virial Expansions for Quantum Plasmas: Fermi-Bose Statistics. *Physical Review E* 53: 5714–5728.

Alastuey, A., V. Ballenegger, and W. Ebeling. 2015. Comment on 'Direct Linear Term in the Equation of State of Plasmas' by Kraeft et al. *Physical Review E* 92: 047101. (see also Kraeft et al. (2015b)).

Arkhipov, Yu.V., F.B. Baimbetov, and A.E. Davletov. 2000. Thermodynamics of Dense High-Temperature Plasmas: Semiclassical Approach. *The European Physical Journal D* 8: 299–304.

Arkhipov, Yu.V., F.B. Baimbetov, and A.E. Davletov. 2011. Self-Consistent Chemical Model of Partially Ionized Plasmas. *Physical Review E* 83: 016405.

Baimbetov, F.B., M.A. Bekenov, and T.S. Ramazanov. 1995. Effective Potential of a Semiclassical Hydrogen Plasma. *Physics Letters A* 197: 157–158.

Barker, A.A. 1968. Monte Carlo Study of a Hydrogenous Plasma Near the Ionization Temperature. *Physical Review* 171: 186–188.

Barker, A.A. 1969. Radial Distribution Functions for a Hydrogenous Plasma in Equilibrium. *Physical Review* 179: 129–134.

Beth, E., and G.E. Uhlenbeck. 1937. The Quantum Theory of the Non-ideal Gas. II. Behaviour at Low Temperatures. *Physica* 4: 915–924.

Bogolyubov, N.N. 1946. Kinetic Equations. *Journal of Physics* 10: 265–274.

Bogolyubov, N.N. 2005–2009. *Collected Papers*. Vols. 1–12. Moscow: Fizmatlit.

Brillouin, L. 1931a. *Die Quantenstatistik und ihre Anwendung auf die Elektronentheorie der Metalle*. Berlin: Springer.

Brillouin, L. 1931b. *Les statistiques quantiques et leurs applications*. Paris: Presses Universitaires de France - PUF.

Brown, L.S., and L.G. Yaffe. 2001. Effective Field Theory of Highly Ionized Plasmas. *Physics Reports* 340: 1–164.

Debye, P., and E. Hückel. 1923. Zur Theorie der Elektrolyte. I. Gefrierpunktserniedrigung und verwandte Erscheinungen. *Physikalische Zeitschrift* 24: 185–206.

Deutsch, C. 1977. Nodal Expansion in a Real Matter Plasma. *Physics Letters A* 60: 317–318.

DeWitt, H.E. 1962. Evaluation of the Quantum-Mechanical Ring Sum with Boltzmann Statistics. *Journal of Mathematical Physics* 3: 1216–1228.

DeWitt, H.E. 1976. Asymptotic Form of the Classical One-Component Plasma Fluid Equation of State. *Physical Review A* 14: 1290–1293.

DeWitt, H.E., M. Schlanges, A.Y. Sakakura, and W.D. Kraeft. 1995. Low Density Expansion of the Equation of State for a Quantum Electron Gas. *Physics Letters A* 197: 326–329.

Ebeling, W. 1967. Statistische Thermodynamik der Bindungszustände in Plasmen. *Annalen der Physik (Berlin)* 19: 104–112.

Ebeling, W. 1968a. Ableitung der freien Energie von Quantenplasmen kleiner Dichte aus den exakten Streuphasen. *Annalen der Physik (Berlin)* 477: 33–39.

Ebeling, W. 1968b. The Exact Free Energy of Low Density Quantum Plasmas. *Physica* 40: 290–292.

Ebeling, W. 1969. Zur Quantenstatistik der Bindungszustände in Plasmen. I Cluster- Entwicklungen. *Annalen der Physik (Berlin)* 22: 383–391.

Ebeling, W. 1974. Statistical Derivation of the Mass Action Law or Interacting Gases and Plasmas. *Physica* 73: 573–584.

Ebeling, W. 2016. The Work of Baimbetov on Nonideal Plasmas and Some Recent Developments. *Contributions to Plasma Physics* 56: 163–175.

Ebeling, W., and G.E. Norman. 2003. Coulombic Phase Transitions in Dense Plasmas. *Journal of Statistical Physics* 110: 861–877.

Ebeling, W., H.J. Hoffmann, and G. Kelbg. 1967. Quantenstatistik des Hochtemperatur-Plasmas im thermodynamischen Gleichgewicht. *Contributions to Plasma Physics* 7: 233–248.

Ebeling, W., G. Kelbg, and K. Rohde. 1968. Binäre SLATER-Summen und Verteilungsfunktionen für quantenstatistische Systeme mit COULOMBWechselwirkung. II. *Annalen der Physik (Berlin)* 476 (5–6): 235–243.

Ebeling, W., W.D. Kraeft, and D. Kremp. 1976. *Theory of Bound States and Ionisation Equilibrium in Plasmas and Solids.* Berlin: Akademie-Verlag.

Ebeling, W., V.E. Fortov, and V.S. Filinov. 2017. *Quantum Statistics of Dense Gases and Nonideal Plasmas. Springer Series in Plasma Science and Technology.* Cham: Springer.

Eggert, J. 1919. Über den Dissoziationzustand der Fixsterngase. *Physikalische Zeitschrift* 20: 570–574.

Falkenhagen, H. 1971. *Theorie der Elektrolyte.* Leipzig: Hirzel.

Falkenhagen, H., and W. Ebeling. 1971. Equilibrium Properties of Ionized Dilute Electrolytes. In *Ionic Interactions*, ed. S. Petrucci, vol. 1, 1–59. New York/London: Academic.

Filinov, V.S., M. Bonitz, W. Ebeling, and V.E. Fortov. 2001. Thermodynamics of Hot Dense H-Plasmas: Path Integral Monte Carlo Simulations and Analytical Approximations. *Plasma Physics and Controlled Fusion* 43 (6): 743–759.

Filinov, A.V., M. Bonitz, and W. Ebeling. 2003. Improved Kelbg Potential for Correlated Coulomb Systems. *Journal of Physics A: Mathematical and General* 36 (22): 5957–5962.

Filinov, V.S., M. Bonitz, P. Levashov, V.E. Fortov, W. Ebeling, M. Schlanges, and S.W. Koch. 2003. Plasma Phase Transition in Dense Hydrogen and Electron–Hole Plasmas. *Journal of Physics A: Mathematical and General* 36 (22): 6069–6076.

Filinov, A.V., V.O. Golubnychiy, M. Bonitz, W. Ebeling, and J.W. Dufty. 2004. Temperature-Dependent Quantum Pair Potentials and Their Application to Dense Partially Ionized Hydrogen Plasmas. *Physical Review E* 70: 046411.

Fortov, V.E. 2009. *Extreme States of Matter (in Russian).* Moskva: Fiz-MatGis.

Fortov, V.E. 2011. *Extreme States of Matter: On Earth and in the Cosmos.* Berlin: Springer.

Friedman, H.L. 1962. *Ionic Solution Theory.* New York: Interscience.

Gombás, P. 1965. Pseudopotentiale. *Fortschritte der Physik* 13: 137–156.

Hansen, J.P., and I.R. McDonald. 1981. Microscopic Simulation of a Strongly Coupled Hydrogen Plasma. *Physical Review A* 23: 2041–2059.

Hellmann, H. 1935. A New Approximation Method in the Problem of Many Electrons. *The Journal of Chemical Physics* 3: 61–61.

Hemmer, P.C., H. Helge, and S. Kjelstrup Ratkje, eds. 1996. *The Collected Works of Lars Onsager.* Singapore: World Scientific.

Hoffmann, H.J., and W. Ebeling. 1968a. On the Equation of State of Fully Ionized Quantum Plasmas. *Physica* 39: 593–598.

Hoffmann, H.J., and W. Ebeling. 1968b. Quantenstatistik des Hochtemperatur-Plasmas im thermodynamischen Gleichgewicht. II. Die freie Energie im Temperaturbereich 10^6 bis $10^8\, o$K. *Contributions to Plasma Physics* 8 (1): 43–56.

Ichimaru, S. 1992. *Statistical Plasma Physics.* Redwood: Addison-Wesley.

Kalman, G.J., J.M. Rommel, and K. Blagoev, eds. 1998. *Strongly Coupled Coulomb Systems.* New York: Springer.

Kelbg, G. 1963a. Quantenstatistik der Gase mit Coulomb-Wechselwirkung. *Annalen der Physik* 467: 354–360.

Kelbg, G. 1963b. Theorie des Quanten-Plasmas. *Annalen der Physik* 467: 219–224.

Kelbg, G. 1964. Klassische statistische Mechanik der Teilchen-Mischungen mit sortenabhängigen weitreichenden zwischenmolekularen Wechselwirkungen. *Annalen der Physik (Leipzig)* 14: 394–403.

Kelbg, G. 1972. *Einige Methoden der statistischen Thermodynamik hochionisierter Plasmen, Ergebnisse der Plasmaphysik und Gaselektronik.* Vol. Bd. III. Berlin: Akademie-Verlag.

Kelbg, G., and H.J. Hoffmann. 1964. Quantenstatistik realer Gase und Plasmen. *Annalen der Physik* 469: 310–318.

Kleinert, H. 1995. *Path Integrals in Quantum Mechanics, Statistics and Polymer Physics.* Berlin: Springer.

Klimontovich, Yu.L. 1982. *Statistical Physics (in Russian)* Moscow: Nauka.

Klimontovich, Yu.L., and W. Ebeling. 1972. Quantum Kinetic Equations for a Nonideal Gas and a Nonideal Plasma. *Journal of Experimental and Theoretical Physics* 36: 476–481.

Kraeft, W.D., and D. Kremp. 1968. Quantum-Statistical Mechanics of a System of Charged Particles at High Temperatures. *Zeitschrift für Physik* 208 (5): 475–485.

Kraeft, W.D., D. Kremp, W. Ebeling, and G. Röpke. 1986. *Quantum Statistics of Charged Particle Systems.* Berlin: Akademie-Verlag.

Kraeft, W.D., D. Kremp, and G. Röpke. 2015a. Direct Linear Term in the Equation of State of Plasmas. *Physical Review E* 91: 013108.

Kraeft, W.D., D. Kremp, and G. Röpke. 2015b. Reply to Alastuey, Ballenegger, and Ebeling 2015. *Physical Review E* 92: 047102.

Kremp, D., and W.D. Kraeft. 1972. Analyticity of the Second Virial Coefficient as a Function of the Interaction Parameter and Compensation Between Bound and Scattering States. *Physical Review A* 38: 167–168.

Kremp, D., M. Schlanges, and W.D. Kraeft. 2005. *Quantum Statistics of Nonideal Plasmas.* Berlin: Springer.

Landau, L.D., and E.M. Lifshitz. 1980. *Statistical Physics.* Oxford: Butterworth-Heinemann.

Larkin, A.I. 1960. Thermodynamic Functions of a Low-Temperature Plasma. *Journal of Experimental and Theoretical Physics* 11: 1363–1364.

Macke, W. 1950. Über die Wechselwirkungen im Fermi-Gas. *Zeitschrift für Naturforschung A* 5a: 192–208.

Mayer, J.E. 1950. The Theory of Ionic Solutions. *The Journal of Chemical Physics* 18: 1426–1436.

Milner, S.R. 1912. The Virial of a Mixture of Ions. *The London, Edinburgh, and Dublin Philosophical Magazine and Journal of Science* 23: 551–578.

Montroll, E., and J. Ward. 1958. Quantum Statistics of Interacting Particles. *The Physics of Fluids* 1: 55–72.

Morita, T. 1959. Equation of State of High Temperature Plasma. *Progress of Theoretical Physics (Kyoto)* 22: 757–774.

Ortner, J. 1999. Equation of States for Classical Coulomb Systems: Use of the Hubbard-Schofield Approach. *Physical Review E* 59: 6312–6327.

Ortner, J., I. Valuev, and W. Ebeling. 1999. Semiclassical Dynamics and Time Correlations in Two-component Plasmas. *Contributions to Plasma Physics* 39 (4): 311–321.

Ortner, J., I. Valuev, and W. Ebeling. 2000. Electric Microfield Distribution in Two-Component Plasmas. Theory and Simulations. *Contributions to Plasma Physics* 40: 555–568.

Pines, D., and P. Nozieres. 1966. *The Theory of Quantum Liquids*. New York: Benjamin.

Planck, M. 1924. Zur Quantenstatistik des Bohrschen Atommodells. *Annalen der Physik* 75: 673–684.

Riewe, K., and R. Rompe. 1938. Über die Besetzungszahlen der Elektronenterme in einem teilweise ionisierten Gas. *Zeitschrift für Physik* 111: 79–94.

Rohde, K., G. Kelbg, and W. Ebeling. 1968. Binäre SLATER-Summen und Verteilungsfunktionen für quantenstatistische Systeme mit COULOMBWechselwirkung. I. *Annalen der Physik (Berlin)* 477: 1–14.

Sadykova, S., and W. Ebeling. 2007. Electric Microfield Distributions in Dense One- and Two-component Plasmas. *Contributions to Plasma Physics* 47: 659–669.

Saha, M.N. 1920. Ionization in the Solar Chromosphere. *Philosophical Magazine Series VI* 40: 472–478.

Schmitz, G., and D. Kremp. 1967. Quantenmechanische Verteilungsfunktion für ein Elektronen-gas. *Zeitschrift für Naturforschung A* 23: 1392–1395.

Starostin, A.N., and V.C. Roerich. 2006. Bound States in Nonideal Plasmas: Formulation of the Partition Function and Application to the Solar Interior. *Plasma Sources Science and Technology* 15: 410–415.

Starostin, A.N., V.C. Roerich, and R.M. More. 2003. How Correct is the EOS of Weakly Nonideal Hydrogen Plasmas? *Contributions to Plasma Physics* 43: 369–372.

Stolzmann, W., and W. Ebeling. 1998. New Padé Approximations for the Free Charges in Two-Component Strongly Coupled Plasmas Based on the Unsöld-Berlin-Montroll Asymptotics. *Physics Letters A* 248: 242–246.

Storer, R.G. 1968a. Path-Integral Calculation of the Quantum-Statistical Density Matrix for Attractive Coulomb Forces. *Journal of Mathematical Physics* 9: 964–970.

Storer, R.G. 1968b. Radial Distribution Function for a Quantum Plasma. *Physical Review* 176: 326–331.

Trigger, S.A., W. Ebeling, V.S. Filinov, V.E. Fortov, and M. Bonitz. 2003. Internal Energy of High Density Hydrogen: Analytic Approximations Compared with Path Integral Monte Carlo Calculations. *Zhurnal Ehksperimental'noj i Teoreticheskoj Fiziki* 123: 527–542.

Trubnikov, B.A., and V.F. Elesin. 1965. Quantum Correlation Functions in a Maxwellian Plasma. *Journal of Experimental and Theoretical Physics* 20: 866–872.

Uhlenbeck, G.E., and E. Beth. 1936. The Quantum Theory of the Non-ideal Gas I. Deviations from the Classical Theory. *Physica* 3: 729–745.

Vedenov, A.A., and A.I. Larkin. 1959. Equation of State of Plasmas (in Russian) *Zhurnal Ehksperimental'noj i Teoreticheskoj Fiziki* 36: 1133.

von Neumann, J. 1932. *Mathematische Grundlagen der Quantenmechanik*. Berlin: Springer.

Wigner, E. 1934. On the Interaction of Electrons in Metals. *Physical Review* 46: 1002–1011.

Wigner, E. 1938. The Transition State Method. *Transactions of the Faraday Society* 34: 29–41.

Zamalin, V.M., G.E. Norman, and V.S. Filinov. 1977. *The Monte Carlo Method in Statistical Thermodynamics (in Russian)*. Moscow: Nauka.

Zelener, B.V., G.E. Norman, and V.S. Filinov. 1981. *Perturbation Theory and Pseudopotential in Statistical Thermodynamics (in Russian)*. Moscow: Nauka.

Chapter 7
Non-ideality and Deep Bound States in Plasmas

7.1 Higher Order Expansions with Respect to Density

7.1.1 Cluster Expansions with Respect to Density

With increasing density of plasmas, non-ideality effects become more and more apparent. This is of particular importance for plasmas with deep bound states, that is, when the thermal energy is smaller than or of the same order as the ground state energies of the atoms in the plasma. For hydrogen this is the case already for $T < 10^5$ K. Then, except for very low density, the convergence of the expansions becomes worse and we have to consider higher order terms. Sometimes even the transition to other expansion parameters, as the fugacity is recommended.

To compute the interaction contributions in higher orders, we start first from the cluster expansion for real quantum gases with respect to density. It is useful to split the free energy into the contribution of a Boltzmann gas and an excess part, which contains all correlations including the correlations due to the symmetry properties of the particles:

$$
F(T, V, N) = F_{\text{id}} + F_{\text{ex}} = N k_{\text{B}} T \left[\ln \left(n \Lambda^3 \right) - 1 \right]
$$
$$
- k_{\text{B}} T V \left[\frac{1}{2} \sum_{ab} n_a n_b \, I \left(S_{ab}^{(2)} \right) + \frac{1}{6} \sum_{abc} n_a n_b n_c \, I \left(S_{abc}^{(3)} \right) + \dots \right],
$$

$$(7.1)$$

© Springer Nature Switzerland AG 2019
W. Ebeling, T. Pöschel, *Lectures on Quantum Statistics*,
Lecture Notes in Physics 953, https://doi.org/10.1007/978-3-030-05734-3_7

with the abbreviations

$$I\left(S_{ab}^{(2)}\right) = \int d\mathbf{r}_1 d\mathbf{r}_2 \left(S_{ab}^{(2)} - 1\right) \tag{7.2}$$

$$I\left(S_{abc}^{(3)}\right) = \int d\mathbf{r}_1 d\mathbf{r}_2 d\mathbf{r}_3 \left(S_{abc}^{(3)} - S_{ab}^{(2)} S_{ac}^{(2)} S_{bc}^{(2)}\right) \tag{7.3}$$

etc.

This way we obtain the screened cluster expansion for the free energy density,

$$\beta f(T, n_a) = \beta f_{\mathrm{id}} - \frac{\kappa^3}{12\pi} - \sum_{ab} n_a n_b B_{ab}(\kappa) - \sum_{abc} n_a n_b n_c B_{abc}(\kappa) + \ldots, \tag{7.4}$$

with $f_{\mathrm{id}} = F_{\mathrm{id}}/V$. Here the screened cluster integrals are constructed in complete analogy to the screened cluster integrals for classical ionic solutions given by Meeron (1962) (see Friedman 1962)

$$B_{ab} = \frac{1}{2} I\left(S_{ab}^{(2)}\right) ; \qquad B_{abc} = \frac{1}{6} I\left(S_{abc}^{(3)}\right). \tag{7.5}$$

The representation is exact, but slowly convergent and the orders in density are

$$B_2(\kappa) = \mathcal{O}\left(\kappa^0\right) ; \qquad\qquad B_3(n) = \mathcal{O}\left(\log \kappa\right) ;$$

$$B_4(n) = \mathcal{O}\left(\kappa^{-2}\right) ; \qquad\qquad B_5(n) = \mathcal{O}\left(\kappa^{-3}\right). \tag{7.6}$$

This estimate is correct only for the case that the third moments are zero, otherwise, if $\mu_3 = \sum_a n_a e_a^3 \neq 0$, the estimate is even worse and gives (Friedman 1962)

$$B_2(n) = \mathcal{O}\left(\log \kappa\right) ; \qquad\qquad B_3(n) = \mathcal{O}\left(\kappa^{-1}\right) ;$$

$$B_4(n) = \mathcal{O}\left(\kappa^{-3/2}\right) ; \qquad\qquad B_5(n) = \mathcal{O}\left(n^{-4}\right). \tag{7.7}$$

In the case of dilute Coulomb plasmas, the second cluster integrals read explicitly (Ebeling 1968, 1969; Kraeft et al. 1986)

$$B_{ab}(\kappa) = \frac{1}{2!} \int d\mathbf{r}_2 \left(S_{ab}(r) e^{g_{ab} + \beta V_{ab}} - 1 - g_{ab} - \frac{1}{2} g_{ab}^2\right). \tag{7.8}$$

7.1.2 Density Expansions of Pressure and Free Energy

We collect the results we obtained so far and order them term by term with respect to density. First, we calculated the higher order terms to the second virial coefficient and obtained:

$$G'' B_{ab}(\kappa) = 2\pi^{3/2} \lambda_{ab}^3 (1 + \beta e_a e_b \kappa) \sum_{m \geq 4} \frac{\xi_{ab}^m}{2^m \Gamma \left(\frac{m}{2} + 1\right)}$$

$$\times \left[\zeta(m-2) \pm \delta_{ab} \frac{1 - 2^{2-m}}{2s_a + 1} \zeta(m-1) \right]. \qquad (7.9)$$

Then we calculated the lower orders also term by term. We included the ideal terms, the Hartree and Fock quantum terms and the screening terms, and with particular effort, the term of order e^6. The result of these calculations are the following lower order contributions

$$G' B_{ab}(\kappa) = 2\pi \lambda_{ab}^3 (1 + \beta e_a e_b \kappa) \left[-\frac{1}{6} \xi_{ab} - \frac{\sqrt{\pi}}{8} \xi_{ab}^2 - \frac{1}{6} \left(\frac{1}{2} C + \ln 3 - \frac{1}{2} \right) \xi_{ab}^3 \right]$$

$$+ \frac{\pi}{3} (\beta e_a e_b)^3 \left[(1 + \beta e_a e_b \kappa) \ln (3\kappa \lambda_{ab}) - \beta e_a e_b \kappa \left(1 - \ln \frac{4}{3} \right) \right]$$

$$\pm \delta_{ab} 2\pi \lambda_{ab}^3 (1 + \beta e_a e_b \kappa) \left(\frac{\sqrt{\pi}}{4} + \frac{\xi_{ab}}{2} + \frac{\sqrt{\pi} \ln 2}{4} \xi_{ab}^2 + \frac{\pi^2}{72} \xi_{ab}^3 \right), \qquad (7.10)$$

with Euler's constant, $C = 0.577$. Note again that most difficulties are related to the term of order ξ_{ab}^3. Still there is a running controversy about the term of order ξ_{ab}^1 which, however, is of no influence on the free energy, since it contributes zero after summation. Finally we obtain the density expansion of the free energy density

$$\beta f(n_a, T) = \sum_a n_a \left[\ln \left(n_a \Lambda_a^3 \right) - 1 \right] - \frac{\kappa^3}{12\pi}$$

$$- \sum_{ab} 2\pi n_a n_b \left[\frac{1}{6} (\beta e_a e_b)^3 \ln (\kappa \lambda_{ab}) + \lambda_{ab}^3 K_0 (\xi_{ab}; s_a) \right]$$

$$+ \mathcal{O} \left(n^{5/2} \ln n \right). \qquad (7.11)$$

The corresponding expression for the pressure is

$$\beta p(n_a, T) = \sum_a n_a - \frac{\kappa^3}{24\pi}$$

$$- \sum_{ab} 2\pi n_a n_b \left[\frac{1}{6} (\beta e_a e_b)^3 \ln (\kappa \lambda_{ab}) + \lambda_{ab}^3 K_0 (\xi_{ab}; s_a) + \frac{1}{6} (\beta e_a e_b)^3 \right]$$

$$+ \mathcal{O}\left(n^{5/2} \ln n\right) . \tag{7.12}$$

The function $K_0(\xi)$ (shown in Fig. 7.1) was introduced in analogy to similar functions introduced earlier in electrolyte theory (Falkenhagen 1971; Falkenhagen and Ebeling 1971) and is defined by

$$K_0 (\xi_{ab}, s_a) = Q (\xi_{ab}) + \frac{(-1)^{2s_a}}{2s_a + 1} \delta_{ab} E (\xi_{ab}) . \tag{7.13}$$

Here we used the $E(\xi)$ function defined above (see Fig. 6.8) and define

$$Q(\xi) = -\frac{\xi}{6} - \frac{\sqrt{\pi}}{8}\xi^2 - \frac{1}{6}\left(\frac{C}{2} + \ln 3 - \frac{1}{2}\right)\xi^3 + \sum_{m \geq 4} \frac{\xi^m}{2^m \Gamma\left(\frac{m}{2} + 1\right)} \zeta (m - 2) . \tag{7.14}$$

Note that the direct quantum contribution, $Q(x)$, increases steeply with increasing argument, that is, exponentially for positive argument. For large argument, the asymptotic behavior is Ebeling (1968)

$$Q(\xi) \simeq -\frac{\xi^3}{6} \left(\ln |3\xi| + 2C - \frac{11}{6} \right) + \Theta(\xi) 2\sqrt{\pi}\, \sigma(T) . \tag{7.15}$$

Fig. 7.1 The virial function, $K_0(\xi; s)$, which describes the deviations from Debye's limiting law of the free energy density

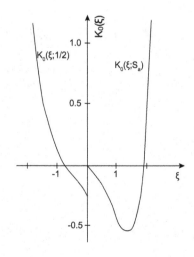

Here, $\sigma(T)$ is the already well studied PBL partition function and Θ is the step-function (one for positive arguments and zero elsewhere). The fact that the PBL-function appears in the asymptotics can be considered as another confirmation of our approach. We mention that the remaining term of order ξ^3 yields the correct classical asymptotics for the electron gas too. Consequently, our result is asymptotically consistent with all results known so far. If we would follow the arguments by Kraeft et al. (2015a), the linear term in $Q(\xi)$ has to be omitted. As an disadvantage, the nice asymptotics of $Q(\xi)$ given by Eq. (7.15) does not hold true but one obtains an additional linear term. Finally we notice that the $K_0(\xi)$ function found here (including the linear term) was also confirmed by at least two independent groups (Alastuey and Ballenegger 2010; Alastuey and Perez 1992, 1996; Alastuey et al. 2008, 2015; Brown and Yaffe 2001; Kahlbaum 2000).

We demonstrate the influence of the second virial term calculated here on the equation of state by means of two examples. Figure 7.2 shows the pressure of hydrogen plasmas relative to the ideal pressure function for $n = 10^{15}\,\mathrm{cm}^{-3}$ as a function of temperature, due to Eq. (7.12), in comparison to the result of a fugacity expansion to be discussed later. We see that for lower temperature, the pressure assumes large negative values. Evidently this is an artefact due to incorrect description of the bound state contribution which increases exponentially at lower temperature. We will show later that this can be corrected by using fugacity expansions or chemical descriptions.

Let us check now the classical limit of our calculations. To this end, we consider an electron gas which is an interesting special case, as there are no bound states at all. In this case, all interaction parameters are zero except the electron-electron

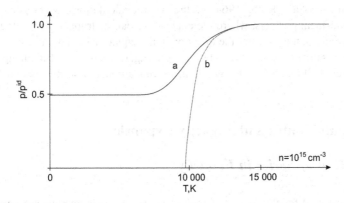

Fig. 7.2 Pressure of hydrogen plasmas relative to the ideal pressure function for $n = 10^{15}\,\mathrm{cm}^{-3}$ as a function of temperature. The dashed line shows the result of a density expansion and the full line shows the corresponding fugacity expansion

parameter, which is negative. The free energy density is

$$\beta f(n, T) = n \left[\ln \left(n \Lambda^3 \right) - 1 \right] - \frac{\kappa^3}{12\pi}$$
$$- 2\pi n^2 \left[\frac{1}{6} \beta^3 e^6 \ln(\kappa \lambda) + \lambda^3 K_0 \left(-\xi; s = \frac{1}{2} \right) \right] + \mathcal{O} \left(n^{5/2} \ln n \right),$$

(7.16)

where $\lambda = \lambda_{ee}$ and $\xi = \xi_{ee}$. The near classical case of the electron gas corresponds to $\xi \to -\infty$ and the virial function has the asymptotic (see also Eq. (7.15))

$$K_0(\xi; s) = -\frac{1}{6} \xi^3 \left(2C - \ln 3 - \frac{11}{6} + \ln |\xi| \right) ; \qquad \xi < 0.$$

(7.17)

Introducing this into Eq. (7.16), we find for the free energy density

$$\beta f(n, T) = n \left[\ln \left(n \Lambda^3 \right) - 1 \right]$$
$$- n k_B T \left[\frac{1}{3} \mu + \frac{1}{12} \mu^2 \left(\ln \mu + 2C + \ln 3 - \frac{11}{6} \right) + \mathcal{O} \left(\mu^3 \right) \right].$$

(7.18)

This result is in full agreement with the classical result for the free energy of the electron gas. Thus, our result for the quantum electron gas is consistent with the classical results given in Sect. 6.3.1 as well as with $\hbar$-expansions (Hoffmann and Ebeling 1968a,b; Kelbg 1964). In the present calculations, we considered only weakly non-ideal plasmas, where the expansions can be truncated after the second virial coefficient, that is, after the order n^2 in the equation of state. For higher order contributions including also the orders $n^{5/2}$ see, e.g. (Alastuey and Ballenegger 2010; Alastuey et al. 2008, 2015; De Witt et al. 1995; Kahlbaum 2000; Riemann 1995).

7.2 Bound States and Fugacity Expansions

7.2.1 Cluster Series in Fugacity

As demonstrated for the theory of real gases, fugacity expansions are very useful alternatives to density expansions, for the case that the systems develop deep bound states. Fugacity expansions are, in principle, completely equivalent to density expansions, they are just an alternative mathematical representation. However, under different physical conditions, both series can have different ranges of convergence. Fugacity expansions are formulated in the grand canonical ensemble and they are representations of the pressure and the density in terms of the fugacities. Derivations of screened fugacity expansions for plasmas were developed by Bartsch and Ebeling (1971), Rogers (1974) and Rogers et al. (1988). Density and fugacity expansions are equivalent as we have seen in the case of gases. Therefore, we write both series and verify the coefficients by comparison term by term. As an ansatz for the plasma pressure as a function of the fugacities, we write

$$\beta p(z_a, T) = \sum_a z_a + \frac{\kappa^3}{12\pi} + \sum_{ab} n_a n_b b_{ab}(\kappa) + \sum_{abc} n_a n_b n_c b_{abc}(\kappa) + \dots, \quad (7.19)$$

where

$$z_a = \frac{2s_a + 1}{\Lambda_a^3} e^{\beta \mu_a}. \quad (7.20)$$

Following the arguments in Chap. 5, the fugacities are normalized in such a way that

$$z_a \to n_a \quad \text{for} \quad n_a \to 0. \quad (7.21)$$

Note that within fugacity expansions the variable κ has a different meaning since in its definition, the densities are replaced by fugacities:

$$\kappa^2 = 4\pi\beta \sum_a z_a e_a^2. \quad (7.22)$$

Further we have the relation between densities and fugacities

$$n_a = z_a \frac{\partial \beta p(z_a, T)}{\partial z_a}. \quad (7.23)$$

Differentiation yields

$$n_a = z_a + \frac{1}{2} z_a \beta e_a^2 \kappa + z_a \sum_b z_b b_{ab}(\kappa) +$$

$$+ 8\pi\beta \, z_a e_a^2 \kappa \sum_b z_b b_{ab}'(\kappa) + \sum_{bc} z_a z_b z_c b_{abc}(\kappa) + \dots. \quad (7.24)$$

By expanding around the limit of small density/fugacity, we obtain for Coulomb plasmas with small fugacity (small density) expressions for the cluster integrals, $b_{ab}(\kappa)$ (Bartsch and Ebeling 1971; Ebeling et al. 1976; Kraeft et al. 1986):

$$b_{ab}(\kappa) = \frac{1}{2!} \int d\mathbf{r}_2 \left[S_{ab}(r) \, e^{G_{ab} + \beta V_{ab}} - 1 - G_{ab} - \frac{1}{2} G_{ab}^2 \right] \quad (7.25)$$

$$b_{abc}(\kappa) = \mathcal{O}(S_{abc}). \quad (7.26)$$

The integral equation determining the screening function in the grand canonical ensemble, $G_{ab}(r)$, is given by

$$G_{ab}(1, 2) + \beta V_{ab}(1, 2) + \sum_c z_c \int d3 \, \beta \, V_{bc}(1, 3) \, G_{bc}(2, 3) = 0 . \qquad (7.27)$$

An explicit calculation of the cluster integrals in the grand canonical ensemble using numerical methods was performed by Rogers (1974) and Rogers et al. (1988).

7.2.2   Series in Powers of Fugacity

The easiest way to find the relations between the cluster coefficients, $B_{ab}(\kappa)$ and $b_{ab}(\kappa)$, is to expand around the low density/low fugacity limit and to compare term by term. Transforming this way the density expansion to a fugacity expansion, we obtain

$$\beta p(z_a, T) = \sum_a z_a + \frac{\kappa^3}{12\pi} + \sum_{ab} z_a z_b \left[\frac{\pi}{3} \, (\beta e_a e_b)^3 \ln (\kappa \lambda_{ab}) \right.$$
$$\left. + \frac{\pi}{2} \beta^3 e_a^2 e_b^4 + 2\pi \lambda_{ab}^3 \, K_0 \, (\xi_{ab}; s_a) \right] + \mathcal{O} \left(z_a^{5/2} \ln z_a \right) . \qquad (7.28)$$

The corresponding expansion of the density follows from

$$n_a = z_a \frac{\partial \beta p(z_a, T)}{\partial z_a} . \qquad (7.29)$$

Again, the fugacities are normalized in a way such that

$$z_a \to n_a \qquad \text{for} \qquad n_a \to 0 . \qquad (7.30)$$

We remember that the fugacities do not necessarily obey the condition $\sum_a z_a e_a = 0$. We have, however, some freedom of scaling the fugacities because of some freedom in the lower bound of the chemical potential. Therefore, it is possible to find a special scaling in which this condition is fulfilled (Alastuey and Perez 1996). For example, for the case of hydrogen plasmas, besides $n_e = n_i$ also $z_e = z_i = z$. Then we find equations for the pressure, p, and the fugacity, z, of hydrogen which can be easily inverted (Bartsch and Ebeling 1971).

Thus, the series in fugacity are similar to the series in density, but the series are not identical as there are additional contributions in particular in relation to the screening terms. We underline that fugacity expansions are conceptually simple and well suited to describe directly special aspects of bound state phenomena in plasmas as shown, e.g., by Bartsch and Ebeling (1971), Ebeling (1974), Hill

(1956), Rogers (1974) and Rogers et al. (1988). We discussed here the advantages and disadvantages of both methods. For mixed representations which interpolate between density and fugacity expansions, based on partial summations (see Ebeling et al. 2012).

7.3 Bound States and Saha Equation

7.3.1 Fugacity Expansion and Ideal Mass Action Laws

The close relation between fugacity expansions and mass action laws for associating or reacting real gases was already discussed in Chap. 5. Here we discuss the problem in relation to Coulomb systems following Ebeling (1974), Ebeling et al. (1976), Friedman and Ebeling (1979), Rogers (1974) and Rogers et al. (1988).

We focus on bound states in hydrogen plasmas. First we compare the density expansions provided by the canonical ensemble with the fugacity expansions of grand canonical representations. Summarizing the most relevant results of Chap. 5 and the previous sections we remind of successful applications of fugacity expansions to gases and plasmas with bound states. Consider a binary plasma with species "+" and "−", disregarding complications like screening and degeneration. The pressure reads

$$\beta p = z_- + z_+ + B_{+-}^{(2)}(T)z_-z_+ . \tag{7.31}$$

Further, virial contributions with indices $++$ and $--$ are neglected since they are irrelevant for bound states. By differentiation we obtain

$$n_- = z_- + B_{+-}^{(2)}(T)z_-z_+ \tag{7.32}$$

$$n_+ = z_+ + B_{+-}^{(2)}(T)z_-z_+ , \tag{7.33}$$

with the virial coefficient

$$B_{+-}(T) = 4\pi^{3/2}\lambda_{+-}^3\,\sigma(T) . \tag{7.34}$$

An equivalent interpretation suggested by our studies in Chap. 5 reads

$$z_- = \text{density of free negative charges}$$

$$z_+ = \text{density of free positive charges} .$$

Following Chap. 5, we understand this as a *chemical picture* and interprete $B_{+-}^{(2)}(T) = K(T)$ as a mass action constant. The quantity

$$z_- z_+ B_{+-}^{(2)}(T) = n_0^* \tag{7.35}$$

has then the meaning of an atomic density. Similar as for ideal gas mixtures studied in Chap. 5, we write

$$\beta p = n_-^* + n_+^* + n_0^* \tag{7.36}$$

$$n_- = n_-^* + n_0^* \tag{7.37}$$

$$n_+ = n_+^* + n_0^* \tag{7.38}$$

and the mass action law

$$\frac{n_0^*}{n_+^* n_-^*} = K(T). \tag{7.39}$$

This mass action law corresponds to the reaction

$$(+) \; + \; (-) \; \rightleftharpoons \; \text{atom}. \tag{7.40}$$

In other words, the formation of deep bound states can be described either by fugacity expansions or by a mass action law. The descriptions given so far correspond to ideal Saha equations.

Let us solve these Saha equations given above for a hydrogen plasma with

$$n_+^* = n_-^* = n^* \qquad\qquad \text{for neutrality}$$

$$n_+ = n_- = n \qquad\qquad \text{total charge neutrality}$$

$$\beta p = 2n^* + n_0^* = 2n^* + K(T)n^{*2} \tag{7.41}$$

$$n = n^* + n_0^* = n^* + Kn^{*2}. \tag{7.42}$$

The quadratic equation (7.42) is solvable and delivers for the density

$$n^* = \frac{1}{K(T)} \left(\sqrt{1 + nK(T)} - 1 \right) \tag{7.43}$$

and for the pressure

$$\beta p = n + \frac{1}{K(T)} \left(\sqrt{1 + nK(T)} - 1 \right). \tag{7.44}$$

The general schema which we developed here allows to include also bound states of more than two charged particles. Consider, for example, a hydrogen plasma at fixed temperature, T, and proton density, n_i. We take into account ionization processes,

$$H \Leftrightarrow p + e^- , \tag{7.45}$$

and dissociation processes,

$$H_2 \Leftrightarrow H + H . \tag{7.46}$$

For simplicity, the formation of other species like H_2^+ and H^- is neglected. The degrees of ionization and dissociation are defined by

$$\alpha = \frac{n_i^*}{n_i^* + n_a^* + 2n_m^*} ; \qquad \beta_d = \frac{n_a^*}{n_a^* + 2n_m^*} ;$$

$$\beta_a = \frac{n_a^*}{n_i^* + n_a^* + 2n_m^*} ; \qquad \beta_m = \frac{2n_m^*}{n_i^* + n_a^* + 2n_m^*} . \tag{7.47}$$

Here all densities denoted by stars shall be understood as densities of free particles. In order to take into account the bound states which appear in the system, the fugacity expansion has to include at least the following terms,

$$\beta p = z_e + z_i + B_{ei}^{(2)}(T) z_e z_i + B_{eiei}^{(4)} z_e z_i z_e z_i + \dots \tag{7.48}$$

Note that we considered here only the key terms, further terms related to screening and interactions are still to be added. With the interpretations

$$z_e = n_e^* ; \quad z_i = n_i^* ; \quad n_a^* = B_{ei}^{(2)}(T) z_e z_i ; \quad n_m^* = B_{eiei}^{(4)} z_e z_i z_e z_i , \tag{7.49}$$

we are able to compute the pressure and the degrees of ionization and dissociation in the system. As an example, Fig. 7.3 shows the degrees of ionization and dissociation for hydrogen at three temperatures $T = (5000, 10000, 15000)$ K. In this context, β_d is the degree of dissociation of molecules into atoms, β_a is the relative amount of protons bound in atoms and β_m is the relative amount bound in molecules.

To solve the system of equations, we take into account that due to the balance relation for the total proton density

$$n_i = n_i^* + n_a^* + 2n_m^* \tag{7.50}$$

which provides the relations

$$\beta_a = \alpha(1 - \beta_d) ; \qquad \beta_m = (1 - \alpha)(1 - \beta_d) ; \qquad \beta_d = \frac{\beta_a}{\beta_a + \beta_m} . \tag{7.51}$$

Fig. 7.3 Degrees of
ionization and dissociation
for hydrogen at $T = 5000\,\mathrm{K}$,
$T = 10000\,\mathrm{K}$, and
$T = 15000\,\mathrm{K}$ as a function of
the total density of protons

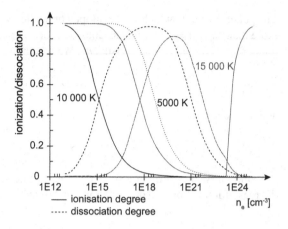

Fig. 7.3 Degrees of ionization and dissociation for hydrogen at $T = 5000\,\mathrm{K}$, $T = 10000\,\mathrm{K}$, and $T = 15000\,\mathrm{K}$ as a function of the total density of protons

Therefore, only one of the β parameters is independent and we have some freedom to make our choice. We prefer here to use α and β_m and will make use of the simplex relation,

$$\alpha + \beta_a + \beta_m = 1. \tag{7.52}$$

In large regions of the density-temperature space, in some approximation we may assume $\beta_a = 0$. Consequently, α remains as the only free parameter. The condition of neutrality requires that electron and ion densities are always equal,

$$n_e = n_i ; \qquad n_e^* = n_i^* . \tag{7.53}$$

The thermodynamic functions depend on 4 independent parameters. The density dependence of the degrees is shown in Fig. 7.3 (Ebeling et al. 2009). We see that atoms appear only in a rather narrow region of the temperature-density plane. We have shown this way that fugacity expansions may be a convenient tool for describing the complicated thermodynamic equilibria in plasmas which is conceptually very simple. Extensive studies of this formalism are due in particular to Rogers (1974).

7.3.2   Ionization and Saha Equation Including Screening

For simplicity, here we consider only binary charge-symmetrical plasmas and use the mean mass approximation (MMA), which—as we have seen—is in good agreement with the strict quantum statistical results for temperature $T < 10^5\,\mathrm{K}$. This follows from the fact that the MMA is a quantum statistical result, stating that in the limit $\xi \to \infty$, that is, $k_B T \ll I$ (see Fig. 6.4), the differences between the relative masses, m_{ab}, do not have significant effect to the thermodynamics, mainly because of the asymptotic of the virial functions (Kraeft et al. 1986). This allows us to transfer the ring function for mass-symmetrical plasmas in an easy way to the ring function for general (not mass-symmetrical) electron-ion plasmas just by replacing the relative wave lengths by an average (Ebeling 2016; Ebeling et al. 1976, 1991). That means, all relative masses may be replaced by the relative electron-ion mass which plays the rôle of a mean mass,

$$m_{ee} \to m_{ie}; \qquad m_{ii} \to m_{ie}; \qquad \frac{1}{m_{ie}} = \frac{1}{m_e} + \frac{1}{m_i}. \tag{7.54}$$

Formally, the plasma behaves in MMA like a mass-symmetrical plasma. Following Ebeling et al. (1976) and Ebeling et al. (1991), the MMA is appropriate for lower temperature $T < 10^5$ K and is asymptotically exact for $\xi \gg 1$ up to the order $\mathcal{O}(\kappa\lambda_{ie})$. As already said, in the MMA all relative thermal wave lengths are replaced by the electron-ion wave length,

$$\lambda_{ab} \to \lambda_{ie} = \frac{\hbar}{\sqrt{2m_{ie}k_B T}}. \tag{7.55}$$

In this approximation, which is not *ad hoc* but based on properties of the virial functions, $Q(x)$, introduced previously, we obtain for the Coulomb energy and the excess chemical potential of the free charges in electron-ion plasmas

$$U_c = -V k_B T \frac{\kappa^3}{8\pi} G(\kappa\lambda_{ie}); \qquad \mu_{ex} = -e^2 \kappa G(\kappa\lambda_{ie}), \tag{7.56}$$

with the G-function

$$G(x) = \frac{1}{1 + \frac{\sqrt{\pi}}{4}x}. \tag{7.57}$$

For two-component systems, κ assumes the value $\kappa^2 = 8\pi\beta n e^2$. This result was originally proven for the quantum Debye–Hückel approximation (QDHA) discussed in Sect. 6.4.2. As shown recently by Ebeling (2016), it can be generalized also to Baimbetov-type approximations, just by using a different G-function,

$$G(x) = \frac{1}{\sqrt{1 + \frac{\sqrt{\pi}}{2}x}}; \qquad x = \kappa\lambda_{ie}. \tag{7.58}$$

For convenience, here we consider only QDHA in combination with MMA. The MMA delivers asymptotically correct results for $k_B T \ll I$. This is the region, where the Saha equation is applied. There are no matrix problems and the expressions are rather simple. Following the above given approximation for the thermodynamic functions and combining it with MMA, the corresponding non-ideal Saha equation for the degree of ionization, α, reads with the ring function of the QDHA (Ebeling et al. 1991)

$$\frac{1-\alpha}{\alpha^2} = n_0 \Lambda_e^3 \, \sigma(T) \, e^{-\beta e^2 \kappa_0 \sqrt{\alpha} \, G(x)} \, ; \qquad x = \kappa_0 \lambda_{ie} \sqrt{\alpha} \, , \qquad (7.59)$$

where $\kappa_0^2 = 8\pi n_0 e^2$ and n_0 is the total atom density. Note that in the QDHA, the excess pressure is given by

$$\beta p = n_0(1+\alpha) - \frac{1}{24\pi} \left(\kappa_0 \sqrt{\alpha}\right)^3 \phi(z) \qquad (7.60)$$

with

$$z = \frac{\sqrt{\pi\alpha} \, \kappa_0 \lambda_{ie}}{4} \, ; \qquad \phi(z) = \frac{3}{z^3} \left[\ln(1+z) - z + \frac{1}{2}z^2\right] . \qquad (7.61)$$

In this approximation, we include the non-ideal terms into the Saha equations in QDHA, however, only Coulomb contributions so far. An interesting property is, that the expressions for thermodynamic functions, in particular the pressure, becomes unstable for higher density. The stability criterion, $\partial p / \partial V < 0$, leads in our approximation to

$$\left(\beta e^2 \kappa\right) \left[G(\kappa\lambda_{ie}) + (\kappa\lambda_{ie}) \, G'(\kappa\lambda_{ie})\right] < 4 \, . \qquad (7.62)$$

This inequality leads in QDHA to instabilities below $T_c \approx 12600\,\text{K}$. The phase transition related to this instability was predicted by Norman and Starostin (1968, 1970). For more detailed investigations see Ebeling et al. (1976, 1991, 2009), Ebeling and Norman (2003), Filinov et al. (2001, 2003), Trigger et al. (2003), and Norman et al. (2015). The critical point obtained here on the basis of our rather elementary QDHA seems to be too high. Recent estimates by Lorenzen et al. (2010) obtained with molecular dynamics and density functional theories lead to a critical point of thermodynamics which is considerably lower. Other ab initio methods were applied to this problem by Norman et al. (2015). By now, several experimental groups found this transition in high-pressure experiments (Fortov 2009, 2011, 2013; Fortov et al. 2007; Knudson et al. 2015). We cannot go in more detail here, since plasma phase transitions are already effects of strong non-ideality, which are beyond the scope of this text.

7.4   Further Problems of Non-ideality in Plasmas

For the case of stronger correlations—*non-ideal plasmas*—which is located in the so-called *corner of correlations*, demonstrated for hydrogen in Fig. 6.4, the physical properties of the plasmas can change drastically in comparison to ideal or weakly non-ideal plasmas. Not only the thermodynamic functions but also the energy levels can change, due to interaction effects. At high density, the bound states can disappear completely. These effects were first studied in early papers by Inglis and Teller (1939). These researchers investigated the merging of levels due to the Stark effect and derived an equation which represents an approximate relationship between an energy level and the electron number density corresponding to the last merging level. Since the Stark effect shifts spectral lines (energy levels), for some energy level n, the splitting is equal to the difference between the adjoining energy level. Beyond this level, spectral lines merge. The equation derived by David R. Inglis and Edward Teller in 1939 for the relation between plasma density and the merging energy level for hydrogen atom is given by

$$\log(n_i + n_e) = 23.491 - 7.5 \log (s_m) . \tag{7.63}$$

Here, s_m shows the merging level, n_i, n_e, the number density of ions and electrons per cm^3. The Inglis-Teller equation is directly related to astrophysics as the electron density of stars is determined using this equation. A different physical effect which also can lead to the disappearance of levels is due to the influence of screening at finite density. This effect was investigated in detail by Nevill Mott (1905–1996) who showed also that this effect plays an essential rôle in semiconductors (Mott 1961; see Fig. 7.4).

The theory of the so-called Mott effect, that is the disappearance of energy levels due to screening, effects is in detail rather complicated (Kremp et al. 2005), and requires the solution of the Bethe-Salpeter equation. The physical idea underlying this theory is, however, easy to understand (Ebeling et al. 1976, 1977; Redmer 1997; Redmer and Röpke 2010). The energy levels in a plasma are only weakly influenced by the density but in first approximation they agree with the unperturbed levels, for instance, for hydrogen

$$E_s' \simeq -\frac{I}{s^2} . \tag{7.64}$$

Fig. 7.4 Nevill Mott, pioneer of screening concepts in solid-state plasma physics. Source: http://www.nndb. com/people/913/000099616/

In a better approximation including screening, the bound state levels correspond to the so-called Ecker-Weizel potential which has unshifted levels (Ecker and Weizel 1956)

$$V^{EW}(r) = -\frac{e^2}{r}\left(e^{-r/r_D} + \frac{1}{r_D}\right).$$ (7.65)

What changes with density, is the series limit which is in first approximation

$$E_\infty = 0 - \frac{e^2}{r_D},$$ (7.66)

where r_D is the Debye radius defined above. We see, that the ground state level disappears according to this estimate at $r_D < 2\,a_B$. Later studies found a more precise value of the Debye radius at which the levels disappear: $r_D < 0.84 a_B$. Mott effects are of different origin (Kremp et al. 2005; Redmer 1997; Redmer and Röpke 2010).

To summarize our findings: The ionization energies in plasma strongly depend on non-ideality effects. In particular the continuum is influenced by density and temperature. With increasing non-ideality the levels merge into the continuum and the discrete states disappear.

The first theories for the thermodynamic properties of weakly correlated quantum plasmas were obtained by Vedenov and Larkin (1959), based on the quantum field theoretical methods developed by Abrikosov et al. (1962, 1965). Quite similar results for the thermodynamic functions including higher approximations were given by De Witt (1961) based on methods developed by Montroll and Ward (1958). The greatest influence on our understanding of non-ideality effects on the properties of plasmas (and gases) had numerical methods as Monte Carlo methods and microscopic simulations pioneered by Alder and Wainwright, see Fig. 7.5. A whole class of algorithms for many particle dynamical simulations was developed

Fig. 7.5 Berni
Alder—pioneer of
microscopic simulations of
gases and plasmas. Source:
https://www.llnl.gov/news/
bernialder-pioneer-times

following the earlier successes of Monte Carlo simulations by Alder and Wainwright in late 1950s and independently by Rahman in the 1960s. Alder and Wainwright (1957) used an IBM 704 computer to simulate perfectly elastic collisions between hard spheres. Rahman (1964) published landmark simulations of liquid argon that utilized a Lennard–Jones potential. Calculations of system properties, such as the coefficient of self-diffusion, compared well with experimental data. DeWitt (1976) presented very precise Monte Carlo simulations for one-component and multiple-component plasmas. Among others we mention also the contributions to the field by Ceperley (1995), Militzer and Pollock (2000) and Zamalin et al. (1977).

An idea, which played an essential rôle in the development of the theory of correlated plasmas is connected with the ideas of collective coordinates (Kelbg 1963; Pines 1961; Pines and Nozieres 1966) and with the methods of effective potentials or pseudopotentials, see Sect. 6.2. This idea was introduced first in quantum chemistry by Hellmann (1935) and Gombás (1965), in solid state physics by Heine (1970) and others and in statistical physics by Morita (1959). Since 1962 Kelbg and co-workers at the Rostock University have developed this method (Kelbg 1963, 1972). The idea was to replace the Coulomb potential by an effective potential which is finite at zero distance due to quantum effects. The Rostock School on Quantum Statistics formed by Günter Kelbg as well as the group of Hugh DeWitt in Livermore (see Figs. 7.6 and 7.7) concentrated on analytical calculations of thermodynamic functions. Other authors developed Monte Carlo calculations using effective potentials (Binder 1979; Deutsch 1977; Zamalin et al. 1977) introduced useful approximations and Hansen and McDonald (1981) and Hansen et al. (1975) gave first applications to molecular dynamics (see Kalman et al. 1998).

Let us discuss now the problems related to stronger coupling, that is, large deviations from non-ideality. Strongly correlated Coulomb systems have been in the focus of recent investigations in many fields, including dense plasmas in space and laboratory (Kraeft et al. 1986; Kremp et al. 2005), in electron-hole plasmas in semiconductors, charged particles confined in traps or storage rings. In these systems, on average the Coulomb interaction energy, U, is larger than the mean kinetic energy, K, i.e. the coupling parameter $\Gamma = |\langle U \rangle|/\langle K \rangle > 1$. In particular, Coulomb and Wigner crystallisation which can occur when $\Gamma \gg 1$ (typically of

Fig. 7.6 Hugh DeWitt, pioneer of analytical and numerical methods for the treatment of non-ideal plasmas in 1977 in Orlean at the first conference on strongly coupled plasmas (author's archive)

Fig. 7.7 Günter Kelbg
founded a school of quantum
statistics at Rostock
University (author's archive)

the order of 100) attracted great attention in recent years. Coulomb crystals have
been observed in ultracold trapped ions, e.g. in dusty plasmas, and in storage
rings. Moreover, there exist many strongly correlated Coulomb systems where
quantum effects are important (Saumon and Chabrier 1989). Examples are dense
astrophysical plasmas in the interior of giant planets or white dwarf stars as well
as electron-hole plasmas in condensed matter, few-particle electron or exciton
clusters in mesoscopic quantum dots (Redmer 1997; Redmer and Röpke 2010). The
formation of Coulomb bound states such as atoms and molecules or excitons and bi-
excitons, of Coulomb liquids and electron-hole droplets are examples of the large
variety of correlation phenomena that exist in these systems (Kremp et al. 2005).
More information on these topics can be found in (Ebeling et al. 2017).

Summarizing this chapter, we provided a complete quantum statistical description of plasmas as Coulomb systems, including bound states, by density and fugacity
expansions and through a chemical approach. The latter approach is based on the
assumption that the system is a gas mixture of chemical species whose composition
is found from fugacity expansions or by solving equivalent mass action laws.

References

Abrikosov, A.A.., L.P. Gor'kov, and I.E. Dzyaloshinskii. 1962. *Quantum Field Theory Techniques
in Statistical Physics* (in Russian). Moscow: Fizmatgiz.
Abrikosov, A.A., L.P. Gor'kov, and I.E. Dzyaloshinskii. 1965. *Methods of Quantum Field Theory
in Statistical Physics*. Oxford: Pergamon Press.
Alastuey, A., and V. Ballenegger. 2010. Pressure of a Partially Ionized Hydrogen Gas: Numerical
results From Exact Low Temperature Expansions. *Contributions to Plasma Physics* 50: 46–53.
Alastuey, A., and A. Perez. 1992. Virial Expansion of the Equation of State of a Quantum Plasma.
Europhysics Letters 20: 19–24.
Alastuey, A., and A. Perez. 1996. Virial Expansions for Quantum Plasmas: Fermi-Bose Statistics.
Physical Review E 53: 5714–5728.
Alder, B., and T. Wainwright. 1957. Phase Transition for a Hard Sphere System. *The Journal of
Chemical Physics* 27: 1208–1209.

Alastuey, A., V. Ballenegger, F. Cornu, and Ph.A. Martin. 2008. Exact Results for Thermodynamics of the Hydrogen Plasma: Low-Temperature Expansions Beyond Saha Theory. *Journal of Statistical Physics* 130: 1119–1176.

Alastuey, A., V. Ballenegger, and W. Ebeling. 2015. Comment on 'Direct Linear Term in the Equation of State of Plasmas' by Kraeft et al. (2015a). *Physical Review E* 92 (see also Kraeft et al. 2015b), 047101.

Bartsch, G.P., and W. Ebeling. 1971. Quantum Statistical Fugacity Expansions for Partially Ionized Plasmas in Equilibrium. *Contributions to Plasma Physics* 11.5: 393–403.

Binder, K., ed. 1979. *Monte Carlo Methods in Statistical Physics*. Berlin: Springer.

Brown, L.S., and L.G. Yaffe. 2001. Effective Field Theory of Highly Ionized Plasmas. *Physics Reports* 340: 1–164.

Ceperley, D.M. 1995. Path Integrals in the Theory of Condensed Helium. *Reviews of Modern Physics* 67: 279–356.

Deutsch, C. 1977. Nodal Expansion in a Real Matter Plasma. *Physics Letters A* 60: 317–318.

DeWitt, H.E. 1961. Thermodynamic Functions of a Partially Degenerate, Fully Ionized Gas. *International Journal of Nuclear Energy Part C: Plasma Physics* 2: 27–45.

DeWitt, H.E. 1976. Asymptotic Form of the Classical One-Component Plasma Fluid Equation of State. *Physical Review A* 14: 1290–1293.

DeWitt, H.E., M. Schlanges, A.Y. Sakakura, and W.D. Kraeft. 1995. Low Density Expansion of the Equation of State for a Quantum Electron Gas. *Physics Letters A* 197: 326–329.

Ebeling, W. 1968. Ableitung der freien Energie von Quantenplasmen kleiner Dichte aus den exakten Streuphasen. *Annals of Physics* 477: 33–39.

Ebeling, W. 1969. Zur Quantenstatistik der Bindungszustände in Plasmen. I Cluster-Entwicklungen. *Annals of Physics* 22: 383–391.

Ebeling, W. 1974. Statistical Derivation of the Mass Action Law or Interacting Gases and Plasmas. *Physica* 73: 573–584.

Ebeling, W. 2016. The Work of Baimbetov on Nonideal Plasmas and Some Recent Developments. *Contributions to Plasma Physics* 56: 163–175.

Ebeling, W., and G.E. Norman. 2003. Coulombic Phase Transitions in Dense Plasmas. *Journal of Statistical Physics* 110: 861–877.

Ebeling, W., W.D. Kraeft, and D. Kremp. 1976. *Theory of Bound States and Ionisation Equilibrium in Plasmas and Solids*. Berlin: Akademie-Verlag.

Ebeling, W., W.D. Kraeft, and D. Kremp. 1977. Nonideal Plasmas. In *Proceedings of the XIIIth International Conference on Phenomena in Ionized Gases*, ed. P. Bachmann, 73–90. Berlin: Physikalische Gesellschaft der DDR.

Ebeling, W., A. Förster, V.F. Fortov, V.K. Gryaznov, and A. Ya. Polishchuk. 1991. *Thermophysical Properties of Hot Dense Plasmas*. Stuttgart: Teubner-Verlag.

Ebeling, W., D. Blaschke, R. Redmer, H. Reinholz, and G. Röpke. 2009. The Influence of Pauli Blocking Effects on the Properties of Dense Hydrogen. *Journal of Physics A: Mathematical and Theoretical* 42: 214033.

Ebeling, W., W.D. Kraeft, and G. Röpke. 2012. On the Quantum Statistics of Bound States within the Rutherford Model of Matter. *Annals of Physics* 524: 311–326.

Ebeling, W., V.E. Fortov, and V.S. Filinov. 2017. *Quantum Statistics of Dense Gases and Nonideal Plasmas*. Springer Series in Plasma Science and Technology. Berlin: Springer.

Ecker, G., and W. Weizel. 1956. Zustandssumme und effektive Ionisierungsspannung eines Atoms im Inneren des Plasmas. *Annals of Physics* 452: 126–140.

Falkenhagen, H. 1971. *Theorie der Elektrolyte*. Leipzig: Hirzel.

Falkenhagen, H., and W. Ebeling. 1971. Equilibrium Properties of Ionized Dilute Electrolytes. *Ionic Interactions*, ed. S. Petrucci, Vol. 1, 1–59. New York, London: Academic Press.

Filinov, V.S., M. Bonitz, W. Ebeling, and V.E. Fortov. 2001. Thermodynamics of Hot Dense H-Plasmas: Path Integral Monte Carlo Simulations and Analytical Approximations. *Plasma Physics and Controlled Fusion* 43 (6): 743–759.

Filinov, A.V., M. Bonitz, and W. Ebeling. 2003. Improved Kelbg Potential for Correlated Coulomb Systems. *Journal of Physics A: Mathematical and General* 36 (22): 5957–5962.

Fortov, V.E. 2009. *Extreme States of Matter* (in Russian). Moskva: Fiz-MatGis.

Fortov, V.E. 2011. *Extreme States of Matter: On Earth and in the Cosmos*. Berlin: Springer.

Fortov, V.E. 2013. *Equation of State From the Ideal Gas to the Quark-Gluon Plasma* (in Russian). Moskva: FizMatGis.

Fortov, V.E., R.I. Ilkaev, V.A. Arinin, V.V. Burtzev, V.A. Golubev, I.L. Iosilevskiy, et al. 2007. Phase Transition in a Strongly Nonideal Deuterium Plasma Generated by Quasi-Isentropical Compression at Megabar Pressures. *Physical Review Letters* 99: 185001.

Friedman, H.L. 1962. *Ionic Solution Theory*. New York: Interscience.

Friedman, H.L., W. Ebeling. 1979. Theory of Interacting and Reacting Particles. *Rostocker Physikalische Manuskripte* 4: 33–48.

Gombás, P. 1965. Pseudopotentiale. *Fortschritte der Physik* 13: 137–156.

Hansen, J.P., and I.R. McDonald. 1981. Microscopic Simulation of a Strongly Coupled Hydrogen Plasma. *Physical Review A* 23: 2041–2059.

Hansen, J.P., I.R. McDonald, and E.L. Pollock. 1975. Statistical Mechanics of Dense Ionized Matter. III. Dynamical Properties of the Classical One-Component Plasma. *Physical Review A* 11: 1025–1039.

Heine, V. 1970. The Pseudopotential Concept. *Solid State Physics* 24: 1–36.

Hellmann, H. 1935. A New Approximation Method in the Problem of Many Electrons. *The Journal of Chemical Physics* 3: 61–61.

Hill, T.L. 1956. *Statistical Mechanics*. New York: McGraw Hill.

Hoffmann, H.J., and W. Ebeling. 1968a. On the Equation of State of Fully Ionized Quantum Plasmas. *Physica* 39: 593–598.

Hoffmann, H.J., and W. Ebeling. 1968b. Quantenstatistik des Hochtemperatur-Plasmas im thermodynamischen Gleichgewicht. II. Die freie Energie im Temperaturbereich 10^6 bis 10^8 °K. *Contributions to Plasma Physics* 8 (1): 43–56.

Inglis, D.R., and E. Teller. 1939. Ionic Depression of Series Limits in One-Electron Spectra. *Astrophysics* 90: 439–448.

Kahlbaum, T. 2000. The Quantum-Diffraction Term in the Free Energy for Coulomb Plasma and the Effective-Potential Approach. *Journal de Physique IV France* 10: 455–459.

Kalman, G.J., J.M. Rommel, and K. Blagoev, eds. 1998. *Strongly Coupled Coulomb Systems*. New York: Springer.

Kelbg, G. 1963. Quantenstatistik der gase mit Coulomb-Wechselwirkung. *Annalen der Physik* 467: 354–360.

Kelbg, G. 1964. Klassische statistische Mechanik der Teilchen-Mischungen mit sortenabhängigen weitreichenden zwischenmolekularen Wechselwirkungen. *Annals of Physics* 14: 394–403.

Kelbg, G. 1972. *Einige Methoden der statistischen Thermodynamik hochionisierter Plasmen, Ergebnisse der Plasmaphysik und Gaselektronik*, Vol. Bd. III. Berlin: Akademie-Verlag.

Knudson, M.D., M.P. Desjarlais, A. Becker, R.W. Lemke, K.R. Cochrane, M.E. Savage, et al. 2015. Direct Observation of an Abrupt Insulator-to-Metal Transition in Dense Liquid Deuterium. *Science* 348(6242): 1455–1460.

Kraeft, W.D., D. Kremp, W. Ebeling, and G. Röpke. 1986. *Quantum Statistics of Charged Particle Systems*. Berlin: Akademie-Verlag.

Kraeft, W.D., D. Kremp, and G. Röpke. 2015a. Direct Linear Term in the Equation of State of Plasmas. *Physical Review E* 91: 013108.

Kraeft, W.D., D. Kremp, and G. Röpke. 2015b. Reply to Alastuey et al. (2015). *Physical Review E* 92: 047102.

Kremp, D., M. Schlanges, and W.D. Kraeft. 2005. *Quantum Statistics of Nonideal Plasmas*. Berlin: Springer.

Lorenzen, W., B. Holst, and R. Redmer. 2010. First-Order Liquid-Liquid Phase Transition in Dense Hydrogen. *Physical Review B* 82: 195107.

Meeron, E. 1962. On Cluster Theory. In *Electrolytes. Proceedings of an International Symposium held in Trieste, June 1959*, 7, ed. B. Pesce. Oxford: Pergamon.

Militzer, B., and E.L. Pollock. 2000. Variational Density Matrix Method for Warm Condensed Matter and Application to Dense Hydrogen. *Physical Review E* 61: 3470–3482.

Montroll, E., and J. Ward. 1958. Quantum Statistics of Interacting Particles. *Physics of Fluids* 1: 55–72.

Morita, T. 1959. Equation of State of High Temperature Plasma. *Progress in Theoretical Physics* 22: 757–774.

Mott, N. 1961. The Transition to the Metallic State. *Philosophical Magazine* 6: 287–309.

Norman, G.E., I.M. Saitov, and V.V. Stegailov. 2015. Plasma-Plasma and Liquid-Liquid First-Order Phase Transitions. *Contributions to Plasma Physics* 55: 215–221.

Norman, G.E., and A.N. Starostin. 1968. Description of Nondegenerate Dense Plasma. *Soviet Physics High Temperature* 6: 394–408.

Norman, G.E., and A.N. Starostin. 1970. Thermodynamics of Strongly Nonideal Plasma. *Soviet Physics High Temperature* 8: 381–395.

Pines, D. 1961. *The Many Body Problem—A Lecture Note*. New York: Benjamin.

Pines, D., P. Nozieres. 1966. *The Theory of Quantum Liquids*. New York: Benjamin.

Rahman, A. 1964. Correlations in the Motion of Atoms in Liquid Argon. *Physics Review* 136: A405–A411.

Redmer, R. 1997. Physical Properties of Dense, Low-Temperature Plasmas. *Physics Reports* 282: 35–157.

Redmer, R., and G. Röpke. 2010. Progress in the Theory of Dense Strongly Coupled Plasmas. *Contributions to Plasma Physics* 50: 970–985.

Riemann, J., M. Schlanges, H.E. DeWitt, and W.D. Kraeft. 1995. Equation of State of the Weakly Degenerate One-Component Plasma. *Physica A* 219: 423–435.

Rogers, F.J. 1974. Statistical Mechanics of Coulomb Gases of Arbitrary Charge. *Physical Review A* 10: 2441–2456.

Rogers, F.J., B.G. Wilson, and C.A. Iglesias. 1988. Parametric potential method for generating atomic data. *Physical Review A* 38: 5007–5020.

Saumon, D., and C. Chabrier. 1989. Fluid Hydrogen at High Density: The Plasma Phase Transition. *Physical Review Letters* 62: 2397–2400.

Trigger, S.A., W. Ebeling, V.S. Filinov, V.E. Fortov, and M. Bonitz. 2003. Internal Energy of High Density Hydrogen: Analytic Approximations Compared with Path Integral Monte Carlo Calculations. *Journal of Experimental and Theoretical Physics* 96: 465–479.

Vedenov, A.A., and A.I. Larkin. 1959. Equation of State of Plasmas. *Zhurnal Eksperimental'noi i Teoreticheskoi Fiziki* 36: 1133.

Zamalin, V.M., G.E. Norman, and V.S. Filinov. 1977. *The Monte Carlo Method in Statistical Thermodynamics* (in Russian). Moscow: Nauka.

Chapter 8
Non-equilibrium: Kinetic Equations

8.1 Development of Classical and Quantum Kinetic Theory

The pioneers of the theory of classical kinetic equations are Rudolf Clausius (1822–1888), James Clerk Maxwell (1831–1879) and Ludwig Boltzmann (1844–1906). Their theories are based on the classical dynamics of particles in the phase space according to Liouville and on detailed studies of the classical mechanics of collisions between neutral molecules. The kinetic equation by Boltzmann was formulated for the classical distribution function, $f(\mathbf{r}, \mathbf{v}, t)$, of molecules and reads:

$$\frac{\partial f}{\partial t} + \mathbf{v} \cdot \frac{\partial f}{\partial \mathbf{r}} = \left(\frac{\partial f}{\partial t}\right)_{\text{coll}}. \tag{8.1}$$

The left-hand side of the Boltzmann equations describes the Liouville-type mechanical changes of the density in coordinate-velocity space and the right-hand side the influence of the collisions, which we gave so far only in symbolic form.

The quantum kinetic theory was developing only half a century later. Essential sources of quantum kinetics are papers by Einstein (1916, 1917a,b) and Pauli (1928). A quite important rôle played also the early studies of diffusion processes by Einstein (1905, 1906), Fokker (1914), Klein (1922), and Planck (1917) as well as the contributions to quantum statistics by von Neumann, Landau, Uhlenbeck, Tolman, Bogolyubov and others (Bogolyubov 1991, 2005–2009; Klimontovich 1982, 1983, 1997; Landau and Lifshitz 1976, 1990; Lifshitz and Pitaevskii 1981; von Neumann 1932). In Chap. 4 some the fundamental tools relevant for this field were already developed.

Here we introduce some basic methods of quantum kinetics which lead to quantum kinetic equations, however, restricted to ideal or weakly non-ideal gases and plasmas. These methods are in fact based on the early papers mentioned above.

The transport theory of plasmas started with adapting the classical kinetic equation by Boltzmann to the specific conditions in plasmas. The Boltzmann-type

© Springer Nature Switzerland AG 2019
W. Ebeling, T. Pöschel, *Lectures on Quantum Statistics*,
Lecture Notes in Physics 953, https://doi.org/10.1007/978-3-030-05734-3_8

kinetic equation for the distribution function, $f(\mathbf{r}, \mathbf{v}, t)$, of plasma particles reads:

$$\frac{\partial f}{\partial t} + \mathbf{v} \cdot \frac{\partial f}{\partial \mathbf{r}} + \frac{e}{m} \left(\mathbf{E} + \mathbf{v} \times \frac{1}{c} \mathbf{H} \right) \cdot \frac{\partial f}{\partial \mathbf{v}} = \left(\frac{\partial f}{\partial t} \right)_{\text{coll}}. \qquad (8.2)$$

The left-hand side describes the Liouville-type mechanical changes, including electrical and magnetic fields and the right-hand side the influence of collisions. The direct application of the Boltzmann-theory leads, however, to divergencies related to the long-range tails of the Coulomb forces. In 1937, Lev D. Landau (1908–1968) showed in a fundamental work (Landau 1936, 1937), how the collision theory for Coulomb forces has to be formulated. It was then Anatoly A. Vlasov (1908–1975), who discussed first in detail, how the electromagnetic fields which modify the left-hand side of the Boltzmann equation, influence the kinetic theory (Vlasov 1950). Vlasov neglected the collision integral at the right-hand side, calculated the fields by using Maxwell equations and studied in detail the solutions of the resulting nonlinear equations, in particular the coupling to the Poisson equation relating the electric field with density. The Vlasov equation, which is related to the Liouville equation and the collisionless Boltzmann equation, is fundamental to plasma physics. Vlasov (1945) showed that this equation, with the collective interaction taken into account, can explain without any additional hypotheses and specifications many plasma effects as e.g. plasma waves. Strictly speaking, the Vlasov equation describes reversible phenomena. Nevertheless damping phenomena were observed in particle beams interacting with the plasma. These effects are based on the energy exchange between an electromagnetic wave with phase velocity, v_{ph}, and particles in the plasma with velocity approximately equal to v_{ph}, which can strongly interact with the wave. The theoretical interpretation was given by Lev Landau, therefore, the effect is called Landau damping. According to Landau's picture, the particles having velocities slightly less than v_{ph} are accelerated by the electric field of the wave to move with the wave phase velocity, while those particles with velocities slightly larger than v_{ph} are decelerated, thus, losing energy to the wave. In the Vlasov-Landau theory, the deviations of the distribution functions from a Maxwell distribution were studied. The development of the kinetic theory of plasmas culminated in the derivation of a more strict kinetic theory of plasmas based on the work of Bogolyubov, Klimontovich, Silin, Rukhadse, Balescu and others (Balescu 1963; Ichimaru 1973, 1992; Klimontovich 1967). Spitzer (1962) wrote the first review of the state of art in kinetic theory including solutions of the kinetic equations; for a more recent review, see Bonitz (1998) and Bonitz and Kraeft (2005) (for photographs of two of the main pioneers see Figs. 8.1 and 8.2).

The following sections are restricted to rarefied gases and dilute plasmas. The reader who is interested in applications to dense gases and strongly coupled plasmas is referred to (Binder 1979; Bonitz 1998; Bonitz and Kraeft 2005; Bonitz and Semkat 2006; Bonzel et al. 1989; Brey et al. 2009; Ebeling et al. 1984, 1991, 2017; Fortov 2009, 2011; Fortov and Yakubov 1994; Fortov et al. 2007; Ichimaru 1992; Kraeft et al. 1986; Kremp et al. 2005; Militzer and Pollock 2000; Nellis et al. 1999; Pöschel and Luding 2001; Redmer 1997; Redmer and Röpke 2010; Ternovoi et al.

Fig. 8.1 Yuri Klimontovich (1924–2002), pioneer of kinetic plasma theory at a visit to Rostock University in the 70th (authors archive)

Fig. 8.2 Radu Balescu (1932–2006), pioneer of plasma kinetic theory, at a Conference on Plasma Theory in Kiev in the 70th (authors archive)

1999; Weir et al. 1996; Zubarev et al. 1996). For extending the theory of dilute gases and plasmas presented here to more dense gases and non-ideal plasmas, numerical methods based on Monte Carlo or Molecular Dynamics algorithms are of increasing importance (Binder 1979; Bonitz and Semkat 2006; Brush et al. 1966; Metropolis et al. 1953; Zamalin et al. 1977). For the fundamental topic of dissipative quantum systems and transport see Dittrich et al. (1998) and Ingold (2002).

8.2 Pauli's Master Equation Approach and H-Theorem

We start the introduction of quantum kinetic theory with a simple method, presented by Pauli (1928) at the occasion of the 60th birthday of Arnold Sommerfeld. We continue with the tight-binding approximation and the Lorentz approximation. We discuss the derivation of quantum Boltzmann equations and, finally, we discuss Klimontovich's kinetic method and fluctuation theory.

In the current section, we introduce Pauli's approach and study a system with a finite spectrum of eigenstates, $i = 1, \ldots, s$. Correspondingly we define the probabilities, $p_i(t)$, of the states of the system. By definition, probabilities have to satisfy some conservation rules such as the normalization,

$$\sum_{i=1}^{s} p_i(t) = 1 . \tag{8.3}$$

The evolution of probabilities in time should conserve the normalization. A rather general kinetic equation which meets this condition is the *Pauli-equation* (Pauli 1928):

$$\frac{d}{dt} p_m(t) = \sum_{m'}^{s} [W_{mm'} p_{m'}(t) - W_{m'm} p_m(t)] . \tag{8.4}$$

We introduced here the transition rates from state m' to state m,

$$W_{m \leftarrow m'} = W_{mm'} . \tag{8.5}$$

According to Pauli's postulate, the transition rates are symmetric

$$W_{mm'} = W_{m'm} , \tag{8.6}$$

which expresses the reversibility of the microscopic quantum mechanics and is in close relation to Onsager's symmetry postulate of nonequilibrium thermodynamics (Ebeling and Sokolov 2005). With the initial conditions for time $t = 0$:

$$p_m = 1 ; \quad p_{m'} = 0 , \tag{8.7}$$

for short time we find with

$$\frac{d}{dt} p_m(t) = -R_m p_m(t) \tag{8.8}$$

and

$$R_m = \sum_{m'}^{s} W_{mm'} \tag{8.9}$$

the solution valid for $t \ll 1/R_m$:

$$p_m(t) = e^{-R_m t} p_m(0) . \tag{8.10}$$

We see that the evolution in time is irreversible and can show further that in the limit of large time, $t \gg 1/R_m$, we obtain the equi-distribution,

$$p_m(t) \rightarrow \frac{1}{s} \qquad \text{for all m.} \tag{8.11}$$

Inserting this into Eq. (8.4), we can check that indeed we found a solution:

$$\frac{d}{dt} p_m(t) = \sum_{m'}^{s} (W_{mm'} - W_{m'm}) \frac{1}{s} = 0, \tag{8.12}$$

that is, the equi-distribution is a stationary solution. As we know, the equi-distribution corresponds to a micro-canonical ensemble. The stationary solution corresponding to a canonical ensemble reads

$$p_m^{(0)} = \frac{1}{Q} e^{-\beta E_m}. \tag{8.13}$$

The requirement that this is a stationary solution leads us to some properties of the transition matrix whose components, evidently, should depend on temperature, $W_{mm'}(\beta)$. In the stationary state, we obtain the condition

$$\sum_{m'} \left[W_{mm'}(\beta) e^{-\beta E_{m'}} - W_{m'm}(\beta) e^{-\beta E_m} \right] = 0. \tag{8.14}$$

Following Pauli we request that this balance is fulfilled not only globally but for each individual step, called the condition of *detailed balance*,

$$W_{mm'}(\beta) e^{-\beta E_{m'}} = W_{m'm}(\beta) e^{-\beta E_m}. \tag{8.15}$$

There exists a whole class of transition matrices which meet these conditions, some of them will be discussed below. The most simple transition matrix which describes the embedding of the system into a heat bath at temperature, T, is

$$W_{mm'}(\beta) = W_{mm'} \begin{cases} 1 & \text{for} \quad E_m < E_{m'} \\ e^{-\beta(E_m - E_{m'})} & \text{for} \quad E_m > E_{m'}, \end{cases} \tag{8.16}$$

where $\beta = 1/k_B T$ is the reciprocal temperature and k_B is the Boltzmann constant. This expression defines the so-called Monte-Carlo process. That is, a quantum mechanical transition taked always place if it is related to a decrease of energy. If the transition would lead to an increase of the energy, it takes place with some smaller probability which contains a Boltzmann factor.

In the next section, we discuss methods to estimate (symmetric and temperature-independent) quantum mechanical transition matrices, $W_{mm'}$. In the context of the

Pauli equation, consider the *Shannon entropy, Kullback information and H-theorem*
The Shannon-Entropy of a distribution, $P = (p_1, p_2, \ldots, p_n)$, is defined as

$$H(P) = -\sum_{i=1}^{s} p_i \log (p_i) . \tag{8.17}$$

In analogy, we define the Kullback-information or relative entropy by

$$K\left(P, P'\right) = \sum_{i=1}^{s} p_i \log \left(\frac{p_i}{p_i'}\right) . \tag{8.18}$$

The Kullback entropy depends on two distributions, P and P'. It is positive definit,
$K(P, P') \geq 0$ and vanishes for the case $P = P'$. The Kullback information can
be considered as a distance between the distributions P and P'. Now we choose
$P' = P^{(0)}$, where $P^{(0)}$ is the stationary distribution which solves the equation

$$\sum_{m'}^{s} \left[W_{mm'} p_{m'}^{(0)} - W_{m'm} p_m^{(0)} \right] = 0 . \tag{8.19}$$

This is a linear equation and the conditions for solvability are well known. In
particular, we can be sure that the solution is unique. For the most simple example
the equi-distribution, we obtain

$$H\left(P^{(0)}\right) = \log s = H_{\max} , \tag{8.20}$$

which correspond to the largest possible value the Shannon entropy can assume. For
the Kullback entropy we find

$$K\left(P, P^{(0)}\right) = \sum_{i=1}^{s} p_i \log (s\, p_i) = \log s + \sum_{i=1}^{s} p_i \log p_i = H_{\max} - H > 0 . \tag{8.21}$$

This means that in the given case the Kullback entropy, $K(P, P')$, is the difference
between the maximal and the actual value of the Shannon entropy. For the case that
the canonical distribution is the stationary distribution,

$$p_i^{(0)} = \frac{1}{Q} e^{-\beta E_i} = \frac{e^{-\beta E_i}}{\sum_{i=1}^{s} e^{-\beta E_i}} \tag{8.22}$$

we obtain the Kullback entropy

$$K\left(P, P^{(0)}\right) = \sum_{i=1}^{s} p_i \log p_i + \sum_{i=1}^{s} \log Q + \beta \sum_{i=1}^{s} p_i E_i$$

$$= \beta \langle E \rangle - H(P) + \log Q, \qquad (8.23)$$

and, therefore,

$$k_B T K\left(P, P^{(0)}\right) = \langle E \rangle - k_B T H(P) - k_B T \log Q = F - F_{eq} > 0. \qquad (8.24)$$

We will prove now a theorem for the evolution of the Kullback information. To study evolution of the function K, we differentiate with respect to time and use the Pauli equation for the evolution of the probabilities:

$$\frac{d}{dt}\left[K\left(P, P^{(0)}\right)\right] = \sum_{i=1}^{s} \dot{p}_i \log \frac{p_i}{p_i^{(0)}}$$

$$= \sum_{\substack{i=1 \\ j=1}}^{s} \left(W_{ij} p_j \log \frac{p_i}{p_i^{(0)}} - W_{ji} p_i \log \frac{p_i}{p_i^{(0)}}\right)$$

$$= \sum_{\substack{i,j=1 \\ i \neq j}}^{s} W_{ij} p_j \log\left(\frac{p_i p_j^{(0)}}{p_i^{(0)} p_j}\right). \qquad (8.25)$$

With the inequality

$$\log\left(\frac{p_i p_j^{(0)}}{p_i^{(0)} p_j}\right) \leq \frac{p_i p_j^{(0)}}{p_i^{(0)} p_j} - 1, \qquad (8.26)$$

we transform Eq. (8.25) to the inequality

$$\frac{d}{dt}\left(K(P, P^{(0)})\right) \leq \sum_{\substack{i=1 \\ j=1 \\ i \neq j}}^{s} W_{ij} p_j \left(\frac{p_i p_j^{(0)}}{p_i^{(0)} p_j} - 1\right)$$

$$= \sum_{\substack{i=1 \\ j=1}}^{s} p_i \sum_{j=1}^{s} W_{ji} - \sum_{\substack{i=1 \\ j=1}}^{s} p_j \sum_{i=1}^{s} W_{ij} = \sum_{i=1}^{s} \dot{p}_i = 0. \qquad (8.27)$$

Consequently, the Kullback functional is either monotonically decreasing or constant. For the case that the stationary distribution is the equi-distribution, corresponding to a micro-canonical ensemble with $p_i^{(0)} = 1/s$, we find

$$- \dot{K}\left(P, P^{(0)}\left(\frac{1}{s}\right)\right) = \dot{H}(P). \tag{8.28}$$

This means that the entropy, H, is monotonically increasing and asymptotically reaching the equilibrium state, corresponding to the equi-distribution.

Quantum Statistical Transition Probabilities
Following Pauli, above we assumed that for a discrete quantum system with a finite number of states, there exists a well-defined probability describing the transitions between the states. Again following Pauli, here we calculate these transition probabilities. We start with the von-Neumann equation (see Chap. 4)

$$i\hbar \frac{\partial}{\partial t}\hat{\rho}(t) = \left[\hat{H}, \hat{\rho}\right]. \tag{8.29}$$

Here $\hat{\rho}(t)$ is the time-dependent density operator and square brackets stand for commutators. The matrix elements are calculated with a complete set of eigenfunctions, $\rho_{mn} = \langle m| \hat{\rho}(t) n\rangle$. In the following, the diagonal elements are denoted by $P_m(t) = \langle m| \hat{\rho}(t) m\rangle$. We will prove that under certain conditions, these probabilities satisfy the Pauli equations. This fundamental property is related to the reversibility of the von-Neumann equation, in contrast to the Pauli equation which is irreversible. These properties are, by far, not trivial (Ebeling and Sokolov 2005; Tolman 1938; Zubarev et al. 1997).

We write the Hamilton operator of the system in the form

$$\hat{H} = \hat{H}_0 + \lambda\hat{V}, \tag{8.30}$$

assuming a basic term which defines the energy eigenfunctions,

$$\hat{H}_0|m\rangle = E_m^0|m\rangle, \tag{8.31}$$

and some perturbation proportional to a small parameter, λ. The matrix elements

$$H_{mm'} = E_m^0\delta_{mm'} + \lambda V_{mm'} \tag{8.32}$$

consist of the basic contributions and the perturbative terms, $\hat{V}$. We assume $\mathrm{Tr}\left(\hat{V}\right) = 0$. We multiply the von Neumann equation from both sides with energy eigenfunctions

$$i\hbar\frac{\partial}{\partial t}\langle m|\hat{\rho} m\rangle = \left\langle m\left|\left[\hat{H}, \hat{\rho}\right] m\right.\right\rangle \tag{8.33}$$

and write

$$\frac{\partial}{\partial t}\rho_{mm'} = -\frac{i}{\hbar}\sum_{m''}\left(H_{mm''}\rho_{m''m'} - \rho_{mm''}H_{m''m'}\right)$$

$$= -\frac{i}{\hbar}\left(E_m^0 - E_{m'}^0\right)\rho_{mm'} - \frac{i}{\hbar}\lambda\sum_{m''}\left(V_{mm''}\rho_{m''m'} - \rho_{mm''}V_{m''m'}\right) .$$

$$(8.34)$$

For the diagonal elements we obtain

$$\frac{\partial}{\partial t}\rho_{mm} = -\frac{i}{\hbar}\lambda\sum_{m''}\left(V_{mm''}\rho_{m''m} - \rho_{mm''}V_{m''m}\right) . \qquad (8.35)$$

From Eq. (8.34) follows that the transitions between different states are due to the interactions. The parameter λ measures the strength of these interactions. For $\lambda \to 0$ no transitions would occur. The derivative, $\partial P_m(t)/\partial t = \mathcal{O}(\lambda)$, depends on diagonal as well as on non-diagonal elements of the density matrix. Let us assume in some approximation that the diagonal terms, $\rho_{mm}(t)$, contain the essential information on the system and the non-diagonal terms, $\rho_{mm'}$, are small and, thus, less relevant perturbations. Selecting in Eq. (8.34) the diagonal elements, we obtain

$$\frac{\partial}{\partial t}\rho_{mm'} = -\frac{i}{\hbar}\left(E_m^0 - E_{m'}^0\right)\rho_{mm'} - \frac{i}{\hbar}\lambda\left(V_{mm'}\rho_{m'm'} - \rho_{mm}V_{mm'}\right) . \qquad (8.36)$$

In this approximation, the neglected terms in Eq. (8.36) are considered as a perturbation which can be treated as a damping term, $\frac{i}{\hbar}\lambda\{\dots\} \to \gamma_{mm'}\rho_{mm'}$. This way, from Eq. (8.36) we find

$$\frac{\partial}{\partial t}\rho_{mm'} = -\left[\gamma_{mm'} + \frac{i}{\hbar}\left(E_m^0 - E_{m'}^0\right)\right]\rho_{mm'} - \frac{i}{\hbar}\lambda\left(V_{mm'}\rho_{m'm'} - \rho_{mm}V_{mm'}\right) . \qquad (8.37)$$

The non-diagonal terms, $\rho_{mm'}(t)$, are assumed to relax quickly to stationary values on a time scale which is much smaller than the time scale of the diagonal elements. In adiabatic approximation, $\frac{\partial \rho_{mm'}(t)}{\partial t} = 0$, we obtain finally

$$\rho_{mm'} = -i\frac{\lambda/\hbar}{\gamma_{mm'} + \frac{i}{\hbar}\left(E_m^0 - E_{m'}^0\right)}V_{m'm}\left(\rho_{m'm'} - \rho_{mm}\right) , \qquad (8.38)$$

were we took into account the symmetry of the perturbation matrix. Now we have expressed all non-diagonal elements by diagonal elements and can derive a closed equation for the diagonal matrix elements. Inserting Eq. (8.38) into (8.35) and replacing the summation indices, $m'' \rightarrow m'$, we find

$$\frac{\partial}{\partial t}\rho_{mm}(t) = -\left(\frac{\lambda}{\hbar}\right)^2 \sum_{m' \neq m} \left\{ |V_{m'm}|^2 \left[\left(\frac{1}{\gamma_{m'm} + \frac{i}{\hbar}\left[E^0_{m'} - E^0_m\right]} \right. \right. \right.$$

$$\left. + \frac{1}{\gamma_{mm'} + \frac{i}{\hbar}\left[E^0_m - E^0_{m'}\right]} \right) \rho_{m'm'} - \left(\frac{1}{\gamma_{m'm} + \frac{i}{\hbar}\left[E^0_{m'} - E^0_m\right]} \right.$$

$$\left. \left. \left. + \frac{1}{\gamma_{mm'} + \frac{i}{\hbar}\left[E^0_m - E^0_{m'}\right]} \right) \rho_{mm} \right] \right\} . \tag{8.39}$$

We Identify $\rho_{mm}(t) = P_m(t)$ and $\rho_{m'm'} = P_{m'}(t)$ and compare Eq. (8.39) with the Pauli equation,

$$\frac{\partial}{\partial t}P_m(t) = \sum_{m' \neq m} [W_{mm'}P_{m'}(t) - W_{m'm}P_m(t)] . \tag{8.40}$$

The comparison yields

$$W_{m'm} = \left(\frac{\lambda}{\hbar}\right)^2 |V_{m'm}|^2$$

$$\times \left(\frac{1}{\gamma_{mm'} + \frac{i}{\hbar}\left(E^0_m - E^0_{m'}\right)} - \frac{1}{\gamma_{m'm} - \frac{i}{\hbar}\left(E^0_{m'} - E^0_m\right)} \right) . \tag{8.41}$$

We note that the transitions rates are proportional to $(\lambda/\hbar)^2$. For a more rigorous notation, we use the identity by Sochotzki and Plemelj (also called Dirac identity) which we explain in a short mathematical excursus:

$$\lim_{\substack{\varepsilon \to 0 \\ z=z'+i\varepsilon}} \int_{-\infty}^{\infty} \frac{f(y)}{z - y}dy = \text{P.V.} \int_{-\infty}^{\infty} \frac{f(y)}{z' - y}dy - i\pi f(z') \tag{8.42}$$

$$\lim_{\substack{\varepsilon \to 0 \\ z=z'+i\varepsilon}} \frac{1}{z' - y + i\varepsilon} = \text{P.V.}\frac{1}{z' - y} - i\pi\delta(z' - y) \tag{8.43}$$

$$\lim_{\substack{\varepsilon \to 0 \\ z=z'+i\varepsilon}} \frac{i}{z' - y + i\varepsilon} = \text{P.V.}\frac{i}{z' - y} + \pi\delta(z' - y) \tag{8.44}$$

$$\int_0^{\infty} dt e^{i(\omega-\omega')-\Delta t} = \frac{i}{\omega - \omega' + i\Delta}, \tag{8.45}$$

where "P.V." stands for the principle value. Symbolically we write

$$\frac{1}{z + i \cdot 0} = P.V. \left(\frac{1}{z}\right) - i \cdot \pi \delta(z), \tag{8.46}$$

Substituting $z = x - y$, multiplying Eq. (8.46) by i and adding the complex conjugate, we obtain

$$\lim_{\varepsilon \to +0} \left\{ \frac{1}{\varepsilon - i \cdot (x - y)} + \frac{1}{\varepsilon + i \cdot (x - y)} \right\} = 2\pi \delta(x - y). \tag{8.47}$$

Inserting this into Eq. (8.47) we find the transition rates in the limit $\gamma \to 0$:

$$W_{m'm} = \left(\frac{\lambda}{\hbar}\right)^2 |V_{m'm}|^2 \pi \delta \left(\frac{E_m^0 - E_{m'}^0}{\hbar}\right) = \lambda^2 \frac{2\pi}{\hbar} |V_{m'm}|^2 \delta \left(E_m^0 - E_{m'}^0\right). \tag{8.48}$$

As the result, we found that the rates for the transition from state (m) to state (m') are proportional to $|V_{m'm}|^2 / \hbar$. Here the δ-function stands for the law of energy conservation during the transition. Our result, Eq. (8.48), is sometimes called *Fermi's Golden Rule*. In conclusion, we write the Pauli equation in the form resulting from perturbation theory:

$$\frac{\partial}{\partial t} P_m(t) = \frac{2\pi}{\hbar} \sum_{m' \neq m} \left\{ |\lambda V_{m'm}|^2 [P_{m'}(t) - P_m(t)] \delta \left(E_m^0 - E_{m'}^0\right) \right\}. \tag{8.49}$$

8.3 Stochastic Dynamics Including a Heat Bath

In the present form, Pauli equation is restricted to processes in the micro-canonical ensemble, due to the existence of factors keeping the energy constant. This is related to the limit $\gamma \to 0$. In order to improve the approximations which in fact leads to unsatisfactory results at higher temperature, let us introduce the heat bath in a more explicit way. For thermal systems at finite temperature, the δ function is replaced by a temperature-dependent function in order to provide the correct stationary distribution. This procedure is not unique since there exist many different heat reservoirs. On the other side, there are several sufficiently general standard methods (Binder 1979; Böttger and Bryksin 1985; Chetverikov et al. 2009, 2011b, 2012). We write the transition probabilities in the form

$$W_{m'm} = \lambda^2 \frac{1}{\hbar} |V_{m'm}|^2 E(m, m'). \tag{8.50}$$

Here, a factors appears which expresses the energy conservation in microscopic transitions,

$$E(n, n') = 2\pi\delta\left(\frac{E_n - E_{n'}}{V_0}\right).$$
(8.51)

This δ-function shall be replaced in a next step by the temperature function (Chetverikov et al. 2009, 2011b; Tolman 1938):

$$E\left(n, n'\right) = e^{-\frac{\beta}{2}(E_n - E'_n)} F(n, n') \quad \text{with} \quad F\left(n, n'\right) = F\left(E_n - E'_n\right),$$
(8.52)

where $F(E_n - E'_n)$ is an even function of the difference of the energy levels. There are several variants for this even function given in the literature. The simplest form is defined by the phenomenological ansatz of the Monte-Carlo process, where energy-descending transitions are weighted by $E = 1$ and ascending transitions by a factor smaller than unity (Binder 1979). This corresponds to

$$F\left(E_n - E_{n'}\right) = e^{-\frac{\beta}{2}|E_n - E'_n|}.$$
(8.53)

Proper statistical derivations of the thermal factors are based on certain microscopic models of the heat bath. An interesting approach was proposed by Böttger and Bryksin (1985). Starting from the von Neumann equation for the density matrix, the authors give the following general expression:

$$F\left(E_n - E_{n'}\right) = \int_{-\infty}^{\infty} e^{\frac{i}{\hbar}\tau|E_n - E_{n'}|} K(|\tau|) d\tau,$$
(8.54)

where $K(|\tau|)$ is a rapidly decaying memory kernel. The decay of these correlations is related to the damping of lattice-particle motion. In the simplest case we assume here an exponential decay with the same damping constant as appears in the above introduced Langevin dynamics. This leads to the Lorentz profile

$$F\left(E_n - E_{n'}\right) = 2\pi \frac{\gamma}{\gamma^2 + \frac{|E_n - E_{n'}|}{\hbar^2}}.$$
(8.55)

In the limit of small damping we recover the δ-function in the Pauli expression for the transition probabilities.

For illustration of the Pauli equation and the H-theorem, we consider an example with Monte Carlo transitions to nearest neighbors only (Chetverikov et al. 2011a),

$$W\left(n, n'\right) = \tau \delta\left(m - n' - 1\right) E(n, m) + \tau \delta\left(m - n' + 1\right) E(n, m).$$
(8.56)

The result of simulations are presented in (Chetverikov et al. 2009, 2011b, 2012).

Lorentz' Kinetic Equations for Electron Gases

As an interesting application we consider now the kinetics of an electron gas. So far, we considered systems with discrete states, as for example atomic or molecular systems with discrete energy levels, and as an interesting application the kinetics of electrons on a lattice in the approximation of tight-binding. Let us consider now another sufficiently simple kinetic equation, the so-called Lorentz approximation. We study an ideal gas which consists of very light particles; we think about electrons which are moving in a force field of heavy scatterers located at random positions. In some sense this is just the opposite case to tight-binding electrons. We assume our electron gas which interacts with unmovable scatterers is enclosed in a rectangular box with the sides, a_i, and the volume, $V = \prod_i a_i$. The momentum eigenfunctions are then given by

$$\langle \mathbf{x} | \mathbf{p} \rangle = \frac{1}{\sqrt{V}} e^{\frac{i}{\hbar} \mathbf{p} \cdot \mathbf{x}}, \tag{8.57}$$

with the quantized momentum eigenvalues

$$\mathbf{p} = (p_1, p_2, p_3) = h \left(\frac{n_1}{a_1}, \frac{n_2}{a_2}, \frac{n_3}{a_3} \right), \qquad n_i = 0, \pm 1, \pm 2, \ldots \tag{8.58}$$

The Hamilton operator has the form

$$\hat{H} = \hat{H}_0 + \lambda \hat{V}, \tag{8.59}$$

consisting of the ideal gas term,

$$\hat{H}_0 = \sum_i \frac{\hat{p}_i^2}{2m}, \tag{8.60}$$

and the interaction term which describes the scattering at heavy scatterers. In the case the particles are charged, we write

$$\hat{V} = V(\mathbf{r} - \mathbf{R}) = \pm \frac{Z\tilde{e}^2}{|\mathbf{r} - \mathbf{R}|}, \tag{8.61}$$

where the positions of the particle and the scatterer are denoted by $\mathbf{r}$ and $\mathbf{R}$, respectively. We define the momentum distribution of the Lorentz gas as

$$P_m(t) = \rho_{mm}(t) = \langle m | \hat{\rho}(t) m \rangle = \langle \mathbf{p} | \hat{\rho}(t) \mathbf{p} \rangle = \text{const.} f(\mathbf{p}, t), \tag{8.62}$$

denoting $f(\mathbf{p}, t)$ as the kinetic distribution function of the electrons. Then the Pauli equation reads

$$\frac{\partial}{\partial t} f(\mathbf{p}, t) = \frac{2\pi}{\hbar} n_s \int \frac{d\mathbf{p}'}{h^3} \left|\lambda V(\mathbf{p}, \mathbf{p}')\right|^2 \left[f(\mathbf{p}', t) - f(\mathbf{p}, t)\right] \delta \left(\frac{\mathbf{p}'^2}{2m} - \frac{\mathbf{p}^2}{2m}\right),$$
(8.63)

where n_s is the density of scatterers and $V(\mathbf{p}', \mathbf{p})$ is the Fourier transform of the potential. By introducing the approximation

$$\delta \left(\frac{\mathbf{p}'^2}{2m} - \frac{\mathbf{p}^2}{2m}\right) \approx \frac{\delta(z - z')}{|f'(z)|}$$
(8.64)

as well as spherical coordinates in momentum space (with the differential $d\mathbf{p} = p^2 \, dp \, d\Omega$), we write

$$\frac{\partial}{\partial t} f(\mathbf{p}, t) = \frac{2\pi}{\hbar} m \, n_s \int d\Omega \int\limits_{p=p'} dp' \, p' \left|\lambda V(\mathbf{p}, \mathbf{p}')\right|^2 \left[f(\mathbf{p}', t) - f(\mathbf{p}, t)\right].$$
(8.65)

Introducing the scattering cross-section,

$$\sigma(\Omega) = \text{const.} \left|\lambda V(\mathbf{p}, \mathbf{p}')\right|^2,$$
(8.66)

we obtain the quantum kinetic equation for the Lorentz gas in the form

$$\frac{\partial}{\partial t} f(\mathbf{p}, t) = n_s \int\limits_{p=p'} d\Omega \, \frac{p}{m} \sigma(\Omega, p) \left[f(\mathbf{p}', t) - f(\mathbf{p}, t)\right].$$
(8.67)

Similar to the Pauli equation, this kinetic equation contains second order terms in the potential, in spite of the fact that it is essentially based on first order perturbation theory.

As a more simple special case we can study *isotropic scatterers*, this means scatterers which are rotationally symmetric.

8.4 Bogolyubov's Kinetic Theory Based on Reduced Density Operators

In the previous section we derived several special kinetic equations for electrons and other particles which all have in common that they are linear with respect to the probabilities or the distributions function, $f(\mathbf{p}, t)$. Prototypes of this type of linear kinetic equations are the Pauli master equation, the Lorentz equation, and the diffusion equations in coordinate and in velocity space. On the other hand it is well known that the Boltzmann equation, derived already at the end of the nineteenth century for classical gases is nonlinear, since it contains collision integrals which are quadratic functionals of the distribution function, $f(\mathbf{p}, t)$. This means that we should expect that more general quantum kinetic equations are nonlinear as well. In order to derive such equations, we follow the method by Bogolyubov (1991, 2005–2009), which we discussed already in brief in Chap. 4.

As we have shown in Chap. 4, the complete information about the quantum statistical state is contained in the density operator, $\hat{\rho}(1, \ldots, N, t)$. The time-dependence of this operator is given by the von Neumann equation,

$$i\hbar \frac{\partial}{\partial t} \hat{\rho}(t) = \left[\hat{H}, \hat{\rho}(t) \right] . \tag{8.68}$$

Further we have shown that for the calculation of many physical quantities, as e.g. the kinetic energy, the potential energy, correlation functions etc., we do not need the complete information on the system. Therefore, several researchers such as Yvon, Bogolyubov, Born, Green, Kirkwood and others proposed to introduce *reduced distribution functions*. In Chap. 4, we introduced reduced density operators following Bogolyubov (1991, 2005–2009) and Bogolyubov and Bogolyubov jr. (1991) by the prescription

$$\hat{F}_s(1, \ldots, s) = V^s \operatorname{Tr}_{(s+1,\ldots,N)} \hat{\rho}(1, \ldots, N) . \tag{8.69}$$

Here s is the order of the reduced operator which is, in general, a small number, e.g. $s = 1$ or $s = 2$, corresponding to first order or second order reduced density operators. Similar to the classical BBGKY hierarchy, Bogolyubov derived the quantum statistical hierarchy for density operators (see Chap. 4)

$$\hat{F}_s(1, \ldots, s) = V^s \hat{\rho}_s(1, \ldots, s) = \operatorname{Tr}_{(s+1,\ldots,N)} \left[V^s \hat{\rho}(1, \ldots, N) \right] . \tag{8.70}$$

Let us consider a gas with only one component and the Hamiltonian

$$\hat{H} = \sum_{k=1}^{N} \hat{T}_k \left(\hat{p}_k^2 \right) + \sum_{i<j}^{N} U_{(ij)} \left(\mathbf{r}_i, \mathbf{r}_j \right) . \tag{8.71}$$

By averaging we find the mean kinetic energy expressed by $\hat{F}_1$ and the mean potential energy expressed by $\hat{F}_2$ (see Chap. 4). This way, the operator $\hat{F}_1$ contains the same information as the distribution function $f(\mathbf{r}, \mathbf{p})$ which we studied in the previous section. Therefore, we will concentrate in the following on the first order density operator, $\hat{F}_1$ or $\hat{\rho}_1$, respectively. In order to find a kinetic equation for the reduced density operators, we carry out the trace, $\operatorname{Tr}_{(s+1,\ldots,N)}$, over $N - s$ particle coordinates in the von Neumann equation (see Chap. 4) to obtain

$$i\hbar \frac{\partial}{\partial t} \hat{F}_s = \text{Tr}_{(s+1,...,N)} \left[\hat{H}, \hat{\rho}_N \right] . \qquad (8.72)$$

By splitting the Hamiltonian, Eq. (8.71), into subgroups $1, \ldots, s$ and $s + 1, \ldots, N$, we take into account that $\hat{H}_s$ acts only in the subspace of the s-particle system. This way we obtain the hierarchy of operator equations

$$i\hbar \frac{\partial}{\partial t} \hat{F}_s = \left[\hat{H}_s, \hat{F}_s \right] + V \sum_{i=1}^{s} (N - s) \text{Tr}_{s+1} \left(\left[U_{i,s+1}, \hat{F}_{s+1} \right] \right) . \qquad (8.73)$$

This is the Bogolyubov hierarchy which determines the reduced density operators, $\hat{F}_s$, by an equation which contains $\hat{F}_{s+1}$. Therefore, our equation for $\hat{F}_s$ is not closed. We can obtain, however, an equation for $\hat{F}_s$, if we have a good approximation for $\hat{F}_{s+1}$. We will exploit this idea by getting $\hat{F}_1$, by means of an approximation for $\hat{F}_2$. Using the formalism by Bogolyubovs F functions, we concentrate now on $\hat{F}_1$. From the definition of $\hat{F}_s(1, \ldots, s, t)$, we found a hierarchy of operator equations which we reformulate now by using the symmetry of particles in the form

$$i\hbar \frac{\partial}{\partial t} \hat{F}_s(1, \ldots, s, t) = \left[\hat{H}_s, \hat{F}_s(1, \ldots, s, t) \right]$$
$$+ n \, \text{Tr}_{(s+1)} \left(\left[V(1, s + 1), \hat{F}_{s+1}(1, \ldots, s + 1, t) \right] \right) . \qquad (8.74)$$

Note the parameter n which is small for dilute gases and plasmas. For the case, $s = 1$, we proceed as follows: We fix one of the N particles in the system and assign it the number 1, all the others form a field of $(N - 1)$ particles. The commutator $\left[\hat{H}_1, \hat{F}_1 \right]$ describes the interaction of the fixed particle with the external field. For $s = 2$ we proceed in a similar way. We discuss now the problem how to find a closing to the equation.

8.5 Bogolyubov's Derivation of the Quantum Boltzmann Equation

8.5.1 Operator Equations

We obtained a hierarchy of operator equations and need methods for decoupling them in order to find a closed equation for $\hat{F}_1(1, t)$. In momentum representation, we have

$$f(\mathbf{p}, t) = \text{const.} \left\langle \mathbf{p}_1 \left| \hat{F}_1(1, t) \, \mathbf{p}_1 \right\rangle . \qquad (8.75)$$

With a *binary collision approximation*, we obtain

$$i\hbar\frac{\partial}{\partial t}\hat{F}_1(1, t) = \left[\hat{H}_1, \hat{F}_1(1, t)\right] + n \operatorname{Tr}_{(2)}\left(\left[V(1, 2), \hat{F}_2(1, 2, t)\right]\right) \quad (8.76)$$

$$i\hbar\frac{\partial}{\partial t}\hat{F}_2(1, 2, t) = \left[\hat{H}_2, \hat{F}_2(1, 2, t)\right] + \{\text{higher orders}\} . \quad (8.77)$$

In the case of dilute gases or plasmas, we neglect higher orders which are $\mathcal{O}(n)$, however, still we have to complete the second equation by initial conditions. To find the initial conditions which lead to binary collisions, first we consider the characteristic times of the problem. The interaction time for a binary collision is of order $\tau_{\text{corr}} \approx 10^{-12}\ldots 10^{-13}$ s. The mean time between two collisions is $\tau_{\text{coll}} \approx 10^{-10}\ldots 10^{-12}$ s. Exploiting $\tau_{\text{coll}} \gg \tau_{\text{corr}}$, we assume that the incoming particles are independent of one another, that is,

$$\lim_{t\to-\infty}\hat{F}_2(1, 2, t) = \hat{F}_1(1, t)\,\hat{F}_1(2, t). \quad (8.78)$$

This is the Bogolyubov condition corresponding to Boltzmann's assumptions about binary collisions. From the physical point of view, the Bogolyubov condition is the source of the irreversibility of the final kinetic equations, that is, the condition (8.78) breaks the time symmetry.

We write the Hamilton-operator in the form

$$\hat{H}_2(1, 2) = \hat{H}_1(1) + \hat{H}_1(2) + V(1, 2) \quad (8.79)$$

and split the two-particle density operator (as previously for classical situations) into two parts

$$\hat{F}_2(1, 2, t) = \hat{F}_1(1, t)\,\hat{F}_1(2, t) + \hat{G}(1, 2, t). \quad (8.80)$$

Here we introduce the quantum statistical correlation operator, $\hat{G}(1, 2, t)$. The Bogolyubov condition reads now

$$\lim_{t_0\to-\infty}\hat{G}(1, 2, t) = 0. \quad (8.81)$$

By introducing this into Eqs. (8.76) and (8.77), we obtain

$$i\hbar\frac{\partial}{\partial t}\hat{F}_1(1, t) = \left[\hat{H}_1, \hat{F}_1(1, t)\right] + n \operatorname{Tr}_{(2,\ldots,N)}\left(\left[V(1, 2), \hat{F}_2(1, 2, t)\right]\right) \quad (8.82)$$

and

$$i\hbar \frac{\partial}{\partial t}\left(\hat{F}_1(1,t)\,\hat{F}_1(2,7) + \hat{G}(1,2,t)\right)$$
$$= \left[\hat{H}_1(1) + \hat{H}_1(2) + V(1,2),\; \hat{F}_1(1,t)\hat{F}_1(2,t) + \hat{G}(1,2,t)\right]. \qquad (8.83)$$

It follows

$$i\hbar \frac{\partial}{\partial t}\hat{G}(1,2,t) = \left[\hat{H}_1(1) + \hat{H}_1(2), \hat{G}(1,2,t)\right] + \left[V(1,2),\hat{F}_1(1,t)\hat{F}_1(2,t)\right]. \qquad (8.84)$$

We neglect higher orders and introduce the short hand notation $\hat{H}_2^0 = \hat{H}_1(1) + \hat{H}_1(2)$. The homogeneous form of Eq. (8.84) reads then

$$i\hbar \frac{\partial}{\partial t}\hat{G}_{\text{hom}}(1,2,t) = \left[\hat{H}_2^0,\; \hat{G}_{\text{hom}}(1,2,t)\right], \qquad (8.85)$$

with the solution

$$\hat{G}_{\text{hom}} = e^{-\frac{i}{\hbar}\hat{H}_2^0 t}\, \hat{G}(1,2,0)\, e^{\frac{i}{\hbar}\hat{H}_2^0 t}. \qquad (8.86)$$

The general solution is

$$\hat{G}(1,2,t) = \hat{G}_{\text{hom}}(1,2,t_0) - \frac{i}{\hbar}\int_0^{t-t_0} dt'\, e^{-\frac{i}{\hbar}\hat{H}_2^0 t'}\left[V(1,2),\, \hat{F}_1(1,t')\hat{F}_1(2,t')\right]e^{\frac{i}{\hbar}\hat{H}_2^0 t'}$$
$$= -\frac{i}{\hbar}\int_0^{\infty} dt'\, e^{-\frac{i}{\hbar}\hat{H}_2^0 t'}\left[V(1,2),\, \hat{F}_1(1,t')\hat{F}_1(2,t')\right]e^{\frac{i}{\hbar}\hat{H}_2^0 t'}, \qquad (8.87)$$

where we used the initial condition, Eq. (8.81). This way, Eq. (8.82) assumes the form

$$i\hbar \frac{\partial}{\partial t}\hat{F}_1(1,t) = \left[\hat{H}_1, \hat{F}_1(1,t)\right] + n\,\text{Tr}_{(2)}\left(\left[V(1,2), \hat{F}_1(1,t)\hat{F}_1(2,t)\right]\right)$$
$$+ n\,\text{Tr}_{(2)}\left(\left[V(1,2), \hat{G}(1,2,t)\right]\right) \qquad (8.88)$$

$$i\hbar \frac{\partial}{\partial t}\hat{F}_1(1,t) = \left[\hat{H}_1, \hat{F}_1(1,t)\right] - \frac{i}{\hbar}n\int_0^{\infty} dt' \times \text{Tr}_{(2)}\left(\left[V(1,2),\right.\right.$$
$$\left.\left.\left\{e^{-\frac{i}{\hbar}\hat{H}_2^0 t'}\left[V(1,2), \hat{F}_1\left(1,t-t'\right)\hat{F}_1\left(2,t-t'\right)\right]e^{\frac{i}{\hbar}\hat{H}_2^0 t'}\right\}\right]\right). \qquad (8.89)$$

We introduced here the Hamilton operator of the self-consistent field, $\hat{\bar{H}}_1$, which fulfils the commutator relation

$$\left[\hat{\bar{H}}_1, \hat{F}_1(1, t)\right] = \left[\hat{H}_1, \hat{F}_1(1, t)\right] + n\,\mathrm{Tr}_{(2)}\left(\left[V(1, 2), \hat{F}_1(1, t)\hat{F}_1(2, t)\right]\right).$$

(8.90)

When suppressing retardation effects, that is, with

$$\hat{F}_1\left(1, t - t'\right)\hat{F}_1\left(2, t - t'\right) = \hat{F}_1(1, t)\,\hat{F}_1(2, t),$$

(8.91)

we will find later the *quantum statistical Boltzmann equation*, see Eq. (8.98). We remember that in the course of this derivation, we used the binary collision approximation and the Bogolyubov condition about the decay of initial correlations, which corresponds to neglecting *memory effects*.

8.5.2   Quantum Boltzmann Equation for Homogeneous Systems

For uniform systems and uniform fields, the operator of the self-consistent field commutes with the 1-particle density operator, $\left[\hat{\bar{H}}_1, \hat{F}_1(1, t)\right] = 0$. In this case we obtain

$$i\hbar\frac{\partial}{\partial t}\left\langle \mathbf{p}_1 \left| \hat{F}_1(1, t)\right| \mathbf{p}_1\right\rangle = n\int d\,\mathbf{p}_2\left\langle \mathbf{p}_1\mathbf{p}_2 \left|\left[V(1, 2), \hat{G}(1, 2, t)\right]\right| \mathbf{p}_1\mathbf{p}_2\right\rangle.$$

(8.92)

Using completeness relations, we obtain

$$i\hbar\frac{\partial}{\partial t}\left\langle \mathbf{p}_1 \left| \hat{F}_1(1, t)\right| \mathbf{p}_1\right\rangle$$

$$= n\int d\mathbf{p}_1'\, d\mathbf{p}_2'\, d\mathbf{p}_2 \left\{\left\langle \mathbf{p}_1\mathbf{p}_2 \left| V(1, 2)\right| \mathbf{p}_1'\mathbf{p}_2'\right\rangle\left\langle \mathbf{p}_1'\mathbf{p}_2' \left| \hat{G}(1, 2, t)\right| \mathbf{p}_1\mathbf{p}_2\right\rangle\right.$$

$$\left. - \left\langle \mathbf{p}_1\mathbf{p}_2 \left| \hat{G}(1, 2, t)\right| \mathbf{p}_1'\mathbf{p}_2'\right\rangle\left\langle \mathbf{p}_1'\mathbf{p}_2' \left| V(1, 2)\right| \mathbf{p}_1\mathbf{p}_2\right\rangle\right\}.$$

(8.93)

We transform the second term in Eq. (8.93):

$$\left\langle \mathbf{p}_1\mathbf{p}_2 \left| \hat{G}(1, 2, t)\right| \mathbf{p}_1'\mathbf{p}_2'\right\rangle$$

$$= -\frac{i}{\hbar}\int_0^\infty dt'\left\{\left\langle \mathbf{p}_1\,\mathbf{p}_2 \left| e^{-\frac{i}{\hbar}\hat{H}_2^0 t'}\left[V(1, 2), \hat{F}_1\left(1, t'\right)\hat{F}_1\left(2, t'\right)\right]e^{\frac{i}{\hbar}\hat{H}_2^0 t'}\right| \mathbf{p}_1'\,\mathbf{p}_2'\right\rangle\right\}.$$

(8.94)

The momentum functions commute with energy. With $E(p) = p^2/2m$, we obtain

$$\left\langle \mathbf{p}_1 \mathbf{p}_2 \left| \hat{G}(1, 2, t) \, \mathbf{p}_1' \mathbf{p}_2' \right. \right\rangle$$

$$= -\frac{i}{\hbar} \int_0^\infty dt' \, d\mathbf{p}_1' \, d\mathbf{p}_2' \, d\mathbf{p}_2 \left\{ e^{-\frac{i}{\hbar}[E(p_1)+E(p_2)-E(p_1')-E(p_2')]t'} \right.$$

$$\times \left(\left\langle \mathbf{p}_1 \mathbf{p}_2 \left| V(1, 2) \, \hat{F}_1(1, t') \, \hat{F}_1(2, t') \, \mathbf{p}_1' \mathbf{p}_2' \right. \right\rangle \right.$$

$$\left. \left. - \left\langle \mathbf{p}_1 \mathbf{p}_2 \left| \hat{F}_1(1, t') \, \hat{F}_1(2, t') \, V(1, 2) \, \mathbf{p}_1' \, \mathbf{p}_2' \right. \right\rangle \right) \right\} \, . \tag{8.95}$$

Using again completeness relations and

$$\left\langle \mathbf{p} \left| \hat{F}_1(1, t) \, \mathbf{p} \right. \right\rangle = \hat{F}_1(\mathbf{p}, t) \, \delta(\mathbf{p} - \mathbf{p}') \, , \tag{8.96}$$

we write

$$\left\langle \mathbf{p}_1 \mathbf{p}_2 \left| \hat{G}(1, 2, t) \, \mathbf{p}_1' \, \mathbf{p}_2' \right. \right\rangle$$

$$= -\frac{i}{\hbar} \int_0^\infty dt' \left\{ e^{-\frac{i}{\hbar}[E(p_1)+E(p_2)-E(p_1')-E(p_2')]t'} \left\langle \mathbf{p}_1 \mathbf{p}_2 \left| V(1, 2) \, \mathbf{p}_1' \mathbf{p}_2' \right. \right\rangle \right.$$

$$\left. \times \left[\hat{F}_1(\mathbf{p}_1', t') \, \hat{F}_1(\mathbf{p}_2', t') - \hat{F}_1(\mathbf{p}_1, t') \hat{F}_1(\mathbf{p}_2, t') \right] \right\} \, . \tag{8.97}$$

Inserting Eq. (8.97) in (8.93), performing the time integration and exploiting symmetry, we obtain the quantum statistical Boltzmann equation

$$\frac{\partial}{\partial t} f(\mathbf{p}, t) = \frac{1}{\hbar} \int d\mathbf{p}_1' \, d\mathbf{p}_2' \, d\mathbf{p}_2 \left\{ |V(p_1 - p_2)|^2 \right.$$

$$\times \delta(\mathbf{p}_1' + \mathbf{p}_2' - \mathbf{p}_1 - \mathbf{p}_2) \, \delta[E(p_1') + E(p_2') - E(p_1) - E(p_2)]$$

$$\left. \times [f(\mathbf{p}_1', t) \, f(\mathbf{p}_2', t) - f(\mathbf{p}_1, t) \, f(\mathbf{p}_2, t)] \right\} \, . \tag{8.98}$$

Here we introduced again $\hat{F}_1(\mathbf{p}, t) = \text{const.} \, f(\mathbf{p}, t)$. Similar to the original Boltzmann equation, our final kinetic equation (8.98) contains quadratic terms in $f(\mathbf{p}, t)$. Therefore, in contrast to the Lorentz equation, Eq. (8.98) is nonlinear. The physical reason is that we released the assumption of the Lorentz equation that the scatterers reside at fixed positions. For the current derivation, thus, we take into account that the scatterers move according to their distribution function and interact with one another. The meaning of the δ-functions in the kinetic equation (8.98) is to enforce the conservation of energy and momentum in scattering processes.

8.6   Theory of Fluctuations and Fluctuation-Dissipation Relations

8.6.1   Basic Einstein-Onsager-Kubo Relations

The method by Einstein and Onsager explained in detail in Chap. 2, is based on the idea that—in principle—any macroscopic quantity, x, can be considered as a fluctuating variable, which is determined by a certain probability distribution, $\omega(x)$. The mean value is given by the first momentum of the probability distribution,

$$\langle x \rangle = \int x \, \omega(x) dx \,. \tag{8.99}$$

We assume that the stationary state, $x_0 = 0$, corresponds to equilibrium and any value of $x(t)$ different from zero corresponds to a non-equilibrium state. According to the Second Law, the equilibrium state is an attractor of the dynamics since it corresponds to the maximum of entropy, S. Due to Onsager's relaxation-fluctuation theory, the derivative of entropy with respect to the variable x plays the rôle of a thermodynamic force,

$$S(x = 0) \rightarrow \max; \qquad \left.\frac{\partial S}{\partial x}\right|_{x=0} = 0; \qquad \left.\frac{\partial^2 S}{\partial x^2}\right|_{x=0} \leq 0 \,. \tag{8.100}$$

Onsager postulates that the relaxation dynamics of the variable x is determined by the first derivative of the entropy, which is different from zero in non-equilibrium. Starting from a deviation from equilibrium (an entropy value below the maximum), spontaneous irreversible processes will drive the entropy to increase,

$$\frac{d}{dt}S(x) = \frac{\partial S}{\partial x}\frac{dx}{dt} \geq 0 \,. \tag{8.101}$$

Onsager interpreted the factors in the following way:

1. The derivative of the entropy,

$$X = -\frac{\partial S}{\partial x} \,, \tag{8.102}$$

is the *driving force* of the relaxation to equilibrium.
2. The derivative of the variable

$$J = -\frac{dx}{dt} \,, \tag{8.103}$$

is the resulting *thermodynamic flux*.

Onsager postulated a linear relation between the thermodynamic force and the corresponding flux. Both are related by a linear cause-effect relation,

$$J = LX . \qquad (8.104)$$

The coefficient L is the *Onsager's phenomenological coefficient*, or *Onsager's kinetic coefficient*. From the Second Law follows that the Onsager-coefficients are strictly positive,

$$P = \frac{d}{dt} S(x) = J X = L X^2 \geq 0 . \qquad (8.105)$$

The linearity of the force–flow relation corresponds to the bi-linearity of the entropy production. This way we find for the neighborhood of the equilibrium state

$$X = -\frac{\partial S}{\partial x} = k_B \beta x \qquad (8.106)$$

and finally

$$\dot{x} = -k_B L \beta x . \qquad (8.107)$$

With the abbreviation

$$\lambda = L k_B \beta \qquad (8.108)$$

being the *relaxation coefficient* of the quantity x we obtain

$$\dot{x} = -\lambda x ; \qquad x(t) = x(0) e^{-\lambda t} . \qquad (8.109)$$

This linear kinetic equation describes the relaxation of a thermodynamic system which is initially out of equilibrium. The time, $t_0 = \lambda^{-1}$, is the decay time of the initial deviation from equilibrium. This coefficient which is responsible for the relaxation to equilibrium is in close relation to the fluctuation properties of the considered system. Indeed, Eq. (8.108) relates a kinetic property, λ, with a fluctuating quantity, β. Relations of this type are called *fluctuation-dissipation relation*.

The first fluctuation-dissipation relation was found in 1905 by Einstein as a relation between the mean square displacement of a Brownian particle and the (spatial) diffusion coefficient, D. Further, in Chap. 2 we discussed the relations between the diffusion coefficient and velocity-velocity correlations found by Taylor (1922) and extended by Kubo (1957a,b, 1959) to a relation between the conductivity, σ, of a system of electrons with density n and the velocity-velocity correlation function

$$D = \int_0^\infty \langle v(t)v(0) \rangle \, dt ; \qquad \sigma = \frac{ne^2}{k_B T} \int_0^\infty \langle v(t)v(0) \rangle \, dt . \qquad (8.110)$$

When dealing with several thermodynamic variables we take the entropy in dependence on x_i ($i = 1 \ldots N$) and obtain the relaxation equations using the convention that summation over double indices is implied

$$S(x_1, \ldots, x_n) = S_{\text{max}} - \frac{1}{2} \beta_{ij} \, x_i x_j \,. \tag{8.111}$$

Following Onsager's idee, we obtain for the thermodynamic forces and fluxes the relations

$$X_i = -k_B \, \beta_{ij} \, x_j \,; \qquad J_i = -\dot{x}_i \,. \tag{8.112}$$

The generalized linear Onsager ansatz and the conditions from the Second Law read

$$J_i = L_{ij} X_j \,; \qquad L_{ij} X_i X_j \ge 0 \,. \tag{8.113}$$

The inequality holds for any value of X_i. The inequality is fulfilled as an equation only for $X_i = 0$, $i = 1, \ldots, n$. This corresponds to the requirement of positive definiteness of the matrix L_{ij}. Inserting Eq. (8.112) into Eq. (8.113), we obtain

$$\dot{x}_i = -k_B L_{ij} X_j \,. \tag{8.114}$$

Introducing the matrix of relaxation coefficients of linear processes near equilibrium, arrive at

$$\dot{x}_i = -\lambda_{ij} \, x_j \,; \qquad \lambda_{ij} = k_B \, L_{ik} \, \beta_{kj} \,. \tag{8.115}$$

The matrix β_{ij} determines the dispersion of the stationary fluctuations, this gives again a close relation between fluctuations and dissipation. Thus, we have obtained a fluctuation-dissipation relation for a matrix of fluctuating and relaxing variables. Silently Onsager assumed that deviations from equilibrium and fluctuations around the equilibrium experience the same kinetics.

8.6.2   Brownian Motion and Onsager–Casimir Relations

In order to illustrate this procedure, we consider Brownian motion, that is, the motion of a heavy particle in a viscous liquid. This problem was studied more than 100 years ago by Einstein. For the velocity of the particle we assume the linear equation of motion

$$\dot{v} = -\gamma_0 v \,. \tag{8.116}$$

For the stationary correlation function we obtain

$$\frac{d}{d\tau} \langle v(t+\tau)\,v(t)\rangle = -\gamma_0 \,\langle v(t+\tau)v(t)\rangle \;; \quad \left\langle v(t)^2\right\rangle = \frac{k_{\mathrm{B}}T}{m}\,, \tag{8.117}$$

with a certain initial condition. Solving these equations, we find the correlation function and the spectrum of Brownian particles,

$$\langle v(t+\tau)\,v(t)\rangle = \frac{k_{\mathrm{B}}T}{m} e^{-\gamma|\tau|}\;; \quad S_{vv}(\omega) = \frac{k_{\mathrm{B}}T}{m}\,\frac{2\gamma_0}{\gamma_0^2 + \omega^2}\,. \tag{8.118}$$

This frequency distribution is sometimes called a red spectrum, since the maximum of intensity is at low ω. A more complex structure of the spectrum is obtained if the particle is additionally under the influence of a harmonic force. Then the dynamical equations read

$$\dot{x} = v\;; \quad \dot{v} = -\gamma_0 v - \omega_0^2 x\,, \tag{8.119}$$

and the corresponding system of equations for the correlation functions is

$$\dot{C}_{xv}(\tau) = C_{vv}(\tau)\;; \quad \dot{C}_{vv} = -\gamma_0 C_{vv} - \omega_0^2 C_{xv}\,. \tag{8.120}$$

Following the approach described above, we find the spectrum

$$S_{vv}(\omega) = \frac{k_{\mathrm{B}}T}{m}\,\frac{2\gamma_0}{\gamma_0^2 + \left(\omega - \dfrac{\omega_0^2}{\omega}\right)^2}\,. \tag{8.121}$$

This distribution has a peak at $\omega \simeq \omega_0$, that is, a resonance.

The symmetry relations of the Onsager matrix, L_{ij}, relate the thermodynamic fluxes and forces near equilibrium. Empirically, it was found that the matrix is symmetrical with respect to exchange of the indices and in some cases one observes antisymmetry

$$L_{ji} = \pm L_{ij}\,. \tag{8.122}$$

This macroscopic property means that if a force, X_j, induces the thermodynamic flux, J_i, we find also a flux, J_j, generated by the force X_i. This property was also described by Onsager. For the proof of this relation, we use properties of the correlation function, in particular, $C_{ij}(\tau) = C_{ji}(-\tau)$. The second fact which has to be taken into account is that the fluctuating quantities are functions of the microscopic variables. Therefore,

$$x_i(t) = \varepsilon_i x_j(-t)\,, \tag{8.123}$$

which expresses the reversibility of the microscopic motion. Here, ε is the parity coefficient which is $+1$ for even variables and -1 for odd ones. Thus,

$$C_{ij}(\tau) = \langle x_i(t)\, x_j(t+\tau)\rangle = \varepsilon_i \varepsilon_j \langle x_i(-t)\, x_j(-t-\tau)\rangle = \varepsilon_i \varepsilon_j C_{ij}(-\tau).$$
(8.124)

Taking into account stationarity, we obtain

$$C_{ij}(\tau) = \varepsilon_i \varepsilon_j C_{ji}(\tau).$$
(8.125)

As a consequence of the general regression dynamics, we find dynamical equations for the correlation functions:

$$\frac{d}{d\tau}C_{ij}(\tau) = -\lambda_{ik}\, C_{kj}(\tau).$$
(8.126)

Hence we obtain

$$\lambda_{ik}\, C_{kj}(\tau) = \varepsilon_i \varepsilon_j\, \lambda_{jk}\, C_{ki}(\tau).$$
(8.127)

For $\tau = 0$, the correlation function reduces to the standard deviation with $C_{ij}(0) = (\beta_{ij})^{-1}$. Using further the definition of the relaxation coefficients, λ_{ij}, we arrive at the *Onsager–Casimir symmetry relation*

$$L_{ji} = \varepsilon_i \varepsilon_j\, L_{ij}.$$
(8.128)

This relation is one of the fundamentals of linear irreversible thermodynamics. It is well confirmed by many experiments.

The approach presented so far is restricted to the dynamics of classical variables near equilibrium. In the next section, we describe the extension to quantum variables. More difficult is the extension situations far from equilibrium. As discussed in Chap. 2, there exist many approaches to solve this problem (see also Ebeling and Sokolov 2005; Klimontovich 1982, 1997). In the following we discuss only a special case of quantum statistical generalization and applications to dilute plasmas.

8.7 Quantum Fluctuation-Dissipation Relations

8.7.1 Nyquist Theorem and Callen–Welton Theorems

In 1928, Johnson studied experimentally thermal noise in an electrical resistor and discovered special relations between mean square fluctuations of voltage and resistance. Soon after these results, Harry Nyquist explained them by assuming a noise, now called Johnson–Nyquist noise (Nyquist 1928). Without applying a current, it was shown that the mean-square voltage is proportional to the resistance, R, the temperature, $k_B T$, and the bandwidth, $\Delta \nu$, over which the voltage is measured:

$$\left\langle V^2 \right\rangle = 4 R \, k_B T \, \Delta \nu .\tag{8.129}$$

Nyquist understood, that this relation is an extension to the classical fluctuation-dissipation theory to classical current-electric field relations. In the same paper, on the basis of Planck's theory Nyquist discussed already a first generalization of a classical fluctuation-dissipation relation to quantum variables. Nyquist considered the fluctuations of the current and voltage in a linear circuit which were detected by J. B. Johnson. He presented a theoretical deduction based on the well known relations for a circuit

$$\frac{dQ}{dt} = J ; \qquad L\frac{dJ}{dt} + RJ + \frac{Q}{C} = V ,\tag{8.130}$$

where Q is the charge of the capacitor, C, J is the current in the circuit with inductivity, L, and resistance, R. In the classical case, Nyquist discussed the electrical relations in the full analogy to the theory of Brownian motion given above. Based on this analogy and in particular on the equipartition theorem, Nyquist found the famous relation

$$(VV)_\omega = 2R \, k_B T .\tag{8.131}$$

For the current-current spectrum follows

$$(JJ)_\omega = \frac{2R}{|Y(\omega)|^2} \, k_B T ,\tag{8.132}$$

where the complex admittance is

$$Y(\omega) = R - i\omega L + \frac{i}{\omega C} .\tag{8.133}$$

The corresponding energy contained in the fluctuating electromagnetic field is

$$U = \frac{1}{2}\left\langle LJ^2 \right\rangle = \int \frac{d\omega}{2\pi} \frac{2R \, k_B T}{|Y|^2} = \frac{k_B T}{2} .\tag{8.134}$$

Further Nyquist argued as follows: "In what precedes the equipartition law has been assumed, assigning a total energy per degree of freedom of $k_B T$. If the energy per degree of freedom be taken

$$\frac{h\nu}{e^{\frac{h\nu}{k_B T}} - 1}\tag{8.135}$$

the expression for the electromotive force becomes" (from now on written in our notation)

$$(VV)_\omega = 2R k_B T \frac{\hbar\omega}{e^{\frac{\hbar\omega}{k_B T}} - 1} . \tag{8.136}$$

Evidently, Nyquist was not aware of the ground state contribution, $\hbar\omega/2$, which was found by Planck (1911). This term was included by Callen and Welton (1951), and the final expression of the Nyquist-Callen-Welton relation for the current-current correlations reads

$$S_{JJ} = (JJ)_\omega = 2\mathrm{Re}\,[Y(\omega)]\, k_B T \left(\frac{\hbar\omega}{e^{\frac{\hbar\omega}{k_B T}} - 1} + \frac{\hbar\omega}{2} \right)$$

$$= 2\mathrm{Re}\,[Y(\omega)]\, \hbar\,\omega\coth\left(\frac{\hbar\omega}{2k_B T} \right) . \tag{8.137}$$

We followed here the original paper by Nyquist (1928) in the more refined version by Klimontovich (1982, 1997, 2001). Introducing a frequency-dependent temperature,

$$T_\omega = \hbar\omega\coth\left(\frac{\hbar\omega}{2k_B T} \right) , \tag{8.138}$$

we formulate the quantum fluctuation-dissipation theorem according to Klimontovich as

$$(JJ)_\omega = 2\mathrm{Re}\,[Y(\omega)]\, k_B T_\omega . \tag{8.139}$$

A general quantum version of the fluctuation-dissipation theorem was given by Callen and Welton (1951) and Kubo (1957a) which is explained in detail by Landau and Lifshitz (1976, 1990) and Lifshitz and Pitaevskii (1981). For some internal vector quantity, $\mathbf{X}$, of the system, the Callen-Welton fluctuation-dissipation theorem reads

$$\left(\delta X_i\,\delta X_j \right)_\omega = \hbar\,\mathrm{Im}\left(\alpha_{ij} \right)\coth\left(\frac{\hbar\omega}{2k_B T} \right) . \tag{8.140}$$

According to Ginzburg and Pitaevskii (1987), this form of the fluctuation-dissipation theorem is one of the exact fundamental relations of quantum statistics. Interestingly enough, there is a controversial discussion about this statement (Klimontovich 1997). Following Klimontovich, the Callen–Welton derivation does not take into account the (weak) irreversibility present in all real processes and is, therefore, of limited validity. The controversy about this statement between Klimontovich (1997) on one hand and Ginzburg and Pitaevskii (1987) on the other hand, in fact, refers to the meaning of the frequency which appears in the relation. In the Callen, Welton and Kubo derivation, there is no deeper discussion of this problem. The frequency appears simply as the running frequency of the Fourier spectral decomposition. In Nyquist's and Klimontovich's derivations, ω is the frequency of the circuit under discussion, which is specific and non-universal. If this is true, the fluctuation-dissipation theorem would not be universal but specific to the circuits or oscillatory phenomena under study. Klimontovich argues that both frequencies agree only under the condition that the resonances are infinitesimally narrow (Klimontovich 2001). This implies the condition of infinitesimally small damping effects.

In our view the problem is at least in part an unsolved and deep mathematical puzzle related to the formula by Sochotzki and Plemelj, Eq. (8.45). As shown by these relations, due to dissipative effects a very small term can change dramatically the analytical properties and the values of integrals. According to Klimontovich, not all versions of the Callen-Welton relations take a proper account of the fact that small irreversible effects are present also in macroscopic processes. We consider this question as so deep and fundamental, that it cannot be decided on purely theoretical arguments. At present time, however, experimental observations which would provide a proof of one or the other version of the quantum fluctuation-dissipation theorem are still not available, as far as we know. Evidently this is one of the important open questions in quantum statistics.

8.7.2   Klimontovich–Silin Theory of Plasma Fluctuations

Let us repeat in brief first the view of Klimontovich and Silin on plasmas, developed also by Pines, Rukhadse, Ichimaru and many others, which is somehow different from the standard view. Remember that we considered plasmas as systems of particles interacting by Coulomb potentials. In the view of the mentioned authors, the central quantity of plasmas are the fluctuating electromagnetic fields, that is, electric and magnetic fields distributed in space and changing rapidly in time. These fields are stochastic, determine the dynamics of the plasma particles, and carry the main part of energy (Ichimaru 1973; Klimontovich 1967, 1983; Klimontovich and Silin 1960; Pines 1961; Silin and Rukhadse 1961). As far as the velocities are non-relativistic, the Coulomb field between the particles is the most relevant of the fields. The electrical field, $\mathbf{E}(\mathbf{r}, t)$, is essentially classically and the local energy density is given by the square of the local field strength. The study of the fluctuating field in the plasma and its correlations is a very fruitful non-traditional view which was worked out in much detail by Bohm and Pines, and Klimontovich (1967, 1983), Silin and Rukhadse (1961), and Sitenko (1982). Here, we will use mainly methods developed by Klimontovich (1967, 1982, 1983, 2001) which we discussed in part already in Chap. 4. In much respect, this approach is also related to the notations of collective excitations or plasma waves. Sometimes, this technique is called *dielectric formalism* since the dielectric function plays a primary rôle (Pines

1961, 1963; Silin and Rukhadse 1961). An attractive feature of these methods is that the classical and the quantum cases are treated in a very similar way. The physical reason for this similarity is, of course, that the electrical field in a plasma is a classical object, and only the dynamics of the particles is determined by quantum mechanics.

In a medium, the vacuum field or the corresponding potential is to be divided by the dielectric constant of the medium, ε_r,

$$\mathbf{E}(\mathbf{r}, t) \rightarrow \frac{\mathbf{E}(\mathbf{r}, t)}{\varepsilon_r} . \tag{8.141}$$

Let us consider now a single charge with a potential and field

$$V(\mathbf{r}) = \frac{e}{r} ; \qquad \mathbf{E}(\mathbf{r}) = -\frac{e\mathbf{r}}{r^3} , \tag{8.142}$$

with the Fourier transforms

$$V(\mathbf{k}) = \frac{4\pi e}{\varepsilon_r k^2} ; \qquad \mathbf{E}(\mathbf{k}) = -\frac{e\mathbf{k}}{\varepsilon_r k^3} . \tag{8.143}$$

The energy density of this field is not integrable, that is, the field energy of a Coulomb charge is infinite. This picture changes if we consider the model of the one-component plasma, that is, the charges are imbedded into a uniformly distributed density of counter charges, neutralising the system. Then, instead of the Poisson equation we obtain the Debye equation for the effective Debye field surrounding our central charge, as shown in Chap. 6:

$$\Delta V^D(\mathbf{r}) = \kappa^2 V_D(\mathbf{r}) ; \qquad \kappa^2 = \frac{8\pi \beta n e^2}{\varepsilon_r} , \tag{8.144}$$

with the solution

$$V^D(r) = \frac{e\, e^{-\kappa r}}{\varepsilon_r r} . \tag{8.145}$$

The prefactor follows by considering the special case, $n = 0$, where again the Coulomb law should be valid. On the level of Fourier transforms, this reads for the potential and the field

$$V^D(k) = \frac{4\pi e}{k^2} \frac{1}{\varepsilon_r \left(1 + \frac{\kappa^2}{k^2}\right)} \tag{8.146}$$

$$\mathbf{E}(\mathbf{k}) = -4\pi \frac{e}{\varepsilon_r} \frac{\mathbf{k}}{k} \frac{1}{k^2 + \kappa^2} . \tag{8.147}$$

Formally, the denominator plays the same rôle as the relative dielectric constant. We denote it, therefore, as the k-dependent dielectric function of the plasma:

$$\varepsilon(k) = 1 + \frac{\kappa^2}{k^2}. \tag{8.148}$$

As previously, we assume mostly $\varepsilon_r = 1$. We can show that the field energy becomes finite due to screening. According to the Parseval theorem, we have the relation

$$\int d\mathbf{r}\, \mathbf{E}(\mathbf{r}) \cdot \mathbf{E}(\mathbf{r}) = \frac{1}{(2\pi)^2} \int d\mathbf{k}\, \mathbf{E}(\mathbf{k}) \cdot \mathbf{E}(\mathbf{k}). \tag{8.149}$$

By integrating the Fourier transform of the Debye field we find for the Coulomb energy per charge

$$u = -\frac{e^2}{2r_D}, \tag{8.150}$$

which is nothing else but Debye's law obtained in an independent way from the field energy. Before generalizing this derivation to the quantum case, let us remind that a static plasma does not exist in the quantum case, as at least the zero-point quantum oscillations are always present. Therefore, it is reasonable to include the time variable in the classical picture. Following Klimontovich, we introduce a time and space dependent electric field and a time and momentum dependent local plasma density and their Fourier representations by

$$\delta\,\mathbf{E}(\mathbf{r}, t) = \int d\mathbf{k}\, d\omega\, e^{i\omega t - i\mathbf{k}\cdot\mathbf{r}}\, \delta\mathbf{E}(\mathbf{k}, \omega); \tag{8.151}$$

$$\delta N(\mathbf{r}, \mathbf{p}, t) = \int d\mathbf{k}\, d\omega\, e^{i\omega t - i\mathbf{k}\cdot\mathbf{r}}\, \delta N(\mathbf{k}, \mathbf{p}, \omega). \tag{8.152}$$

In this Fourier picture, the Poisson equation reads

$$\delta\mathbf{E}(\mathbf{k}, \omega) = -i\frac{\mathbf{k}}{k^2}\, 4\pi e \int d\mathbf{p}\, \delta N\, (\mathbf{k}, \mathbf{p}, \omega). \tag{8.153}$$

Together with the Vlasov equation for the densities which is kind of analog of an random phase approximation (RPA), this leads to the dielectric function

$$\varepsilon(\omega, k) = 1 - \frac{4\pi e^2}{k^2}\, \Pi(\omega, k); \qquad \Pi(\omega, k) = -n \int d\mathbf{p}\, \frac{\mathbf{k} \cdot \frac{\partial f(\mathbf{p})}{\partial \mathbf{p}}}{\omega - kv + i\Delta}. \tag{8.154}$$

Here Π denotes the dielectric response function or polarization function and $i\Delta$ provides a small imaginary increment. For the density of the field energy we obtain in Fourier space the fluctuation-dissipation theorem for plasmas:

$$(\delta E \, \delta E)_{\omega,k} = \text{Im} \left[\frac{1}{\varepsilon(\omega, \mathbf{k})} \right] \frac{8\pi \, k_B T}{\omega} . \tag{8.155}$$

By integrating over the frequencies we recover the previous relations for the k-dependent dielectric function and the energy density (Klimontovich 1967). The advantage of the new formulae is, however, that they are also valid in the quantum case, up to small details. This is, of course, again due to the fact that the electrical field is essentially a classical quantity. Thus, as stated already above, large part of the quantum derivation is in full analogy to the classical case, when Klimontovich's form of the Wigner representation discussed in Chap. 4 is used. Referring to Klimontovich's textbook (see Chapter 17 of Klimontovich 1982, 1986), the Fourier components of the electrical field and the potential can be again represented by

$$\delta E(\omega, k) = -i \, k \, \delta\varphi(\omega, k) ; \qquad \delta U(\omega, k) = e \, \delta\varphi(\omega, k) , \tag{8.156}$$

and the correlations satisfy the fluctuation-dissipation theorem for quantum plasmas (which replaces Eq. (8.155)),

$$(\delta E \, \delta E)_{\omega,k} = \text{Im} \left[\frac{1}{\varepsilon(\omega, \mathbf{k})} \right] \frac{8\pi}{\omega} F_\omega , \tag{8.157}$$

where

$$F_\omega = \frac{\hbar\omega}{2} + \frac{\hbar\omega}{e^{\beta\hbar\omega} - 1} . \tag{8.158}$$

The dielectric function has the same form as in the classical case but the dielectric response function is largely modified due to quantum effects (Klimontovich 1967, 1983, 2001; Kraeft et al. 1986),

$$\varepsilon(\omega, k) = 1 - \frac{4\pi e^2}{k^2} \Pi(\omega, \mathbf{k}) , \tag{8.159}$$

with

$$\Pi(\omega, \mathbf{k}) = -(2s + 1) \int \frac{d\mathbf{p}'}{(2\pi)^3} \frac{f\left(p' + \frac{p}{2}\right) - f\left(p' - \frac{p}{2}\right)}{\hbar\omega + E(p', p)} \tag{8.160}$$

$$E(p', p) = \frac{\hbar^2}{2m} \left[\left(p' - \frac{p}{2} \right)^2 - \left(p' + \frac{p}{2} \right)^2 \right] . \tag{8.161}$$

The theory of fluctuations in plasmas is by now an extended field of statistical physics. For further details, we refer to the special literature (Ichimaru 1973; Klimontovich 1967, 1983; Silin and Rukhadse 1961; Sitenko 1982).

References

Balescu, R. 1963. *Statistical Mechanics of Charged Particles*. London: Wiley.
Binder, K., ed. 1979. *Monte Carlo Methods in Statistical Physics*. Berlin: Springer.
Bogolyubov, N.N. 1991. *Selected Works*. Vol. II. Quantum and Classical Statistical Mechanics. New York: Gordon and Breach.
Bogolyubov, N.N. 2005–2009. *Collected Papers*. Vol. 1–12. Moscow: Fizmatlit.
Bogolyubov, N.N., and N.N. Bogolyubov, Jr. 1991. *Introduction to Quantum Statistical Mechanics*. New York: Gordon and Breach.
Bonitz, M. 1998. *Quantum Kinetic Theory*. Stuttgart: B. G. Teubner.
Bonitz, M., and W.D. Kraeft, eds. 2005. *Kinetic Theory of Nonideal Plasmas*. Vol. 11. Berlin: Springer.
Bonitz, M., and D. Semkat, eds. 2006. *Introduction to Computational Methods for Many Body Systems*. Princeton: Rinton Press.
Bonzel, H.P., A.M. Bradshaw, and G. Ertl, eds. 1989. *Physics and Chemistry of Alkali Metal Adsorption*. Vol. 57. Materials Science Monographs. Amsterdam: Elsevier.
Böttger, H.B., and V.V. Bryksin. 1985. *Hopping Conduction in Solids*. Berlin: Akademie-Verlag.
Brey, J.J., I. Goldhirsch, and T. Pöschel, eds. 2009. *Granular Gases: Beyond the Dilute Limit*. Vol. 179. European Physical Journal Special Topics. Berlin: Springer.
Brush, S.G., H.L. Sahlin, and E. Teller. 1966. Monte Carlo Study of a One-Component Plasma I. *The Journal of Chemical Physics* 45: 2102–2117.
Callen, H.B., and T.A. Welton. 1951. Irreversibility and Generalized Noise. *Physical Review* 83: 34–40.
Chetverikov, A.P., W. Ebeling, and M.G. Velarde. 2009. Local Electron Distributions and Diffusion in Anharmonic Lattices Mediated by Thermally Excited Solitons. *The European Physical Journal B* 70: 217–227.
Chetverikov, A.P., W. Ebeling, G. Röpke, and M.G. Velarde. 2011. Hopping Transport and Stochastic Dynamics of Electrons in Plasma Layers. *Contributions to Plasma Physics* 51: 814–829.
Chetverikov, A.P., W. Ebeling, and M.G. Velarde. 2011. Soliton-Like Excitations and Solectrons in Two-Dimensional Nonlinear Lattice. *The European Physical Journal B* 80: 137–145.
Chetverikov, A.P., W. Ebeling, and M.G. Velarde. 2012. Controlling Fast Electron Transfer at the Nano-Scale by Solitonic Excitations Along Crystallographic Axes. *The European Physical Journal B* 85: 291.
Dittrich, T., P. Hänggi, G.-L. Ingold, B. Kramer, G. Schön, and W. Zwerger. 1998. *Quantum Transport and Dissipation*. Weinheim: Wiley-VCH.
Ebeling, W., and I. Sokolov. 2005. *Statistical Thermodynamics and Stochastic Theory of Nonequilibrium Systems*. Singapore: World Scientific.
Ebeling, W., V.E. Fortov, Yu. L. Klimontovich, N.P. Kovalenko, W.D. Kraeft, et al., eds. 1984. *Transport Properties of Dense Plasmas*. Boston: Birkhäuser.
Ebeling, W., A. Förster, V.F. Fortov, V.K. Gryaznov, and A. Ya. Polishchuk. 1991. *Thermophysical Properties of Hot Dense Plasmas*. Stuttgart: Teubner-Verlag.
Ebeling, W., V.E. Fortov, and V.S. Filinov. 2017. *Quantum Statistics of Dense Gases and Nonideal Plasmas*. Springer Series in Plasma Science and Technology. Berlin: Springer.
Einstein, A. 1905. Über die von der molekularkinetischen Theorie der Wärme geforderte Bewegung von in ruhenden Flüssigkeiten suspendierten Teilchen. *Annalen der Physik* 17: 549–560.

Einstein, A. 1906. Zur Theorie der Lichterzeugung und Lichtabsorption. *Annalen der Physik* 20: 199–206.

Einstein, A. 1916. Strahlungs-Emission und -Absorption nach der Quantentheorie. *Verhandlungen der Deutschen Physikalischen Gesellschaft* 18: 318–323.

Einstein, A. 1917a. Quantentheorie der Strahlung. *Mitteilungen der Physikalischen Gesellschaft Zürich* 16: 47–62.

Einstein, A. 1917b. Zur Quantentheorie der Strahlung. *Physikalishce Zeitschrift* 18: 121–128.

Fokker, A.D. 1914. Die mittlere Energie rotierender elektrischer Dipole im Strahlungsfeld. *Annalen der Physik* 43: 810–820.

Fortov, V.E. 2009. *Extreme States of Matter*. Moskva: FizMatGis.

Fortov, V.E. 2011. *Extreme States of Matter: On Earth and in the Cosmos*. Berlin: Springer.

Fortov, V.E., and I.T. Yakubov. 1994. *Nonideal Plasmas* (in Russian). Moskva: Énergoatomizdat.

Fortov, V.E., R.I. Ilkaev, V.A. Arinin, V.V. Burtzev, V.A. Golubev, I.L. Iosilevskiy, et al. 2007. Phase Transition in a Strongly Nonideal Deuterium Plasma Generated by Quasi-Isentropical Compression at Megabar Pressures. *Physical Review Letters* 99: 185001.

Ginzburg, V.L., and L.P. Pitaevskii. 1987. Quantum Nyquist Formula and the Applicability Ranges of the Callen–Welton Formula. *Soviet Physics Uspekhi* 30: 168–171.

Ichimaru, S. 1973. *Basic Principles of Plasma Physics*. Reading: Benjamin.

Ichimaru, S. 1992. *Statistical Plasma Physics*. Redwood: Addison-Wesley.

Ingold, G.-L. 2002. Path Integrals and Their Application to Dissipative Quantum Systems. In *Coherent Evolution in Noisy Environments*, ed. A. Buchleitner and K. Hornberger, vol. 611, 1–53. Lecture Notes in Physics. Berlin: Springer.

Johnson, J.B. 1928. Thermal Agitation of Electricity in Conductors. *Physics Review* 32: 97–109.

Klein, O. 1922. Zur statistischen Theorie der Suspensionen und Lösungen. *Arkiv För Matematik, Astronomi Och Fysik* 16: 1–51.

Klimontovich, Y.L. 1967. *Statistical Theory of Nonequilibrium Processes in Plasmas*. Oxford: Pergamon.

Klimontovich, Y.L. 1982. *Statistical Physics* (in Russian). Moscow: Nauka.

Klimontovich, Y.L. 1983. *The Kinietic Theory of Electromagnetic Processes*. Berlin: Springer.

Klimontovich, Y.L. 1986. *Statistical Physics*. New York: Harwood.

Klimontovich, Y.L. 1997. *Statistical Theory of Open Systems*. Amsterdam: Kluwer.

Klimontovich, Y.L. 2001. *Statistical Theory of Open Systems* (in Russian). Vol. III. Physics of Open Quantum Systems. Moskva: Janus.

Klimontovich, Y.L., and V.P. Silin. 1960. The Spectra of Systems of Interacting Particles and Collective Losses During Passage of Charged Particles Through Matter. *Soviet Physics Uspekhi* 3: 84–114.

Kraeft, W.D., D. Kremp, W. Ebeling, and G. Röpke. 1986. *Quantum Statistics of Charged Particle Systems*. Berlin: Akademie-Verlag.

Kremp, D., M. Schlanges, and W.D. Kraeft. 2005. *Quantum Statistics of Nonideal Plasmas*. Berlin: Springer.

Kubo, R. 1957a. Statistical-Mechanical Theory of Irreversible Processes. I. General Theory and Simple Applications to Magnetic and Conduction Problems. *Journal of the Physical Society of Japan* 12: 570–586.

Kubo, R. 1957b. Statistical-Mechanical Theory of Irreversible Processes. II. Response to Thermal Disturbance. *Journal of the Physical Society of Japan* 12: 1203–1211.

Kubo, R. 1959. Some Aspects of the Statistical Mechanical Theory of irreversible Processes. In *Lectures on Theoretical Physics*, ed. E. Brittin and L.G. Dunham, 120–203. Lectures on Theoretical Physics. New York: Interscience.

Landau, L.D. 1936. The Kinetic Equation for the Case of Coulomb Interaction. *Physikalische Zeitschrift der Sowjetunion* 10: 154.

Landau, L.D. 1937. The Kinetic Equation for the Case of Coulomb Interaction. *Zhurnal Eksperimental'noi i Teoreticheskoi Fiziki* 7: 203.

Landau, L.D., and E.M. Lifshitz. 1976. *Statistical Physics (Part I)*. Moscow: Nauka.

Landau, L.D., and E.M. Lifshitz. 1990. *Statistical Physics*. New York: Pergamon.

Lifshitz, E.M., and L.P. Pitaevskii. 1981. *Physical Kinetics*. Vol. 10. Course of Theoretical Physics. New York: Pergamon.

Metropolis, N., A.W. Rosenbluth, M.N. Rosenbluth, A.H. Teller, and E. Teller. 1953. Equation of State Calculations by Fast Computing Machines Equation of State Calculations by Fast Computing Machines. *The Journal of Chemical Physics* 21: 1087–1092.

Militzer, B., and E.L. Pollock. 2000. Variational Density Matrix Method for Warm Condensed Matter and Application to Dense Hydrogen. *Physical Review E* 61: 3470–3482.

Nellis, W.J., S.T. Weir, and A.C. Mitchell. 1999. Minimum Metallic Conductivity of fluid Hydrogen at 140 GPa. *Physical Review B* 59: 3434–3449.

Nyquist, H. 1928. Thermal Agitation of Electric Charge in Conductors. *Physical Review* 32: 110–113.

Pauli, W. 1928. Über das H-theorem vom Anwachsen der Entropie vom Standpunkt der neuen Quantenmechanik. In *Probleme der Modernen Physik—Festschrift zum 60. Geburtstage A. Sommerfelds*, ed. P. Debye, 30–45. Leipzig: Hirzel.

Pines, D. 1961. *The Many Body Problem—A Lecture Note*. New York: Benjamin.

Pines, D. 1963. *Elementary Excitations in Solids*. Lecture Notes and Supplements in Physics. New York: Benjamin.

Planck, M. 1911. Eine neue Strahlungshypothese. *Verhandlungen der Deutschen Physikalischen Gesellschaft* 13: 138–148.

Planck, M. 1917. Über einen Satz der statistischen Dynamik und seine Erweiterung in der Quantentheorie. *Sitzungsberichte der Preussischen Akademie der Wissenschaften* 24: 324–341.

Pöschel, T., and S. Luding, eds. 2001. *Granular Gases*. Berlin: Springer.

Redmer, R. 1997. Physical Properties of Dense, Low-Temperature Plasmas. *Physics Reports* 282: 35–157.

Redmer, R., and G. Röpke. 2010. Progress in the Theory of Dense Strongly Coupled Plasmas. *Contributions to Plasma Physics* 50: 970–985.

Silin, V.P., and A.A. Rukhadse. 1961. *Electromagnetic Properties of Plasmas and Plasma-Like Matter* (in Russian). Moskva: Atomizdat.

Sitenko, O.G. 1982. *Fluctuations and Nonlinear Wave Interactions in Plasmas*. Oxford: Pergamon Press.

Spitzer, L. 1962. *Physics of Fully Ionized Plasmas*. New York: Wiley.

Taylor, G.I. 1922. Diffusion by Continuous Movements. *Proceedings of the London Mathematical Society Series* 2 (20): 196–212.

Ternovoi, V. Ya., A.S. Filimonov, V.E. Fortov, S.V. Kvitov, D.D. Nikolaev, et al. 1999. Thermodynamic Properties and Electrical Conductivity of Hydrogen Under Multiple Shock Compression to 150 GPa. *Physica B* 265: 6–11.

Tolman, R. 1938. *The Principles of Statistical Physics*. Oxford: Oxford University Press.

Vlasov, A.A. 1945. On the Kinetic Theory of an Assembly of Particles with Collective Interaction. *Russian Physics Journal* 9: 25–40.

Vlasov, A.A. 1950. *Many-Particle Theory*. Moscow: Gostekhizdat.

von Neumann, J. 1932. *Mathematische Grundlagen der Quantenmechanik*. Berlin: Springer.

Weir, S.T., A.C. Mitchell, and W.J. Nellis. 1996. Metallization of fluid Molecular Hydrogen at 140 GPa (1.4 Mbar). *Physical Review Letters* 76: 1860–1863.

Zamalin, V.M., G.E. Norman, and V.S. Filinov. 1977. *The Monte Carlo Method in Statistical Thermodynamics* (in Russian). Moscow: Nauka.

Zubarev, D.N., V. Morozov, and G. Röpke, eds. 1996. *Statistical Mechanics of Nonequilibrium Processes*. Vol. 1. Basic Concepts, Kinetic Theory. Weinheim: Wiley-VCH.

Zubarev, D.N., V. Morozov, and G. Röpke, eds. 1997. *Statistical Mechanics of Nonequilibrium Processes*. Vol. 2. Relaxation and Hydrodynamic Processes. Weinheim: Wiley-VCH.

Index

© Springer Nature Switzerland AG 2019

W. Ebeling, T. Pöschel, *Lectures on Quantum Statistics*,
Lecture Notes in Physics 953, https://doi.org/10.1007/978-3-030-05734-3

Printed in the United States
By Bookmasters